People Managing Forests

The Links between Human Well-Being and Sustainability

EDITED BY
Carol J. Pierce Colfer
and Yvonne Byron

RESOURCES FOR THE FUTURE
WASHINGTON, DC, USA

CENTER FOR INTERNATIONAL
FORESTRY RESEARCH
BOGOR, INDONESIA

©2001 Resources for the Future

All rights reserved. No part of this publication may be reproduced by any means, either electronic or mechanical, without permission in writing from the publisher, except under the conditions given in the following paragraph:

Authorization to photocopy items for internal or personal use, the internal or personal use of specific clients, and for educational classroom use is granted by Resources for the Future, provided that the appropriate fee is paid directly to Copyright Clearance Center, 222 Rosewood Drive, Danvers, MA 01923, USA. Telephone (978) 750–8400; fax (978) 750–4470.

Printed in the United States of America

An RFF Press book
Published by Resources for the Future
1616 P Street, NW, Washington, DC 20036–1400

A copublication of Resources for the Future (www.rff.org) and the Center for International Forestry Research (www.cifor.org)

Library of Congress Cataloging-in-Publication Data

People managing forests : the links between human well-being and sustainability / edited by Carol J. Pierce Colfer and Yvonne Byron.
    p. cm.
    Includes bibliographical references.
    ISBN 1–891853–05–8 (lib.bdg) — ISBN 1–891853–06–6 (pbk.: alk. paper)
    1. Forest management—social aspects—Tropics. 2. Forestry and community—Tropics. 3. Sustainable forestry—Tropics. I. Colfer, Carol J. Pierce. II. Byron, Yvonne, 1950–
SD247 P46   2001
333.75′16′0913—dc21   2001019059

f e d c b a

The paper in this book meets the guidelines for permanence and durability of the Committee on Production Guidelines for Book Longevity of the Council on Library Resources.

This book was designed and typeset in Bembo and Gill Sans by Betsy Kulamer. It was copyedited by Pamela Angulo. The cover was designed by Debra Naylor Design. The art on the front cover is by Tessie R. Salazar Lozano, Pucallpa, Peru.

ISBN 1–891853–05–8 (cloth)
ISBN 1–891853–06–6 (paper)

About Resources for the Future

Founded in 1952, Resources for the Future (RFF) contributes to environmental and natural resource policymaking worldwide by performing independent social science research.

RFF pioneered the application of economics as a tool to develop more effective policy about the use and conservation of natural resources. Its scholars continue to employ social science methods to analyze critical issues concerning pollution control, energy policy, land and water use, hazardous waste, climate change, biodiversity, and the environmental challenges of developing countries.

About the Center for International Forestry Research

The Center for International Forestry Research (CIFOR) was established in 1993 as part of the Consultative Group on International Agricultural Research (CGIAR) in response to global concerns about the social, environmental, and economic consequences of forest loss and degradation. CIFOR research produces knowledge and methods needed to improve the well-being of forest-dependent people and to help tropical countries manage their forests wisely for sustained benefits. This research is done in more than two dozen countries, in cooperation with numerous partners. Since it was founded, CIFOR has also played a central role in influencing global and national forestry policies.

Contents

CONTENTS

SECTION FOUR
Rights and Responsibilities to Manage Cooperatively and Equitably

SECTION FIVE
Comparisons: Geographical and Temporal

CONTENTS

Contributors

Mary Ann Brocklesby is a lecturer in development studies at the University of Wales, Swansea. She worked previously as a social development practitioner in Southeast Asia and West Africa, and as a community development advisor on a biodiversity conservation project in Cameroon.

Katrina Brown is senior lecturer in natural resources and development at the University of East Anglia and senior research fellow at the Centre for Social and Economic Research on the Global Environment.

Yvonne Byron, a geographer, is an editor for Lonely Planet Publications in Australia. Previously, she was an editor for the Center for International Forestry Research (CIFOR) and a researcher at Australian National University, focusing on environmental change in Southeast Asia.

Carol J. Pierce Colfer, an anthropologist, is team leader of the CIFOR program on Local People, Devolution, and Adaptive Collaborative Management of Forests. Prior to that, she was a member of the CIFOR project from which the research reported in this book largely derives. She has spent approximately ten years living in villages in Kalimantan and Sumatra, Indonesia, and has field experience in Oman, the United States, Cameroon, Côte d'Ivoire, Brazil, Iran, and Thailand.

Rona A. Dennis is an expert in remote sensing and geographical information systems. Dennis recently joined CIFOR to study the underlying causes and impacts of Indonesia's 1997–1998 fires. Previously, she worked on a project on sustainable forest management jointly sponsored by Indonesia and the U.K. Department for International Development.

Norbert Gami is an anthropologist with extensive experience in Central Africa.

Mario Günter recently completed his doctoral degree at the Geography Department of the University of Heidelberg, Germany.

Emily Harwell, a social ecologist, is a doctoral candidate at the School of Forestry and Environmental Studies of Yale University. Harwell has conducted long-term research in the Danau Sentarum Wildlife Reserve, and she has worked as a consultant for Wetlands International and for the U.N. Development Programme.

Sandrine Lapuyade, a socioeconomist with a focus on natural resource management, is developing a research project studying the impact of conservation interventions on livelihoods and gender roles. Previously, she worked for The Environment and Development Group, a consultancy company based in the United Kingdom. She also has designed and evaluated conservation and development projects in the forest zone of Central Africa.

Cynthia L. McDougall is a social science research fellow at CIFOR in Indonesia, where she is currently a member of the program team of Local People, Devolution, and Adaptive Collaborative Management of Forests, working primarily in Nepal and Indonesia.

Robert Nasi, a silviculturist with strong ecological leanings, is the team leader of CIFOR's biodiversity program. Prior to that, he worked on two other CIFOR projects, studying sustainable forest management in central Africa. He has worked in Cameroon, Mali, Malaysia, and New Caledonia.

Noemi Miyasaka Porro is a doctoral candidate at the Department of Anthropology of the University of Florida.

Roberto Porro is a doctoral candidate at the Department of Anthropology of the University of Florida.

Ravi Prabhu, a tropical forester, was leader of CIFOR's project, Assessing Sustainable Forest Management: Testing Criteria and Indicators, from which the research reported in this book was conceived and conducted. He is currently participating in the CIFOR program, Local People, Devolution, and Adaptive Collaborative Management of Forests.

Atie Puntodewo is a specialist in geographical information systems. She works at CIFOR, where she supports research relating to mapping.

Diane Russell, an anthropologist, works for the U.S. Agency for International Development Mission in the Democratic Republic of the Congo and for the Central African Regional Program for the Environment. Previously, she spent two years in Cameroon as a Rockefeller Fellow at the International Institute for Tropical Agriculture.

Agus Salim is a statistician at CIFOR and is continuing his graduate studies in statistics at University College Cork, Ireland.

Ismayadi Samsoedin is a forester at FORDA, Indonesia's Forestry Research Agency in Bogor. He is currently seconded to CIFOR to coordinate the field center at the Bulungan Research Forest in East Kalimantan.

Mustofa Agung Sardjono is a faculty member at Universitas Mulawarman (UnMul) in Samarinda, East Kalimantan, Indonesia, where he has been active in the development of a Center for Social Forestry.

Joseph A. Tainter is project leader of Cultural Heritage Research in the U.S. Department of Agriculture Forest Service, Rocky Mountain Research Station. He is currently pursuing fieldwork in the Sahel of Mali.

Nicodème Tchamou, a botanist, heads the focal-point office of the Central African Regional Program for the Environment (CARPE), a 20-year effort by the U.S. Agency for International Development to mitigate global climate change by reducing the rate of tropical forest destruction and biodiversity loss in Central Africa. Previously, Tchamou worked as a research associate at the International Institute for Tropical Agriculture.

Bertin Tchikangwa, a sociologist, works in southeast Cameroon as community conservation officer with the World Wide Fund for Nature. Previously, he worked as an anthropologist for ECOFAC (Conservation et Utilisation Rationnelle des Ecosystèmes Forestiers en Afrique Centrale) and as a team member of APFT (Avenir des Peuples des Forêts Tropicales).

Anne Marie Tiani, an ecologist and botanist, serves as an advisor to CAPED, a national nongovernmental organization recently created in Mbalmayo, Cameroon, which conducts research on the sustainable management of natural resources and proposes development projects in rural areas of the forest zone. Tiana has collaborated with CIFOR since 1996 and has been a consultant to several nongovernmental organizations concerned with rural development projects.

Reed L. Wadley, an anthropologist, is a research fellow at the International Institute for Asian Studies, Leiden, The Netherlands, where he is working on the environmental history of the Iban in the Danau Sentarum area of West Kalimantan, Indonesia. Previously, he has been a consultant with the Asian Wetland Bureau (now Wetlands International) and with CIFOR.

Joseph Woelfel is a professor of communication at the State University of New York, Buffalo, and the principal originator of the Galileo method discussed in this book.

Eva (Lini) Wollenberg is a community-based management specialist at CIFOR. Wollenberg was involved in the research reported in this book since its inception, and she currently works with the CIFOR program, Local People, Devolution and Adaptive Collaborative Management of Forests.

Acknowledgements

Although each chapter includes individual acknowledgements that pertain to the work reported therein, we feel the need to emphasize the support of some partners and institutions that contributed to large parts of the endeavor.

The Center for International Forestry Research (CIFOR) provided a supportive and congenial "home base" for planning and coordinating much of the research reported in this book. We thank our many CIFOR colleagues who read drafts, critiqued ideas, and served as valuable sounding boards throughout the process. Special thanks go to Rahayu Koesnadi, Linda Yuliani, and Atie Puntodewo at CIFOR, to Gina Armento and Don Reisman at Resources for the Future, and to Betsy Kulamer and Pamela Angulo for their assistance throughout the production process.

We are grateful to the European Union, U.S. Agency for International Development (AID), and the International Development and Research Center (IDRC) for funds that supported the data collection for this book and to the Asian Development Bank and CIFOR for supporting parts of the analysis and writing.

We thank the individual authors for their patience, good humor, and perseverance throughout a process that took longer than we had anticipated.

We are also grateful to the governments of Brazil, Cameroon, Gabon, Indonesia, Trinidad and Tobago, and the United States, which facilitated our work in various ways. Among the most important contributors are our partners—both direct and indirect—who enhanced our understanding and smoothed the way in many instances in all these countries. But the most important debt of gratitude is reserved for the women, men, and children who live in the communities we visited. Without their patience and forbearance with our many questions, surveys, interviews, and observations, this research would have been impossible. We hope that the findings we present justify their faith in us.

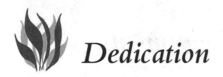 *Dedication*

Carol dedicates this book to her mother,
Gwendolyn Marie Harris Pierce,
for instilling in her a concern for human well-being,
and to her daughter,
Megan Melina Colfer Steinert,
for showing that concern daily as a loving, caring
mother herself.

Yvonne dedicates this book to her family—
Neil, Kathryn, and Andrew—
for encouragement and support in this project
and all else that she attempts.

People
Managing
Forests

INTRODUCTION

History and Conceptual Framework

Carol J. Pierce Colfer and Yvonne Byron, with Ravi Prabhu and Eva Wollenberg

Confusion and dismay are rampant among those concerned about human and environmental issues in the tropics, and with good reason. Forests are being degraded at apparently ever-increasing rates, and human welfare in forested areas is remaining at a constant level at best, more often deteriorating. Many people—researchers, environmentalists, and policymakers of various hues—are trying to address these problems. This book represents the evolution of one cooperative effort to understand and develop mechanisms for dealing with these interrelated problems, and the authors propose some suggestions for improving our future efforts.

In our research, we have asked ourselves one fundamental question: how can we create conditions that allow local people who live in and around forests to maintain the valued aspects of their own way of life and to prosper while still protecting those forests on which they, and perhaps the rest of us, depend? To answer that question, we needed first to identify the conditions that contribute to sustainable forest management (SFM) in general and to the well-being of forest-dwelling people in particular. Satisfied that we had a good grasp of the most important conditions (see later), we set out to examine their relationship to sustainability. This examination is the central theme of this book. Central issues of concern include the identification and roles of relevant stakeholders (including gender and diversity, discussed in Section 1, and the relevance of a "conservation ethic," discussed in Section 2), security of intergenerational access to forest resources (Section 3), and rights and responsibilities to manage forests cooperatively and equitably (Section 4).

1

We began this exploration looking at criteria and indicators (C&I) for SFM. The primary purpose of C&I is as a tool to assess the sustainability of particular systems quickly, easily, and reliably. Initially conceived as tools for use by external evaluators, the C&I concept has evolved. Some individuals and projects are now using modified C&I in cooperation with local communities as monitoring instruments to make management more adaptive (see Concluding Remarks).

The C&I approach is built on a hierarchical framework in which principles, criteria, indicators, and verifiers are identified, each level more concrete than the previous (see Prabhu and others 1996; Lammerts van Bueren and Blom 1997).

From the beginning, we were convinced that human well-being (HWB) played a part in SFM and were anxious to clarify the links between these concepts. Through a complex global process (described later), we identified relevant C&I and then set out to test the causal links between HWB and SFM, using those C&I.

We started with a series of assumptions, one of which we have since concluded was not an assumption at all but a testable hypothesis:[1] "that human systems are complex adaptive systems, intimately connected with each other and with biological systems, in a self-organizing process of coadaptation" (Colfer and others 1995). In retrospect and technically, we should have identified the null hypothesis, that human systems are *not* said "complex adaptive systems...." Such a null hypothesis suggests that simple cause-and-effect relationships can in fact be found, that clear and consistent links exist between, for instance, HWB and SFM.

Links were indeed what we were initially seeking. Building on the best science we could bring to bear, within the state-of-the-art C&I framework, we sought evidence of such links. We applied the C&I framework to define and refine concepts and to test specific links in several locations, drawing on the long-term experience of various researchers. However, the results, though rich in insights, provided little conclusive evidence of such links. Concepts such as gender and other kinds of diversity, views about nature, secure access to resources, equitable sharing of benefits, and participation were found to be important everywhere, but in different ways in different places. Marshalling clear evidence to link these issues to SFM in the direct way demanded by reductionist science proved impossible.

Ultimately, after giving it our best effort, we concluded that we must reject the null hypothesis. We cannot thereby prove our hypothesis that human systems are in fact the complex and adaptive systems we think they are. But our findings certainly tend to support that view. Our findings have serious implications for our usual research methods. If we are indeed dealing with complex, interrelated, and adaptive systems, new research paradigms, approaches, and methods are vitally needed. We suggest a few ways of approaching this issue in our concluding chapters.

This introduction is composed of three major parts. The first part is a chronological treatment of the six years of research on which this book is based, divided in a way that reflects the evolving, iterative nature of our research approach. We describe the series of field tests used to evaluate C&I for SFM; lay out the conceptual framework with which we began this research by providing definitions of terms, assumptions, and three conceptual issues relating to our scientific "worldview"; discuss the social principles that were identified in the multinational field tests of C&I (the results of previous tests); and introduce the themes and hypotheses (that resulted from the C&I research) that are the focus of this book. These ideas discussed in this chapter were our foundation when we began the C&I tests in 1995.

In the second part, we describe the relevant tools and approaches used in the analyses. The methods, tested in the social science methods tests undertaken between 1996 and 1998, inform much of the research reported here. In most cases, these methods are supplemented by longer-term, more qualitative methods.

In the final part of this chapter, we examine our findings, our methodological shortcomings, and draw some conclusions about the nature of scientific inquiry that focuses on dynamic and interdependent systems.

History and Context for This Research

The Past

In 1994, the Center for International Forestry Research (CIFOR) in Bogor, Indonesia, initiated a project to assess existing sets of C&I for SFM for timber in several locations—initially, in Germany, Indonesia, Côte d'Ivoire, and Brazil.[2] For this work, we considered principles as abstract, "motherhood" statements; criteria as desirable conditions, somewhat more specific; indicators as ideally measurable, observable, and directly linked to the criteria under which they fall; and verifiers—the subject of some controversy in the literature—to be similar but even more specific than indicators (they can be threshold levels or means of verification). The term C&I refers to this conceptual framework that helped to guide much of the research reported here.

On the basis of the CIFOR team's experience in Germany, where no social scientists were included in the interdisciplinary test team, project leader Ravi Prabhu concluded that a conceptual framework was needed to deal with social issues. Colfer joined the group to develop a conceptual framework for dealing with social issues (Colfer and others 1995), to help future team members address social issues more systematically. We gave the conceptual framework to all CIFOR test teams, stressing their freedom to accept, adapt, or reject it in their evaluation of the sets of social C&I. In the initial testing process, which eventually expanded to include Austria, Cameroon, the United States, and Gabon, the social scientist team members typically found the conceptual framework useful but still were dissatisfied with the assessment tools available to them. Although lingering doubts remained, a "generic template" of C&I gradually evolved (Prabhu and others 1996, 1998; see Annex 1 for the most recent version), primarily based on materials from humid tropical forests managed for timber. At the same time, parallel activities were under way that looked at C&I in forests managed by communities (coordinated by Nicolette Burford de Oliveira with Cynthia McDougall) (Burford de Oliveira 1997, 1999; Burford de Oliveira and others 1998, 2000; Ritchie and others 2000) and in plantations (coordinated by Christian Cossalter at CIFOR) (Muhtaman and others 2000; Sankar and others 2000).

By clearly demonstrating the importance of social issues in SFM, the first five field tests (in forests managed for timber) convinced us to mount a subsidiary effort focused specifically on the social C&I. In 1996, Wadley and Colfer tested eight social science methods in West Kalimantan as possible mechanisms to make quick and reliable assessments of HWB issues (see Chapters 5, 8, and 12; Colfer and others 1997c). After that experience, a selection of 12 methods was systematically tested by teams in Cameroon, Indonesia, and Brazil in 1997 and 1998; in most cases, teams were led by social scientists.[3] Based on the results of these tests, the methods were revised again and then published (Colfer and others 1999a, 1999b, 1999c; Salim and others 1999). The major themes stressed in the first four sections of this book were selected based on the results of these various C&I tests.

The testing of these methods, however, was only half of the task we had set ourselves. We also wanted to gain a better understanding of the causal relationships between HWB and sustainability. In the unanimous judgement of our interdisciplinary and multicultural test teams, HWB was an important concern in SFM. Yet the links remained clouded; some evidence seemed contradictory.

At that stage, we hoped that by more carefully examining the research results from our methods tests across sites—particularly in comparison with long-term, qualitative studies in the area and with forest quality—we might be able to shed some light on the relationships between HWB and SFM. Authors compared methodological test results with their long-term knowledge of study sites to draw conclusions about these issues. As part of our initial approach (which reflected some reductionist influences that have in fact strengthened the clarity with which we can reject the null hypothesis, discussed earlier), we also placed study sites on a loose continuum from "forest rich" to "forest poor"; we indicate the forest quality, using this rough guideline, for each research site (see Annex 2).

The chapters in this book reflect a range of studies that have come out of this research effort, plus a few others that offer related insights. Figure 1 is a

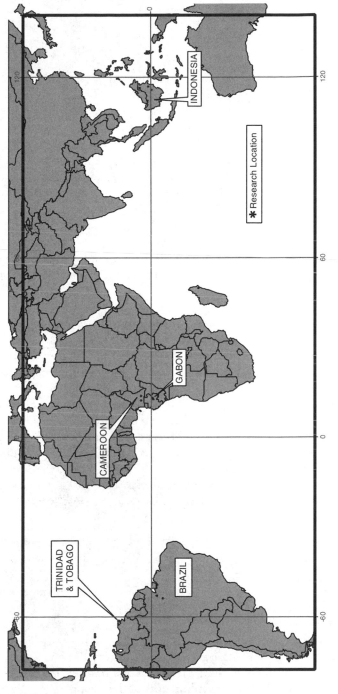

Figure 1. Map of the World Showing the Research Locations Covered in This Book

world map on which we have indicated all the research locations covered in this book.

Conceptual Framework

In this section, we define our key concepts and assumptions and discuss some of the issues that have recurred in our work. The discussion draws heavily on the conceptual framework with which we began much of the research reported in this book.[4]

One of the problems in coping adequately with human issues in sustainability has been the lack of a shared vocabulary and a common conceptual framework. Gale and Cordray (1994), for instance, identify nine different views on what should be sustained. The concept of sustainability itself is inherently value-laden (see Chapter 15). The following sustainability criteria reflect our values as professionals concerned with sound resource management practices and our assessment of values that are widely held among other stakeholders. We describe our shared vocabulary and put forth our perceptions of what needs to be addressed most fundamentally—always striving for a minimalist approach.

Definitions. *Sustainability.* With only slight alteration, we can use the definition of sustainable development accepted at the 1992 Earth Summit in Rio de Janeiro, Brazil, for SFM: "Sustainable forest management aims to meet the needs of the present without compromising the ability of future generations to meet their own needs." We followed Prabhu (1995) in considering that the satisfaction of two conditions would be sufficient to indicate sustainability in the context of forest ecosystem management: Ecosystem integrity is ensured or maintained, and people's well-being is maintained or enhanced. In these definitions, we need to specify what is meant by *well-being* and *needs* as well as who the relevant *people* are.

Well-Being or Needs. The fundamental needs that we considered contribute to people's well-being, now and in the foreseeable future, are the following:[5]

- *Security and sufficiency of access to resources now and in the future.* Ultimately, all human life depends on this element; therefore, it plays a crucial role in human–forest interactions.
- *Economic opportunity.* Forest activities should maintain or enhance people's livelihood opportunities.
- *Decisionmaking opportunity.* People have a right to participate meaningfully in decisions affecting their lives.
- *Heritage and identity.* People's rights to their values, behavior, networks, land use, and material goods should be respected, both for the present and as a necessary context for the enculturation of the young.

- *Justice.* Conflict and distribution of benefits, rights, responsibilities, and incentives should be resolved fairly (recognizing varying interpretations of "fair").[6]
- *Health and safety.* Employment in, residence in, or use of a forest should not endanger people's safety or health (physical or mental).

Although this list was compiled with forest-dwelling people in mind, we believe that they do not have very different needs from those of other human beings.

People. We recognize the ultimate interdependence of all people in our assumptions (see the next section): A forest dweller may be dependent on the forest for his daily fare; a settler in a nearby village may need forest-dependent environmental services; a consumer in the nation's capital may suffer if wood prices rise due to deforestation; a farmer in a distant country may depend on the forest for the rains that water her crops or for a stable climate.

For the purposes of effective forest management, the population of people who must be directly considered in daily management needs to be limited in some way. Formal forest managers, for instance, are not omnipotent and cannot be given the responsibility for ensuring the well-being of all humanity, nor can local community managers typically enforce their version of management on outsiders. Even within the forests they manage, these stakeholders are unlikely to be able to affect all the important variables that determine the sustainability of forest management.

It is therefore important first to define who has some interest or rights in forest management, that is, who has a "stake" in the forest (see Chapters 1–6). The most common word used in the SFM literature to designate these people (though inconsistent with the dictionary definition) is "stakeholder." Behan's (1988) discussion of a forest's "constituency" is also quite similar, defined as "the people who know about and care about" that forest.

Once the stakeholders have been defined, it is necessary to ascertain the varying rights and responsibilities among them. Recognition that forest dwellers have been disadvantaged in interactions with outsiders who come ostensibly to manage local forests has been widespread and increasing.[7] Reasons for resolving this human problem, in pursuit of SFM, are both ethical and pragmatic. Ethically, the "well-being" of these people, according to the earlier definition, has in many cases been adversely affected. Pragmatically, when people's well-being is thus affected, potential for conflict, forest and landscape degradation, marginalization, and cultural disintegration is increased. Ultimately, in the worst-case scenario, if *forest actors*—people who have resided in and managed an area for long periods of time and have pre-existing claims and responsibilities in that area, both for themselves and for their descendants—feel their situations are unacceptable, no forest may be left for any would-be claimants.

To identify these forest actors, we developed a simple technique for differentiating among stakeholders (Colfer and others 1999c). First, stakeholders are identified; then, the central forest actors are differentiated by their "scores" on seven dimensions: proximity, preexisting rights, dependency, poverty, indigenous knowledge, the integration of their culture with the forest, and power deficit vis-à-vis other stakeholders.

This method is a convenient mechanism for defining which stakeholders have the most pressing rights (with corresponding responsibilities) and thus constitute a sort of bottom line for stakeholder satisfaction; however, it is not a carte blanche for ignoring the rest. Sustainable forestry will ultimately, and probably inevitably, involve continuing negotiation and conflict management among stakeholders. Some progress is being made toward accomplishing this task in a constructive manner (see Resolve 1994, for examples from Ecuador, Bolivia, and Brazil; Ramirez 1999; Engel and others in press) (see also Concluding Remarks); however, much remains to be done. Some of the most difficult issues revolve around the extreme differences in power among stakeholders (for one perspective on this, with a summary of other views, see Edmunds and Wollenberg 1999; Wollenberg and others in press). One issue not adequately foreseen at this early stage of our research was the importance of intragroup differences, which are highlighted in Sections 1 and 2.

Fundamental Assumptions. Given the complexity of interactions between people and the forest, we acknowledge the probability that numerous unrecognized assumptions will need clarification as research continues. However, outlining two basic assumptions seems useful at this stage:

- *The landscape, where we are evaluating sustainability, is intended to remain largely natural forest in the foreseeable future.* The *natural forest* as discussed here can include logged forest as well as areas in various stages of regrowth (from spontaneous, natural, or planned human causes), or small areas that have been cleared. This assumption derives from a global perception that protecting some forested areas is in the best interests of humans. If, in particular areas, people do not want to protect the forest—as long as the global perception is that forests need protection—we must devise mechanisms whereby sufficient forest benefits accrue to those who live there. Trying to force forest protection has generally been shown to be ineffective (and/ or prohibitively expensive). The principles and criteria presented later (see *The Generic C&I Template*) reflect our view that forest protection must also be perceived by local people to be in their best interests.
- *Sustainable natural forest management locally will contribute to sustainable natural forest management nationally and regionally.* Nations and regions are made up of smaller parts that, by definition, include local forests. Although sustainable, local natural forest management is possible without national and regional SFM, the reverse is generally impossible (Lele 1993).

As noted in the opening paragraphs, we initially considered human systems to be complex adaptive systems, intimately connected with each other and with biological systems, in a self-organizing process of coadaptation. We drew this conclusion from the huge body of anthropological literature that showed the changing and interdependent nature of human systems. A cornerstone of cultural anthropology is the holistic nature of culture, and adaptation has been key to human ecological theories.

We decided to re-examine the idea that human systems are complex and adaptive as a hypothesis to be tested, in light of such features as networks of interconnected nodes, self-organization and emergence, self-organized criticality, dynamism between order and chaos, increasing returns, prediction, and feedback (see Waldrop 1992 for a readable exposition of complexity theory). We have explicitly rejected, from the beginning, the ideas that culture change is problematic and that cultural stability is not. Stability and change are aspects of cultural systems that vary in space and time. We have sought to better understand how such changes occur and how they are linked.

Conceptual Issues. Two conceptual issues colored our initial thoughts on principles and criteria for SFM: the nature of the interactions between people and forests, and the role of diversity of human systems in the sustainability of human life on the planet. Although these issues do not form the organizational framework for this book, we continue to consider them important issues related to people and forest management.

Role of People in Relation to the Forest. Most fundamentally, we viewed local people as part of the forest, in recognition of humanity's biological basis and their place in forest ecosystems. People—particularly those we have defined as forest actors—have a relationship of mutual dependence with the forest; they both contribute to and benefit from the forest. In this sense, forest actors constitute a resource, such as biophysical forest resources, available for the benefit of people (themselves and others) and of forests. This interaction between people and their environment means that people living in the forest both depend on it and act on it (Vayda and others 1980; Vayda 1983). Over the past decade, the documentation of long-standing, two-way interaction between human systems and forest ecosystems has been increasing.[8]

Debates about the nature of the human–forest relationship are ongoing. The role of poverty and wealth in affecting people's relationships to the forest is one example. That poor people sometimes constitute a threat to SFM is widely believed, and may be true [though Banuri and Marglin (1993) and Dove (1993) skillfully argue to the contrary]. The degree to which the poor can contribute to SFM has only recently begun to be widely acknowledged (Clay 1988; Posey 1992; Savyasaachi 1993; Colfer and others 1997a), though evidence for this has been around for much longer (for example, Conklin 1957).

One such potential contribution is knowledge. Banuri and Marglin (1993) argue that many indigenous systems of knowledge are available to us based on indigenous people's experience living with and learning from the environment. Those systems, if recognized and allowed to flourish, would have potentially more benign, nurturing implications for the ecosystem than the dominant system of scientific knowledge does. We suspect that a synthesis of kinds of knowledge—indigenous and otherwise—is more likely needed. But whichever view is true, a growing body of evidence suggests that attention to the voices and perceptions of forest actors may be in both humanity's and the forests' best interests.

Maintenance of Cultural Diversity. Cultures and ecosystems represent storehouses of both complex systems not yet fully understood and creative potential that we have argued should be maintained and nurtured. The destruction, or homogenization, of these diverse systems may seriously reduce the human capacity for sustaining itself.

Diversity in itself is of value for reducing risk, expanding the breadth of human potential, and increasing human knowledge and understanding. But human cultural diversity also represents differing solutions to survival in differing contexts (WRI/IUCN/UNEP 1992; Colfer 2000b); it serves as a dynamic global heritage from which future as well as current generations can benefit. Just as we do not now know which plant species may contain the properties needed to overcome an existing or future disease, neither do we know what human cultural characteristics (knowledge, values, social organization) may be needed in the future to sustain the human species. Enhancing the capability of various cultures to flourish, changing in directions selected and monitored by their adherents, constitutes a kind of "insurance policy" for the human species (as Barbier and others 1994 suggest with regard to biodiversity; see also Smith 1994). The availability of multiple cultures on Earth means that the failure or loss of any one is less likely to threaten the viability of the species.

The Generic C&I Template

This section is an outline of the kinds of social issues considered important for SFM by the CIFOR test teams that have visited forests in numerous countries over the past six years. The C&I that came out of the tests discussed in the previous section (*The Past*) formed an initial element of the research reported here. Because of their central role, we comment at some length on the meanings of the social principles and criteria listed in the *CIFOR Generic C&I Template* (Annex 1).

These hierarchically organized concepts are widely used in the literature on SFM, certification, and ecolabeling of timber (ITTO 1992b; Rainforest Alliance 1993; FSC 1994a; Heuveldop 1994; Soil Association 1994). We have followed the *Oxford Dictionary of Current English* (1987) and defined a *principle* as "a fundamental truth or law as the basis of reasoning or action." Principles, then, are stated as imperatives. We also use the dictionary definition of *criterion:* "a principle or standard that a thing is judged by." The FAO (1995a) defines *criterion* with a focus on forest management, consistent with our usage: "identified elements of sustainability against which forest management can be assessed." *Criteria* are phrased as conditions that must be met for a forest to be judged as "sustainably managed."

Three Principles and Nine Criteria. We identified three social principles as fundamental to SFM:[9]

- *Principle 3:* Forest management maintains or enhances fair intergenerational access to resources and economic benefits.
- *Principle 4:* Concerned stakeholders have acknowledged rights and means to manage forests cooperatively and equitably.
- *Principle 5:* The health of forest actors, cultures, and the forest is acceptable to all stakeholders.

These principles recognize in *forest resources* the importance of the physical and economic basis of human life as well as the cognitive, normative, and symbolic elements. Social scientists have debated for decades the priority of one or the other of these two aspects of the human condition (Harris 1968 is a somewhat dated but comprehensive review of this literature from a "techno-environmental" perspective). The view here is that both "hard" and "soft" elements are important for HWB and thus for the sustainability of forests.

Additionally, the proposed principles, criteria, and indicators are built on the assumptions listed earlier and must be taken as a whole. The criteria are interdependent such that, for instance, forest actors' access to resources must be balanced by appropriate mechanisms for monitoring and control. Participation in forest management is likely to be a parody if forest actors do not have secure access to the resources in question.

Principle 3. This principle addresses the issue of maintenance and fair apportionment of goods and services among stakeholders. If adhered to, it guarantees forest actors' security and sufficiency of access to resources over time; enhances their access to health, safety, cultural integrity, and other elements of HWB; and provides a power base for dealing with other stakeholders. Our site visits (and the literature) provide ample evidence that many forest actors—people with the greatest opportunity and potential to degrade and/or sustainably manage the forest—have not been fairly treated with regard to access to forest resources.

Other stakeholders also have legitimate claims that must be negotiated. This principle recognizes the claims of other stakeholders—such as government, private industry, and environmentalists—to resource access that they consider fair as well. The existence of multiple stakeholders with legitimate and varying claims obviously implies a process of communication, negotiation, and conflict resolution for forests to be sustainably managed (our Principle 4 addresses this issue).

Principle 3 (and its related C&I) is based on two pragmatic suppositions: that people are more likely to manifest stewardship toward forests from which they derive benefit, and that people tend to be more willing to sacrifice immediate gain from activities that may result in forest degradation when they are certain their children will benefit (see Palmer 1993 on Maine fisherfolk). The research reported in Section 3 was intended to test aspects of the causal links implied here.

An ethical consideration, based on justice, reinforces the importance of this principle. Although the claims of forest actors are not absolute,[10] justice demands that they should have some priority over the claims of other stakeholders.

In this discussion, we avoid specifying any particular kind of tenure system[11] because various systems could fulfil the central requirement of this principle (that people feel secure and comfortable that they and their children can continue to use the resources that have been available to them and in which they have a personal investment). We explicitly make no assumption that the claims of the state necessarily supersede those of local communities. Instead, we argue that conflicting claims will have to be clarified by a process of negotiation and conflict resolution.

The concern with economic benefits deriving from forest use evolves from our perception that inadequacy of resources can force people to degrade forest resources. Perceptions of unfair distribution of benefits can stimulate purposeful, retaliatory degradation of forest resources as well as other undesirable conflict. From a more positive perspective, people who have adequate access to resources are likely to be able to fulfil their other needs in accordance with their wishes, thus enhancing their well-being in terms of health, education, and other desired goods and services. Again, an ethical element pertains to justice among stakeholders.

Principle 4. This principle supports the rights of those concerned about and making use of the forest to be actively involved in forest management (see Behan 1988). It is important for several reasons. In many areas, forest actors particularly have had few opportunities to be heard or to integrate their views into formal forest management. Having a legitimized voice provides them with a mechanism for

- enunciating traditional rights and responsibilities and existing systems of forest management;
- protecting the rights identified;
- gaining access to a share in the benefits of forest exploitation;
- integrating their own knowledge, experience, and preferences into overall forest management, thus reducing marginalization (van Haaften 1995); and
- protecting their children's futures by all these means.

Such acknowledged rights also are important for other stakeholders. In the United States, for instance, environmentalists from New York City on the East Coast may have strong opinions and attachment to the Olympic National Forest in the northwestern state of Washington, thousands of kilometers away; similarly, Jakarta-based environmentalists have strong views on forest management in distant Borneo. The respective forestry agencies obviously have pertinent input regarding forest management. National citizens may have legitimate concerns about how their taxes are being spent and how forest revenues are being collected. Without the acknowledgement of such varying rights, no widely applicable mechanism exists by which the legitimate forest uses of various stakeholders can be integrated into SFM.

The importance of cultural systems for people's well-being, combined with the nearly infinite diversity of such systems in time and space, makes cooperation a crucial part of SFM. To be able to address stakeholders' concerns, many kinds of forest managers must know each other's concerns. The absence of such feedback to formal forest managers has been most obvious in the case of forest actors. Without the active participation[12] of forest actors in forest management, no viable mechanism has been identified for communicating the relevant aspects of their cultures to other stakeholders (and, to a lesser extent, vice versa).

One of the most important functions of participation is in providing a means for forest-based people to control the speed and direction of changes in their lifestyles. Supporting their rights in forest management can help people protect their existing ways of life (by enhancing cultural diversity and protecting cultural and natural resource integrity), insofar as they want to, and alter these lifestyles in ways they consider desirable (see Oksa 1993). Real participation also can reduce such adverse psychological consequences as stress, marginalization, and related physical health problems (van Haaften 1995). Active stakeholder participation in forest management provides a mechanism for dealing with cultural diversity and with the continually changing interface between people and forests.

The call for active efforts to understand and assimilate differing models in the management of a particular forest is built on the increasing recognition that forest actors often have natural resource management systems that are—or have been—viable. The sense that conventional science can learn from indigenous systems is growing. Proactive attempts to integrate indigenous systems with more conventional management models also may be helpful in minimizing conflicts and lead to better overall management.

The other side of this coin pertains to the well-being of forest actors. Insofar as forests are managed cooperatively with other stakeholders, meshing management systems in mutually beneficial ways, the activities of stakeholders who are not forest dwellers will be less disruptive to forest actors and their existing systems.

Without the support of stakeholders, efforts to control access to resources are unlikely to succeed. Forestry officials in charge of forest protection may not support existing mechanisms for controlling access (for example, by allocating forest concessions based on cronyism or failing to enforce forestry regulations). Forest actors may continue to harvest forbidden species or harvest in protected areas, feeling that their own rights have been usurped. SFM requires that these kinds of problems be resolved in such a way that stakeholders support existing mechanisms for control or help develop new, more viable ones.

Principle 5. We have not examined this principle in a systematic way (beyond its initial selection as a principle) simply because of lack of personnel and funding. But we do argue that maintaining the flow of benefits from resources requires that forest health be maintained. We see a strong interdependence among the well-being of forest actors, their cultures, and the forest. Because people depend on the forest, the forest's health is important to them at some level. The health of the forest, in turn, depends on HWB, because poor and unhealthy human beings (or too many human beings) may need to ravage the forest to survive.

Similarly, human culture affects human action, which can enhance or degrade forest health via such mechanisms as sustainable management systems or useful indigenous knowledge on one hand, or exploitative attitudes and practices on the other. Forest actors, who by definition have a strong forest–culture link, long-term rights in the area, and considerable knowledge of and dependence on the forest, are likely to have important elements in their forest management systems that sustain those systems.[13] But changing circumstances (such as access to markets, opportunities for medical care and education, desire for consumer goods, in-migration, gender roles, and technology) can have dramatic effects on cultures. For this reason, the degree to which and the conditions under which forest actors practice SFM merit additional investigation.

Cultures also affect HWB in other ways;[14] thus, "cultural health" needs monitoring in its own right. Culture, as a dynamic mode of adaptation, provides human beings with (malleable) patterns for communication, subsistence, division of labor, inheritance patterns, enculturation of the young, old age security, and values—all critical to HWB. Indeed, even the meaning of *health of people and forests* is defined culturally. In contributing to HWB, these functions contribute to SFM.[15]

Hypotheses or Themes

In this section, we document the evolution of our research process, which did not follow the same order as the chapters of this book. As noted earlier, we began the research reported here with the idea of tracing the causal links between the relevant C&I and SFM in a somewhat reductionist fashion. Indeed, our initial idea was to test how Principle 3 (Section 3 of this book) and Principle 4 (Section 4 of this book) related to SFM. We used techniques determined during our methods tests to reflect aspects of these principles and then compared the results of those studies across sites. Using long-term, in-depth knowledge of their areas, researchers evaluated the appropriateness of the results obtained from CIFOR's quick assessment methods, which were used in the comparisons. The analyses that most faithfully adhere to this plan are reported in Chapters 11 and 14; we compared results across several sites and tested for differences related to the two principles according to forest quality (the proxy for SFM). The nature of our results supports our view that new and different research strategies are required to reflect the reality of complex, adaptive systems.

We also were interested in testing the importance of two other issues: gender and diversity, and a conservation ethic. Gender and diversity issues (Section 1 of this book) emerged both in the identification of relevant forest actors (or stakeholders) and within the context of all three social principles (Principles 3–5; see Annex 1). In the course of our methods tests, we determined that access to women was both important and difficult in the attempt to assess HWB (see Colfer and others 1997c). Their significance for HWB—by numbers alone, if nothing else!—is obvious. Similar problems were identified with other marginalized groups. We also were convinced of the importance of such people as actors, with existing roles, and their potential contributions in improving both forest management and HWB. Even though Chapter 1 expands on the difficulties of gaining access to marginalized groups, including gender-based groups, Chapters 2 and 3 were initially focused on access to resources. The shift in emphasis reflects the dynamic and systemic nature of the issues we examined as well as the improbability of establishing simple, direct, cross-cultural causal links.

The question of a conservation ethic (Section 2 of this book) has been widely discussed in the West, and environmentalists in particular are interested in its role in SFM. Parties to the debate from various disciplines disagree not only about the degree to which forest dwellers may or may not have a conservation ethic but also about the role of a conservation ethic in enhancing SFM. Our research was designed primarily to test whether a conservation ethic could be identified and to what degree it was correlated with SFM, again using current forest quality as the proxy. As with our other efforts to make systematic comparisons across sites differentiated by forest quality, the results were interesting but not conclusive.

Section 5 of this book covers two more general issues. First, in Chapter 15, we compare SFM in the developed world with the developing-world contexts that dominate this book. Besides providing another view of sustainability, we list the potential differences between developed and developing countries and highlight the ways that context can alter the potential measures of HWB.

The second issue, discussed in Chapter 16, more carefully pertains to our proxy for sustainability. Recognizing that using *forest rich* as a proxy for sustainably managed forests was "iffy," we studied differing management systems in one comparatively forest-rich environment over time, using geographical information systems (GIS) and remote-sensing tools. This approach, although labor intensive, provides a good sense of biophysical sustainability within different human systems in the same area.

Tools and Approaches

Some of our methodological tools were used in several sites (and thus in several chapters). Here we provide an overview of the seven most common methods.

Galileo and CATPAC[16]

The Galileo program (Terra Research and Computing), used in Chapters 5 and 6, is a multidimensional scaling method. (See the introduction to Section 2 for a discussion of our rationale for using this method.) We conducted conventional Galileo studies[17] (Woelfel and Fink 1980) as a possible means to assess three conditions identified in previous research as relevant in establishing people's roles in SFM: the presence or absence of a conservation ethic, a feeling of closeness to the forest, and an intimate link between local culture and the forest.

The Galileo study begins with the identification of locally appropriate concepts pertaining to the domain of study (in this case, forest–people interactions). This method makes no assumptions about congruence between the researchers' and local people's definitions of these concepts. Such locally relevant concepts can be determined through experience or obtained in an unfamiliar area by content analysis of open-ended interviews on the topic of interest.[18] These concepts are then paired in a questionnaire format in the local language.

A *criterion pair* (often the distance between "black" and "white," as seen in the respondents' own minds) is selected as a measuring stick for comparing each of the pairs of study concepts. Literate villagers can fill in the forms themselves; others are interviewed and asked each measurement. The process typically takes about 20–30 minutes for 20 concepts. Data were entered into the Galileo program and analyzed in Bogor, Indonesia, with assistance from Joseph Woelfel (the principal developer of the software) (Woelfel and Fink 1980; Foldy and Woelfel 1990) and Agus Salim.

The most fundamental output of a Galileo is a *means matrix,* in which the mean response (from all the respondents) is computed for every pair of concepts. Put another way, the means matrix reflects the mean distances perceived by the community in question between every concept and every other concept. The program provides extensive descriptive and inferential statistics, including standard deviations; standard errors; indices of skewness and kurtosis; sample size; maximum and minimum values; and other, more global statistics. We have been satisfied with fairly simple analyses.

The results of this procedure made it possible to represent the respondents' attitudes and beliefs in a three-dimensional graph or space. This space provides a precise and holistic picture of the respondents' beliefs and attitudes about forests. Locally defined concepts that are closely related are close together in this space, whereas those that are unrelated are far apart. If people think the forest is good, for example, the concept **forest** will lie close to **good** in the Galileo space.[19] One advantage of this model is that dozens or even hundreds of attitudes and beliefs can be displayed simultaneously in a single picture, which makes it possible to see the interrelationships among the beliefs and attitudes. Seeing the "big picture" is important because changing one attitude or belief often changes others. If forest managers are aware of such indirect consequences of change, then their methods of forest management (as it relates to human involvement in the forest) may become more sensitive.

In recent years, Terra Research and Computing has been developing several relevant new programs built on the idea of neural networks (for example, CATPAC and Oresme; see Chapter 2). They represent a kind of artificial intelligence that may allow us to obtain some of the same information we can generate with the Galileo program more simply. CATPAC can analyze text—in much the same way that open-ended interviews could be analyzed—to identify frequencies and clustering of concepts that recur within that text. The important difference is that CATPAC can do it much more quickly. These programs all can be run on an ordinary PC or laptop computer.

In our use of CATPAC, we asked representative individuals about their views on human–forest interactions, trying not to say anything after the initial, very broad question was asked. We taped their responses and then typed them into the computer. The CATPAC program analyzed the content of the responses in seconds. The results are clusters of concepts that tend to occur together in the respondents' speech, reflecting the cognitive patterns of the interviewees. We tried to interview about ten individuals from any given group to be able to make an accurate statement about their views.

The final component of this group of software—the Automatic Strategy Generator (ASG) and its predecessor, the Automatic Message Generator (AMG)—is of a more general interest to researchers, beyond assessment per se. They identify which concepts should be emphasized in an effort such as planned change (for example, encouraging a conservation ethic, or encouraging people to consider forests in a more positive light). These concepts can then be used in extension or "advertising" to affect people's views of the forest. Insofar as the interviewees' views reflect their behavior (Woelfel and Danes 1980; Cary and Holmes 1982; Woelfel and others 1988a, 1988b; Barnett and Woelfel 1998), such changes could have important impacts on forests.

Biplot Analysis

Biplot analysis (Gabriel 1971; Jolliffe 1986) has been used to compare quantitative results across sites (see Chapters 11 and 14). In this kind of analysis, each variable (such as "benefits being shared" or "rights to manage") is represented by an arrow, whereas each point represents a stakeholder. The length of an arrow indicates the amount of variation within that variable. If, in an example pertaining to "rights and means to manage forests," the arrow for "defining borders" is very long compared with the others, then imbalance among the stakeholders is greater for this right. A short arrow implies that the right is fairly equally shared. Inequalities in sharing of rights can thus be determined quickly from this type of graph.

The position of a point shows the level of rights or benefits (depending on the study) for a particular stakeholder, compared with others. If the point lies in the same direction as the arrow, then the stakeholder has more rights or benefits than the average; if the point is in the direction opposite the arrow, then the stakeholder has fewer rights or benefits than average.

The relationship between two variables (say, "defining boundaries" and "assessing fines") can be determined from the angle formed by the two arrows. If the angle is less than 90° (an acute angle), then the two variables are positively correlated. In this hypothetical example, a greater right to define boundaries would also imply a greater right to assess fines. If the angle is greater than 90°, then the two variables are negatively correlated, that is, a group with greater rights to define boundaries would have fewer rights to assess fines.

Gender and Diversity Analysis

Although our emphasis in this book has been on gender (Section 1), many of the same methods that are used to understand gender apply to diversity (ethnicity, class, caste, economic level, and so forth). Chapter 1 provides the most comprehensive discussion of these methods, but we have made extensive use of several tools, which include

- *participant observation,* a long-term, qualitative, anthropological method that involves setting aside personal assumptions, insofar as possible, and using one's self as a methodological tool to understand the workings of local human systems;
- *rapid rural appraisal techniques,* which are quick methods that typically involve female (in the gender case) team members, attention to both genders in data collection, and assessments of contexts in which both genders function (some of these techniques are described in more detail later); and
- *process methods,* the use of focus groups and other interactive methods designed to draw out community members that would not otherwise be

heard (for example, structuring separate meetings so that marginalized groups have a chance to express their views, helping the illiterate or those with poor national language skills to contribute in creative ways, and holding separate meetings for men and women).

Pebble-Sorting Methods

Sharing of Benefits. The pebble-sorting methods proved to be among the simplest to compare across sites (see Chapter 11). For the studies focused on sharing of benefits, stakeholders and benefits were initially identified by Colfer for Bornean sites (based on long-term ethnographic research experience there) and then adapted by the other researchers for their own sites. An attempt was made to keep the categories as comparable as possible, without misrepresenting local stakeholders or benefits. We selected a sample of 12–15 participants from each of the most important stakeholder groups in each area, trying to represent men and women relatively equally, and to attend to other locally important social differences (for example, age and ethnicity). We conducted the method with individuals and with fairly homogeneous groups, collecting relevant demographic data (age, gender distribution, ethnicity, and occupation) for subsequent analysis.

Necessary materials were revised for local conditions (that is, with locally relevant stakeholders and forest resources, in local languages). We limited the number of stakeholder groups to as few as possible (three to ten), with each researcher determining the minimum number that would allow us to maintain the accuracy and integrity of our analyses. Some researchers used a large matrix, for group use; others used plates representing stakeholders or resources in which participants distributed pebbles or seeds. In the Brazilian tests, we used plates with drawings picturing situations related to each stakeholder. Although we had a preestablished set of benefits and stakeholders for each test, we did include additional plates for other stakeholders eventually suggested by individual interviewees. The researchers transferred the quantity of pebbles in the respective plates to the appropriate matrix cells.

Wherever possible, we used the local language in our interviews. The main benefits from the forest, including subsistence products, were listed. The relevant stakeholders or user groups among whom respondents perceived forest benefits to be divided were also identified. Each participant or group of participants allocated 100 pebbles among the stakeholders. We asked the participants to consider the forests in their area and indicate their perceptions of the division of the listed forest benefits.

Intergenerational Access to Resources. This method is very similar to the previous one in terms of sample selection and process (Chapters 7, 11, and 13). In most cases, we asked each participant or group of participants to allocate 100 pebbles among the generations, with each row equaling 100, explaining to participants that we wanted to understand how local access to resources is changing over time and what they think about the future. We then asked respondents to imagine all the forest resources over time (from the time of one's grandparents through the present to the time of one's grandchildren) and to allocate those resources proportionally among the generations. (Specific adaptations needed for different countries are described in the chapter discussions.)

For a given group of 12–15 group or individual interviews, we computed a mean for each generation. We have been satisfied, based on longer-term familiarity with the areas, that these results provide a succinct and relatively accurate representation of that group's perceptions of changes in access to resources over time.

Rights to Manage Forests. This pebble-sorting method was designed to gain access to local stakeholders' perceptions about the division of management rights and responsibilities among significant stakeholders (Chapters 2, 13, and 14). We initially selected six management functions to reflect overall management:

- defining and protecting boundaries,
- developing and applying rules and regulations,
- monitoring compliance,
- resolving conflict,
- providing leadership or organization, and
- assessing fines and sanctions.

Again, we selected samples of 12–15 respondents from each important stakeholder, user group, or social category in each research locale. They included at least men and women; different ethnic groups; and different occupations. In Cameroon, it was important to differentiate by age as well. All groups selected had a clear relationship with forest management. We conducted interviews in fairly homogenous groups (5–15 people) and individually, collecting relevant demographic data about each respondent (age, sex, ethnic group, and so forth) for use in subsequent analysis.

We used 100 pebbles (or beans, buttons, corn kernels, or nuts, depending on local availability and preference) and a matrix with large enough cells so that people could allocate the pebbles along the rows of the matrix. The rows listed the functions of forest management (earlier), and the columns listed the most important stakeholders.[20] Smaller, paper copies of the matrices were used for recording the data.

We explained to each respondent or group that we were interested in understanding who they considered responsible for managing the forest in the area. We explained that the rows represent different rights and responsibilities in forest management and asked them to allocate the 100 pebbles among the stakeholders listed across the top (once for each row). The results were then analyzed using cluster analysis.

The initial selection of forest management functions may have had a gender bias in some areas, relating predominantly to the male domain, which may in turn have affected the identification of relevant stakeholders. The Brazil team felt that, despite our efforts at gender equity, greater emphasis on management functions that include the participation of women would have added to the explanatory power of the method.

In areas characterized by commercial extraction of nontimber forest products (such as Brazil nut, *babassu*, and *açai*), for instance, women participate in these activities, which could be explicitly incorporated into the method. Additionally, management functions linked to the domestic domain (or to the reproduction of the household) might usefully be included. In Pará, a specific example is related to protection of water sources (for drinking and for washing clothes and kitchen utensils). In the Transamazon forest-poor site (Transiriri), the conversion to pasture resulted in the local stream drying up, a matter of continuing complaint by local women. The same was not observed in the forest-rich area. The role of women in protecting these water sources would be a valuable issue to incorporate for future research. Management functions for the reproduction of the household varied considerably across sites. Female-dominated management functions include firewood collection or making charcoal, obtaining manure (from dead palm trees), fertilizing vegetable gardens, and collecting medicinal plants.

The Iterative Continuum Method (ICM)

This experimental, qualitative method (results reported in Chapters 8 and 12) was designed to provide a framework within which to organize thoughts about and emerging understanding of site conditions, over the course of necessarily brief fieldwork. We devised forms with a continuum—a horizontal line that represented different values on our topic of interest—at the top, and space below for writing. Researchers filled in one form on each day of the fieldwork, assessing where the community (or subgroups within the community) should be placed, based on the researcher's understanding, as of that day.[21] Placement was accompanied by an arrow to show the researcher's perception of the direction of change. The pages were then filled with evidence to support the conclusions marked on the continuum. The process of filling in these forms was iterative, whereby a researcher's growing understanding is reflected in changes in daily assessments.

To gain the kind of understanding needed to estimate the placement of a community or subgroup along the continuum, we spent days with representatives of the various stakeholders and subgroups—discussing, observing, inquiring—using elements from Vayda's contextual analysis approach (Vayda and others 1980; Vayda 1983). This approach strives to trace the links among significant human actions (such as felling timber, monitoring concessionaires, or contributing ideas about forest management to conservation project personnel) in the research setting. The emphasis in this research was on tracing causal links[22] to demonstrate the relevance (or irrelevance) of particular kinds of human actions to SFM.

Researchers supported their initial assessments with cases and evidence. New cases and evidence that accounted for the changes in the researchers' perceptions were documented. By the end of the fieldwork, the state and direction of change along the continua for the locations studied were thereby fine-tuned, and the factors affecting forest management better understood.

All researchers felt the need for some defined points along the continuum (from secure to insecure access to resources [Chapter 8] or from significant to insignificant levels of participation [Chapter 12]) to help "anchor" observations from day to day. Colfer constructed a series of steps (for example, from "very insecure" to "very secure" tenure) that have been systematized by Salim and others (1999).

This method helped qualitative researchers focus on the issue, record what was learned, and think about the implications thereof. It also resulted in a wealth of case material relating to the topic of interest.

Participatory Card-Sorting Method

One prerequisite for effective participation is regular communication. This reasoning led us to develop the participatory card-sorting method (see Chapters 12 and 13). We used a form with a specified number of locally relevant stakeholders. In Danau Sentarum Wildlife Reserve, for instance, stakeholders included the local community, other communities, the government, the timber companies, the Conservation Project, and traders. Each stakeholder was listed on a different colored card, ideally with locally meaningful colors representing each stakeholder. The form also posed four questions, each a concrete example of a component of forest management. The questions in Danau Sentarum Wildlife Reserve pertained to seeking information about fish, looking for rattan, looking for valuable wood, and problems between timber concessionaires and other stakeholders. These questions were designed to reflect local forest management by identifying who had knowledge, who controlled and made use of resources, and who was involved in conflict resolution. We sampled 12–15 respondents in each community, evenly divided (wherever possible) by gender and representing whatever diversity we found. Respondents could be individuals or groups.

People were asked first to rank the stakeholders by importance (when necessary, this term was further explained as involving "rights" or "status" in forest management) for each of these four topics. It was necessary to rank all stakeholders (for analysis purposes), even if their role was quite unimportant. The people were then asked to allocate 100 points among these stakeholders, depending on frequency of interaction, for each topic. Zero was an acceptable value for frequency of interaction.

The results are a simple average of ranking by importance and by frequency of interaction—both important issues in assessing people's involvement in managing forests. Disaggregating the responses by gender, occupation, location, or some other dimension is straightforward.

Assessment of Findings

As described at the outset of this introduction, our broad purpose has been to contribute to creating conditions that would allow people who live in and around forests to prosper while protecting their environment. As part of that process, this book was initially intended to clarify the causal links between HWB and SFM in a fairly reductionist mode. We hoped that by examining some widely accepted aspects of HWB in forests that varied in their apparent sustainability, we would be able to say with some certainty, "Yes, security of intergenerational access to resources is a critical factor in maintaining forest quality" or "No, security of intergenerational access to resources is unrelated to forest quality." We hoped to establish cause. That goal has been elusive—and for very good reasons, as we argue later. Instead, we found that human cultural patterns (behavioral and cognitive) relating to natural resources tend to vary by area and cultural region rather than by forest quality.

It might be tempting to focus on specific shortcomings of the research. From this perspective our first and simplest mistake was using good forest quality as our best available proxy for sustainable management. Our initial fears about this proxy (along with the continued absence of any other straightforward, inexpensive proxy) proved correct. Excellent quality forest can exist, in the short run, in a context of completely unsustainable management (as in many areas of southern and eastern Cameroon and in central Borneo). In fact, currently sustainable practices also can characterize degraded forest areas, not to mention the multitudinous concepts and differences of opinion about the meaning of "degraded." Current forest quality does not suffice as a proxy for good forest management.

Another straightforward problem involved finding sites that differed along the HWB continua. None of our sites in the developing world (Indonesia, Cameroon, Trinidad, Gabon, or Brazil) was characterized, for instance, by secure intergenerational access to resources. In forest-rich West Kalimantan, local people felt reasonably secure,[23] but we had significant reasons to suspect that their rights were in danger. Indeed, in all our sites, the security of intergenerational access to resources was clearly in jeopardy. Even at the U.S. site in Boise, Idaho, where land tenure is comparatively clearly defined, people's timber-related jobs were on the line—forests were being closed to logging, and the mills were downsizing. We could not get the spread of HWB values (principles and criteria) that we had hoped for in our test sites.

One can argue, as we have from time to time (Prabhu 1995; Colfer and others 1999c; see also Chapter 15), that the concept of SFM is inherently value-laden and subject to continual redefinition. We defined SFM as including the maintenance or improvement of both ecological integrity and HWB. We also defined critical aspects of HWB in the *CIFOR Generic C&I Template* (CIFOR 1999; Annex 1). However, the concepts of HWB and of SFM are broad, and specifications are necessarily subject to local interpretation and variation if they are to be useful and widely applicable concepts.

But we would argue that any attempt to make global cross-cultural and interregional comparisons of this sort will be plagued by just such problems. One problem that we managed to avoid but is common in comparative cross-cultural research is the temptation to warp the data to fit a predefined conceptual or analytic framework. We went great distances to ensure that questions and definitions made sense in the local context, to ensure that translations were as comparable as possible, to throw out inappropriate methods, and to add new issues as identified in new contexts. We are pleased that some common issues remained that we could compare across sites (Chapters 6, 11, and 14) but are not surprised that some sites and some issues had to be discarded due to incomparability.

Through a process based on the professional judgements of many researchers, we sifted through the incredible variety presented by the real-world circumstances we were able to examine and extracted a few nuggets (the C&I) that have been widely accepted as important for HWB everywhere. But in trying to make the causal links between SFM and the respective manifestations of HWB in different locales, we found that particular historical sequences of events were more informative about these causal relationships than one-to-one correlations between aspects of HWB and SFM (see also Vayda 1996).

We are convinced that the problem lies not with the research process, which was unusually well funded, systematic, global, interdisciplinary, multicultural, and based on state-of-the-art conceptual frameworks and methods; the problem lies in the approach. We are looking at complex, adaptive systems that are dynamic and fluid. The search for straight-line, cause-and-effect links is simply a chimera, a holy grail that we must stop seeking.

HWB and SFM are, in our view, both bundles of complex and interrelated ideas and practices. They are not clear, monolithic concepts subject to tidy dissection as implied by the hierarchy of principles, criteria, indicators, and verifiers. C&I are very useful for ease of communication and as a conceptual and organizational device. Criteria are goal statements, linked to higher-order goal statements. Although a fair amount of commonality in top-order goal statements is to be expected, culture and context will necessarily dictate that all these goals will not be identical or, if identical, will reflect different values in different places.

CIFOR developed CIMAT (Criteria and Indicators Modification and Adaptation Tool), a software package designed to aid users in adapting C&I to particular contexts. The first version precluded the user from modifying the principles on the assumption that all would agree about their importance; however, feedback from users suggests that even the principles should be subject to modification and adaptation (Ravi Prabhu, unpublished data, February 2000).

Prabhu and Colfer (Prabhu and others in press) have believed for some time that a network was a more appropriate metaphor for the linkages between the various components of HWB and SFM. Testing sets of C&I in the field, we were struck by the varying prominence of one criterion or another in different contexts. Although eventually included in the U.S. C&I set (Woodley and others 2000), the generic template Indicator 3.1.1, "Ownership and use rights to resources (inter- and intragenerational) are clear and respect preexisting claims," was initially rejected—not because it wasn't deemed important but because the issue was felt to have been resolved. In Côte d'Ivoire, health took a much more prominent role in the C&I set selected than in other locales, partly because of the devastating impact of diseases such as malaria and AIDS and partly because of people's emotional stress levels related to dramatic competition for land; environmental degradation; spiraling population growth; and influxes of economic, climatic, and political refugees from neighboring countries (van Haaften 1995).

Instead of a hierarchy, Colfer began to imagine the C&I as hills in a landscape, which she likened to the lumps formed by a child's appendages under a blanket. In one locale, lump A, "a knee," would practically disappear (a criterion demoted to a verifier), and lump B, "an arm," would grow dramatically (an indicator rises to a principle). Lump C, "the head," (a principle) might remain the same. The "topography" of the "blanket" thus takes on a whole new configuration of shapes with every movement of the "child" or in each new locale while maintaining some inherent unity.[24]

So, although we continue to use the hierarchical metaphor of principles, criteria, and indicators because of its utility as a communication and organizational device and as an aid in the practical problem of assessment, we are skeptical of the degree to which this structure represents reality.[25] Because of changing field realities and changing human perceptions and values, we suspect that the kinds of globally mandated values represented in the C&I (and discussed in this book) will remain in continuous oscillation with a self-conscious learning process of location-specific testing and adaptation (see Concluding Remarks and Next Steps).

Throughout our research, we have been increasingly convinced that good forest management—that is attentive to human and ecological needs—will require iterative attention to the systems in which people and forests interact. It implies much more creative approaches to these issues, building on methods that can deal with evolving and interconnected systems, whether participant observation, system dynamics, network analysis, or participatory action research. A wide range of methods are available, waiting to be picked up and applied to the problems of forests and the people who inhabit them by those of us who are concerned about such issues. Our concept of "good forest management" is a far cry from conventional forest management, but we are convinced that without such an iterative process, both our forests and the cultural diversity that so many of us (including forest-dwelling people) value will perish.

Annex 1. Social Criteria and Indicators (C&I) from CIFOR's Generic Template

Principle 3: Forest management maintains or enhances fair intergenerational access to resources and economic benefits.

Criterion 3.1: Local management is effective in controlling maintenance of, and access to, the resource.

Indicator 3.1.1: Ownership and use rights to resources (inter- and intragenerational) are clear and respect preexisting claims.

Indicator 3.1.2: Rules and norms of resource use are monitored and successfully enforced.

Indicator 3.1.3: Means of conflict resolution function without violence.

Indicator 3.1.4: Access to forest resources is perceived locally to be fair.

Verifier 3.1.4.1: Access of small timber operators to timber concessions

Verifier 3.1.4.2: Access of nontimber users to nontimber forest products (NTFPs)

Indicator 3.1.5: Local people feel secure about access to resources.

Criterion 3.2: Forest actors have a reasonable share in the economic benefits derived from forest use.

Indicator 3.2.1: Mechanisms for sharing benefits are seen as fair by local communities.

Indicator 3.2.2: Opportunities exist for local and forest-dependent people to receive employment and training from forest companies.

Verifier 3.2.2.1: The number of local people employed in forest management (disaggregated, for example, by gender and ethnicity)

Indicator 3.2.3: Wages and other benefits conform to national and/or International Labor Organization (ILO) standards.

Indicator 3.2.4: Damages are compensated in a fair manner.

Verifier 3.2.4.1: Number of people affected by off-site impacts, without compensation

Indicator 3.2.5: The various forest products are used in an optimal and equitable way.

Criterion 3.3: People link their and their children's future with management of forest resources.

Indicator 3.3.1: People invest in their surroundings (that is, time, effort, and money).

Indicator 3.3.2: Out-migration levels are low.

Indicator 3.3.3: People recognize the need to balance number of people with natural resource use.

Indicator 3.3.4: Children are educated (formally and informally) about natural resource management.

Indicator 3.3.5: Destruction of natural resources by local communities is rare.

Indicator 3.3.6: People maintain spiritual or emotional links to the land.

Principle 4: Concerned stakeholders have acknowledged rights and means to manage forests cooperatively and equitably.

Criterion 4.1: Effective mechanisms exist for two-way communication related to forest management among stakeholders.

Indicator 4.1.1: More than 50% of timber company personnel and forestry officials speak one or more local languages, or more than 50% of local women speak the national language used by the timber company in local interactions.

Indicator 4.1.2: Local stakeholders meet with satisfactory frequency, representation of local diversity, and quality of interaction.

Indicator 4.1.3: Contributions made by all stakeholders are mutually respected and valued at a generally satisfactory level.

Criterion 4.2: Local stakeholders have detailed, reciprocal knowledge pertaining to forest resource use (including user groups and gender roles) as well as forest management plans prior to implementation.

Indicator 4.2.1: Plans/maps showing integration of uses by different stakeholders exist.

Indicator 4.2.2: Updated plans, baseline studies, and maps are widely available, outlining logging details such as cutting areas and road construction, and include temporal aspects.

Indicator 4.2.3: Baseline studies of local human systems are available and consulted.

Indicator 4.2.4: Management staff recognizes the legitimate interests and rights of other stakeholders.

Indicator 4.2.5: Management of NTFPs reflects the interests and rights of local stakeholders.

Criterion 4.3: Agreement exists on rights and responsibilities of relevant stake-holders.

Indicator 4.3.1: Level of conflict is acceptable to stakeholders.

Principle 5: The health of the forest actors, cultures, and the forest is acceptable to all stakeholders.

Criterion 5.1: There is a recognizable balance between human activities and environmental conditions.

Indicator 5.1.1: Environmental conditions affected by human uses are stable or improving.

Indicator 5.1.2: In-migration and/or natural population increases are in harmony with maintaining the forest.

Criterion 5.2: The relationship between forest management and human health is recognized.

Indicator 5.2.1: Forest managers cooperate with public health authorities regarding illnesses related to forest management.

Indicator 5.2.2: Nutritional status is adequate among local populations.

Indicator 5.2.3: Forest employers follow ILO work and safety regulations and take responsibility for the forest–related health risks of workers.

Criterion 5.3: The relationship between forest maintenance and human culture is acknowledged as important.

Indicator 5.3.1: Forest managers can explain links between relevant human cultures and the local forest.

Indicator 5.3.2: Forest management plans reflect care in handling human cultural issues.

Indicator 5.3.3: There is no significant increase in signs of cultural disinte-gration.

Annex 2. Descriptions of Comparative Research Sites (Chapters 6, 11, and 14)

One aspect of trying to understand the relationship between forest health and human well-being has been to identify and examine patterned differ-ences in forest sites on the basis of their quality. Our first step was to divide research sites into categories characterized as "forest rich" or "forest poor." This qualitative differentiation was suggested by Center for International Forestry Research (CIFOR) silviculturist (and project leader) Ravi Prabhu. For our purposes, *forest rich* refers to a landscape resembling a "sea of forest with islands of people" and *forest poor* refers to "a sea of people with islands of forest." We used input from various biophysical scientists and complementary studies in helping us to place our sites on this continuum.

Here, we simply divide the locations where this methods test was conducted[26] into two groups. The forest-rich sites are

- Long Paking, Bulungan, East Kalimantan, Indonesia;
- the Dja Reserve and Mbongo, Cameroon; and
- Trairão and São João, Pará, Brazil.

The forest-poor sites are

- Long Segar, Kutai, East Kalimantan, Indonesia;
- Mbalmayo, Cameroon; and
- Transiriri and Bom Jesus, Pará, Brazil.

Descriptions of the sites follow.

Forest Rich

Long Paking, Bulungan, East Kalimantan, Indonesia.[27] CIFOR's Bulungan Research Forest area, directly adjacent to this community, is considered by the World Resources Institute to be the most intact remaining forest in Southeast Asia (Figure A1). Varying from lowland dipterocarp to montane forest, the area is a biodiversity treasure. The area immediately surrounding Long Paking has been commercially logged and subject to swidden (more pejoratively called "slash-and-burn") cultivation for several years, but its remoteness and low population density have prevented serious environmental degradation.

Long Paking is a riverine community of Lundaye and Abay Dayak swidden cultivators, recently joined by the inhabitants of five Punan hamlets. The Punan are the hunter-gatherers of Kalimantan. The Indonesian government has unsuccessfully tried to "settle" such people in villages for decades. The Punan rarely remain in the village, spending much of their time upriver in their traditional areas. The Dayak are swidden cultivators who supplement their incomes by seeking forest products, both for subsistence (ferns, medicinal plants, fibers, and timber) and for sale (for example, *gaharu,* a rare and fragrant heartwood used for incense); by fishing and hunting; and by periodic wage labor, particularly with the nearby timber company. The village itself is surrounded by the varying stages of forest regrowth that characterize Kalimantan's swidden cultivation systems, supplying the community with the variety of domestic, semidomestic, and wild products that flourish in the different stages. As with almost all Kalimantan communities, land ownership follows traditional rules and is not officially recognized by the government. Limited logging activity has occurred in the area since the 1970s, but significant and accessible forest areas that are almost undisturbed remain. Transport

29

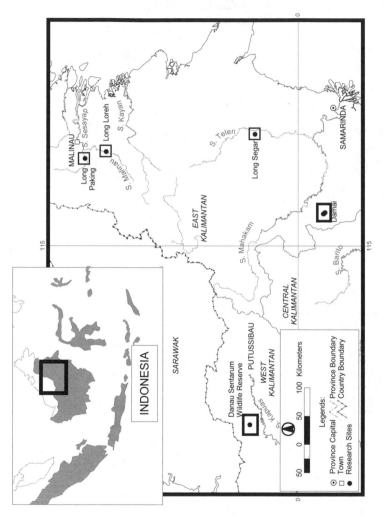

Figure A1. *Map of Indonesia, Focused on Our Research Locations*

is still primarily by river. In the mid-1990s, CIFOR was given access to the adjacent Bulungan Research Forest, where the Indonesian government encourages research, although existing stakeholders (local people, timber companies, mining companies, and plantation companies—most operating to the east of Long Paking) retain rights to continue their activities there.

Dja Reserve, Cameroon.[28] The Dja Reserve (Figure A2) is located in the northeastern corner of the Congo Basin and covers 526,000 hectares. The northern and western boundaries of the reserve can be reached by road from Yaoundé, but access remains difficult in much of the area. Several conservation organizations are active in the area, including Conservation et Utilisation Rationnelle des Ecosystèmes Forestiers d'Afrique Centrale (ECOFAC) in the west, north, and south of the reserve and the International Union for the Conservation of Nature (IUCN) and Soutient au Developpement Durable de la Region de Lomié (SDDL) in the east. ECOFAC and IUCN hope to develop participatory management plans for the reserve. Commercial hunting and logging are seen as the primary threats to the ecosystem, and the projects hope to develop economic alternatives for local people. SDDL operates in a region about 20 kilometers from the reserve boundaries; one of its important goals is the implementation of community forestry, particularly in the *communes* (administrative structures similar to counties) of Lomié and Messok. Another important actor in the region, Société du Littoral pour l'Exploitation et le Transport (SOLET), was the only logging company working in the area (88° N to 89° N) at the time of the test. The main species logged include *Entandrophragma cylindricum* (*sapelli*), *Triplochiton scleroxylon* (*ayous*), *Lovoa trichilioides* (*bibolo*), and *Pericopsis elata* (*asamela*). These species were primarily exported to France, but some of the *sapelli* was sold to PALISCO, another logging company, which had a sawmill at Mindourou to the north of the study zone.

The method reported here was tested in four communities: Messok/Mbaya, Baréko, Pohempoum, and Sembé.

- Messok/Mbaya includes camps of Baka pygmies, selected to reflect the insecurity of tenure that characterizes this group. The Baka inhabitants of Messok, for instance, have been required to move twice since Messok was formed by administrative decrees in 1994.
- Baréko's inhabitants consist of four lineages, which fall into two unequal categories (Grand Baréko and Petit Baréko), both of which were involved in the study reported here. Their relevance in this case derives from their involvement with SOLET, the logging company. One of SOLET's logging areas is completely within the community's forest territory, and another one is shared between Baréko and its neighbor, Messok. About 120 people are employed by the company, the majority from the local area, and the company provides significant contributions to the community.

31

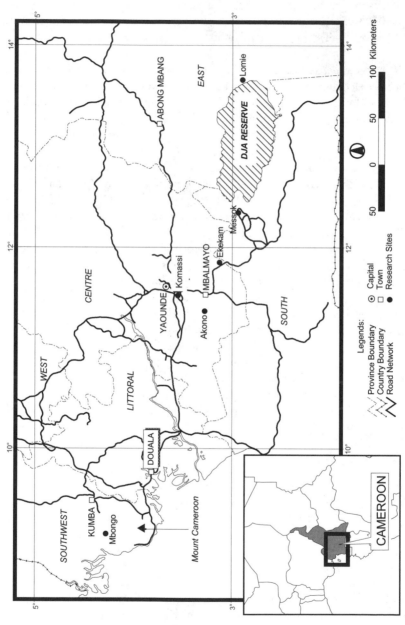

Figure A2. *Map of Cameroon, Focused on Our Research Locations*

- Pohempoum, one of the oldest population resettlement areas in the region, has an ethnically diverse population. Under the Germans, this village was the principal commercial center in the area; Europeans bought rubber and then cocoa and sold manufactured products. This area also is on the periphery of the reserve, which generates considerable concern from conservationists about the hunting proclivities (within the reserve) of the population. Situated close to Sembé, this population has not been concerned with logging and therefore provides an interesting comparison.
- Sembé has a population of newcomers. The first occupants were the Nzime from the clan *babil*, but none remain. The current population is divided among several different ethnic groups. The Kako, who come from the dividing line between the forest and the savanna to the north, are the majority. They were given rights to the forest by the Nzime around 1930. The other groups are the Njem, the Nzime from the clan *balamine*, and the Badjoué, who come from neighboring areas to the north, west, and south, respectively.

Mbongo, Mount Cameroon, Cameroon.[29] Mount Cameroon (Figure A2), located within comparatively easy driving distance from principal population centers, is the site of a major U.K. Department for International Development (DfID)–Cameroonian conservation effort. It is the highest mountain in West Africa (4,095 meters) and the only active volcano. This equatorial forest is Atlantic Biafran evergreen, rich in Caesalpiniaceae. Rainfall ranges from 2,000 to 10,000 millimeters/year (average 3,000 millimeters/year). The area is considered a conservation "hot spot" because of its many rare and endemic species.

Mbongo village, which has the most diverse forest in the Mount Cameroon Project/Limbe area, was the primary location for the study reported here. It is a Balondo village in the Bamusso Subdivision of the Ndian Division in the Southwest Province. It is bordered to the south by the Mokoko River Forest Reserve, to the north by the Atlantic Ocean, to the west by Bonjare village (with which it has close ties), and to the east by Dikome village. This site was selected because of its heterogeneous and comparatively large population and because of its experience with logging companies (over the past ten years). The forest remains in reasonably good condition.

Of Mbongo's people, 55% are native Balondo, about 30% are Ibibios and Ibos from neighboring Nigeria, and about 15% are other Cameroonians (from the Southwest and Northwest Provinces, primarily). Shifting cultivation is the economic base, supplemented by hunting, fishing, and the collection of nontimber forest products. The matrilineal inheritance system was changed to a patrilineal system by a decision of the Balondo Cultural and Development Association (BACUDA) at a meeting in 1980. Interestingly, women were not allowed to own land under the matrilineal system but are now allowed under the patrilineal system. In one sense, the community owns all primary forest land, and individuals own parcels that are being cultivated or have been recently cultivated by themselves or family members. In another sense, all forest land belongs to the state. Land rental has been common since 1990, particularly for the in-migrant Nigerians.

Trairão, Pará, Brazil.[30] Uruará (3°43′S, 53°44′W; Figure A3) is characterized by upland primary forest on terra firma along with secondary forests in areas of older colonist occupation (beginning in the early 1970s) closer to roads. *Attalea* sp. (*babassu*) and *inajá* palms are increasingly dominant in mature secondary succession, with *Cecropia* sp. (*imbauba*) occurring in recent openings. Selective timber extraction began in Trairão around 1993. The access to the site where the tests were conducted is 25 kilometers west and 50 kilometers north of the town of Uruará, on one of the secondary roads of the Transamazon highway, between the Xingu and Tapajos Rivers. Fertile soils made Uruará one of Transamazon's agricultural poles, attracting colonists from the south and northeast of Brazil to invest in subsistence and commercial agriculture (cocoa, pepper, and coffee) and livestock. Beyond the northern edge of lands occupied since the 1970s, Trairão is a small river that gives its name to a settlement project recently established by the Instituto Nacional de Colonização e Reforma Agrária (INCRA), the Brazilian agrarian agency. The river runs through an area divided into 100-hectare plots for 130 families. These settlers, who arrived in the late 1980s, are mostly children or relatives of Transamazon's early colonists, who arrived in nearby areas closer to the road in the early 1970s. Those colonists—predominantly from the Brazilian northeast, with a mixed indigenous, African, and European ancestry—were attracted to the frontier in hopes of obtaining secure land tenure. They encountered forested land with very little human disturbance, a condition that is gradually changing with opening and burning for swidden agriculture and pasture formation. In general, however, disturbance in the area is still limited.

Commercial logging started in Trairão when a timber company called Marajoara began to operate through the demarcation and selective exploration of tracts of land beyond the colonization schemes. Local industry currently operates with only six commercial species [*jatobá, Cedrella odorata* L. (*cedro*), *Tabebuia* sp. (*ipê*), *cumarú, freijó,* and *pau-amarelo*], allegedly because of isolation and transportation costs. Because of the lack of presence of the local and state governments, the operation of the timber company in the area is viewed by the local residents as a substitute for the state, even though operation is only seasonal. As in most areas of recent occupation, small landholders participate in timber extraction mainly by selling standing trees to the timber company. Even the reduced payment received (equivalent to $10–50 per tree, depending on the species) is considered advantageous to a settler in need of cash for initial establishment. In 1998, the prospect of receiving rural credit and land titling (associated with the arrival of hydro-

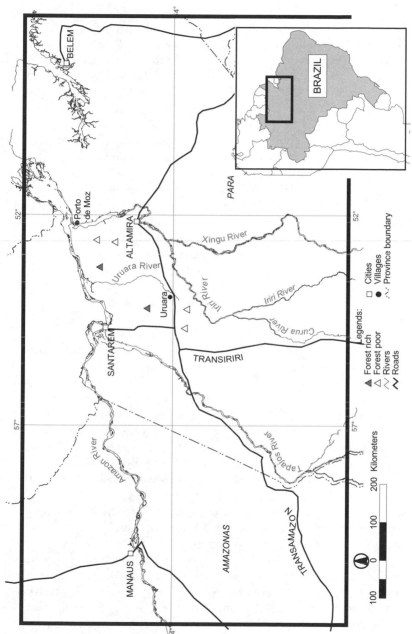

Figure A3. *Map of Brazil, Focused on Our Research Locations*

electric power to Uruará) significantly changed the livelihoods of peasant farmers and the future of the local logging industry.

São João, Pará, Brazil. Porto de Moz (1°45'S, 52°15'W; Figure A3), the area in which São João is found, is located near the mouth of the Xingu River in the Amazonian floodplain. The area is characterized by a very flat landscape with a predominance of alluvial soils and oxisols. Although annual average precipitation is 2,000 millimeters, it is concentrated in the months of December to June. Rainfall from July to November may be less than 60 millimeters/month. The floodplain (the *várzeas*) consists of both seasonally flooded forests and grassy vegetation, whereas upland forests as well as primary and mature secondary growth occur at higher elevations on terra firma.

Technically, all floodplain land belongs to the federal government, under the navy's administration, despite longstanding and culturally accepted resource use patterns by local people. The predominant activities are, in order of importance, timber extraction, fishing, buffalo husbandry, and manioc production. Timber extraction and subsistence agriculture are more important on the terra firma, whereas buffalo raising and fishing take place on the *várzeas*. Human settlements in Porto de Moz are located near the rivers and tiny streams (*igarapés*).

São João is a community 70 kilometers west of the town, or 12 hours by boat. Dwellings are scattered along the margins of the Cupari, an extremely rich fishery and tributary of the Xingu River. A Portuguese merchant first occupied the area early this century, and most of today's 40 families are *ribeirinhos*, or *caboclos*, descending from his family. In contrast to most sites in the vicinity, no timber was commercially extracted in São João. Local forests thus remain almost undisturbed. Lowland resources were traditionally used communally, especially the floodplain. Nearby areas in the terra firma began to be used only a few years ago for agriculture, because the first dwellers focused more on fishing. The Cupari is ideal for catching the more marketable fish species, freshwater turtles, and caimans. Although difficult, it is still possible for the fisherfolk to find *pirarucu* and manatees. Better-off families own small buffalo herds.

New economic opportunities and the prospect of fisheries' depletion are generating friction within the group. Most of the friction has to do with the attempts of some residents to benefit from timber extraction by making individual contracts with Porto de Moz "loggers" who operate in the vicinity. Commercial timber extractors have entered lands in São João twice but were promptly expelled by the majority of the community. Rapid resource depletion without economic compensation, as occurred in neighboring villages, contributed to their opinion: that timber extraction should be carried out only under community managed projects for which they seek government funding. In this regard, a legal entity was recently created and is currently submitting proposals for sustainable forest management activities.

Forest Poor

Long Segar, Kutai, East Kalimantan, Indonesia.[31] Long Segar is situated within a timber concession, where roads have been developed in the last few years, but much travel remains on the river (Figure A1). The area was previously lowland tropical rainforest, and the primary commercial tree species were dipterocarps (most of which have been removed by logging, then plantation development, and most recently, devastating fires related to El Niño in 1997–1998). This equatorial region has two main seasons, rainy (October–May) and dry (June–September), with an additional brief dry spell in January or February.

Long Segar is populated by a community of Uma' Jalan Kenyah Dayak who moved from their remote homeland in the Apo Kayan in the early 1960s. They were attracted by plenty of old-growth forest; abundant pigs; an excellent view of the Telen River and surrounding forest; and more accessible education, medical care, markets, and consumer goods. The people are riverine, swidden cultivators whose animist beliefs were augmented by conversion to Christianity in the Apo Kayan. As with almost all Kalimantan communities, land ownership follows traditional rules and is not officially recognized by the government. Long Segar was part of the U.S.-based Georgia-Pacific timber company's concession in the 1970s and early 1980s. It was technically labeled a "resettlement village" in 1972, giving the inhabitants five years of government assistance of various kinds. In the 1980s, the Muara Wahau transmigration site, covering hundreds of thousands of hectares, was settled by thousands of Indonesian families from other islands a short distance to the north. After the 1983 fires induced by El Niño, Georgia-Pacific relinquished its share of the timber concession to P.T. Kiani Lestari (owned by ex-timber tycoon Mohammad "Bob" Hasan). It later became the site of considerable industrial timber plantation activity, including several other transmigrant settlements designed to supply labor for the timber plantations, in the 1990s. Whereas logging and transmigration represent some competition for forest resources, the industrial timber plantations appear to spell disaster for the local way of life, removing the people's access to most local land. Fires related to El Niño of 1997–1998 seriously degraded the forest in the area.

Mbalmayo, Cameroon.[32] Research took place in an area surrounding the town of Mbalmayo (Figure A2), from Akono (30 kilometers west of Mbalmayo) to Ekekam (60 kilometers southeast of Mbalmayo) along the Sangmelima road. Mbalmayo is an industrial center 45 kilometers south of Yaoundé, the capital of Cameroon. With half a dozen logging industries and many family-run logging units, this town can be considered one of the country's most industrialized areas. Its proximity to the metropolis of Yaoundé on one hand, and industrial growth on the other, have caused a significant population influx to the town. High urban and rural population density directly influences the vegetation; a degraded forest has replaced the original semi-deciduous tropical dense forest based on an equatorial climate with four seasons. Nevertheless, a few areas of virgin forest still remain in remote villages and in areas without ready access.

The Mbalmayo Subdivision has two forest reserves: the Zamakoe Forest Reserve (4,200 hectares) and the Mbalmayo Forest Reserve (9,700 hectares). The Mbalmayo Forest Reserve, very close to town, is subject to a lot of pressure from the surrounding population. Declassification of part of this forest for the purpose of urban expansion is under way. The Mbalmayo Forest Reserve sustains

- the Forestry School (unique in the whole of Central Africa), for field experimentation; and
- the laboratories and experimental farms of the International Institute of Tropical Agriculture (IITA).

The people of Mbalmayo town are made up of many ethnic groups; the following are the most dominant:

- the Beti, most of whom are civil servants or work for logging companies;
- the Bamileke, immigrants from the west of the country who do manual jobs (such as carpentry, bricklaying, or mechanics) and trade activities; and
- Muslims of diverse origins who make up religious-based communities.

The rural populations all belong to the greater Beti tribe, but clans differ from one village to the next. Thus, we have the Bënë at Metet and Nkout, the Mvog Manze at Yop, the Yanda at Akono, and the Etenga at Mendong. The common language spoken in the area is Ewondo. As a general rule, marriage is forbidden within the same clan; consequently, most adult women found in the villages have come there as a result of marriage.

In the villages, the main activity is agriculture: men involve themselves in growing cocoa and, at times, in tapping palm wine; women grow food crops and harvest nontimber forest products. Illegal sawmill operators are particularly active in this area, despite repressive measures taken by the forest administration. This activity is partly due to the area's proximity to Yaoundé and partly to the high cost of sawn lumber since the last currency devaluation. Land ownership follows traditional rules that are officially recognized by—but sometimes in contradiction to—national law. The local land ownership system is complex; acquiring land may be by inheritance, by cutting down community virgin forest, or by gifts under certain conditions.

Transiriri, Pará, Brazil.[33] Transiriri (Figure A3) is the name of one of the most important secondary roads of the Transamazon highway, starting 10 kilometers west of Uruará and continuing south for about 100 kilometers, where it reaches the Iriri River. Until 1982, the road was only 20 kilometers long and forests were relatively undisturbed. By 1995, most land near the

Transiriri was depleted of commercially valuable timber species. The mahogany in the forests south of Uruará and Altamira were the major resources targeted by timber companies, who extended the road. A few other species were considered commercial (the same ones as in Trairão) and also were extracted.

The area selected for the social science methods tests is between kilometers 50 and 60 of the Transiriri road, on lands beyond the official colonization scheme. Its occupation began with the arrival of families of northeastern Brazilians, mostly from the state of Maranhão, who were guided and transported by timber companies after the road was constructed. Half of the plots were deforested by 1998, mostly for cocoa and pepper plantings and for pasture development. In the two years after the road was constructed, two companies—Peracchi and Bannach—removed most of the mahogany. Bannach built a large processing unit near the margins of the river. The peak of "prosperity" and goods circulation in the Transiriri was in the mid- to late 1980s. The main reason for this prosperity was the operation of the timber processing unit at the Iriri and the related truck traffic. Additional features of this period included the frequent crossings of gold miners coming from upriver and the periodic movement of cattle brought from ranches in Mato Grosso and left on local properties for fattening. Today, the situation is quite different. Gold and beef prices are much lower. The tenure security of 450 households beyond kilometer 40 of the Transiriri road was compromised when a 1993 disposition of the Justice Ministry designated a 760,000-hectare area centered on the Transiriri as indigenous (*arara*) land. The timber company was considered illegal and shut down after the territory was demarcated, and timber extraction in the area has since been prohibited. Local residents say that they have lost economic opportunities such as wage labor, a secure market for their annual crops, and free transportation to the city.

Bom Jesus, Pará, Brazil. In contrast to São João, the community leaders of Bom Jesus—the forest-poor floodplain site in Porto de Moz—have a completely different approach to commercial logging. Bom Jesus (Figure A3), at the margins of the Quati River (a tributary of the Xingu that receives waters from the Cupari, where São João is located), was occupied by *caboclos* in the early nineteenth century, but houses were relocated to their current site in 1991, when a record flood destroyed most of the homes. A larger Catholic and a smaller Evangelical settlement constitute the relocated site of Bom Jesus, where *ribeirinhos* allowed and stimulated timber extraction. Whereas the floodplain extends to the northern margin of the Quati River, the two villages are strategically positioned because areas of upland forest start on their boundaries. As a result, Bom Jesus is one of the most suitable locations for timber extraction in Porto de Moz. Commercial logging has been in place since the 1970s, but on a smaller scale during the first decade. Some 20 species are extracted, including *cedro*, *Virola* sp. (*virola*), *sucurba*, *jabutirana, cambará, freijó, Hymenolobium* sp. (*angelim*), *Vochysia maxima* Duck (*quaruba*), and *ipê*. After the village was relocated, ferries brought trucks and tractors to Bom Jesus and left with loads of timber. Small sawmills were installed in the hinterlands. Economic opportunities brought by logging operations enacted a process of demarcation and appropriation of individual tracts of land in the terra firma, mainly for claiming property rights over trees that were sold directly to timber companies operating in the area, or more often to truck drivers (*caminhoneiros*). Today, only a few trees of commercial size are left in Bom Jesus' terra firma. The exhaustion of forest resources along with the depletion of local fisheries raises serious questions about future survival strategies in Bom Jesus.

We have not included descriptions of the sites in Trinidad, Gabon, and West Kalimantan because we felt the case study approaches of the chapters provided enough detail and because the sites were not included in any of our cross-site analyses. The Trinidad sites are shown in Figure A4, the Gabon sites in Figure A5; and the West Kalimantan sites (DSWR) in Figure A1.

Endnotes

1. We are grateful to Jack Ruitenbeek for helping us state this conclusion in this way.

2. Each team produced a report: Burgess and others (1995) for Indonesia; Mengin-Lecreulx and others (1995) for Côte d'Ivoire; Zweede and others (1997) for Brazil; and later, the Federal Ministry of Environment, Youth, and Family (1996) for Austria; Prabhu and others (1998) for Cameroon; Woodley and others (2000) for the United States; and Nasi and others (1998) for Gabon.

3. These results are reported in Brocklesby and others 1997, Sardjono and others 1997, Tiani and others 1997, Diaw and others 1998, McDougall 1998, Oyono and others 1998, Porro and Miyasaka Porro 1998, and Tchikangwa and others 1998.

4. Besides authors in this book, the following individuals had a significant impact on the evolution of these ideas: Neil Byron, previous program leader for CIFOR's policy work; Heleen van Haaften, an agricultural sociologist with expertise in cross-cultural psychology, working with the Tropenbos Foundation in Wageningen, the Netherlands; Ahui Anvo, a sociologist working with SODEFOR in Abidjan (Côte d'Ivoire team); Laksono, a professor of Anthropology at Gadjah Mada University in Yogyakarta, Indonesia (Kalimantan team); and Jan Kressen, an independent consultant from Germany who specializes in sociology (Brazil team).

5. This set has been modified slightly from the original (see Wollenberg and Colfer 1996).

6. Prakash and Thompson (1994) identified four quite different ways to interpret "fairness," for instance: proportionality (to each according to contribution), parity (equal distribution of outputs), priority (inherent rights, like rank/station), and pot-luck (equal chance, like a lottery).

7. For theoretical discussions of the ubiquity (and perhaps inevitability) of this kind of problem, see Smith and Steel 1995 and Dove 1996.

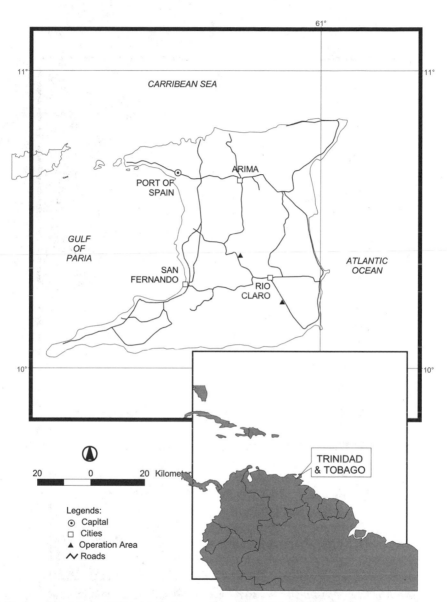

Figure A4. *Map of Trinidad, Focused on Our Research Locations*

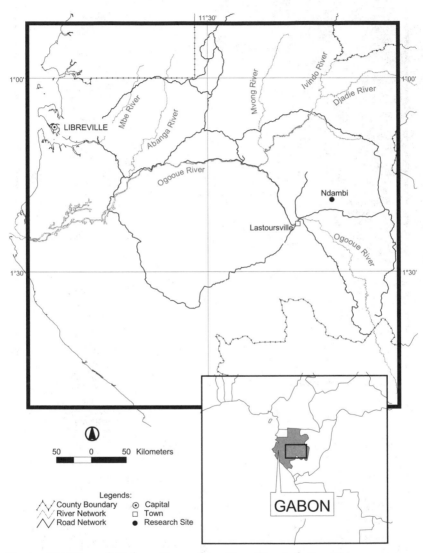

Figure A5. *Map of Gabon, Focused on Our Research Locations*

8. See Fairhead and Leach 1994/95 for examples in Africa; see Roosevelt 1989, Balee 1992, or Salick 1992 for examples in South America.

9. We use the numbering system used in the *CIFOR Generic C&I Template,* for ease of reference. Principles 1 and 2 refer to policy and ecological issues, respectively.

10. This concept is not uncommon. Conditional rights to land are familiar to forest actors in Borneo, for instance, where many communities retain some residual rights, even in otherwise privately held resources (Appell 1986; Colfer with Dudley 1993; Peluso 1994; Ngo 1996). In Côte d'Ivoire, we found a willingness to accept needy outsiders and give them access to land (in return for labor), even when such pressure resulted in environmental degradation (Riezebos and others 1994; SODE-FOR 1994; van Haaften 1995).

11. *Tenure* is defined as "the act, right, manner, or term of holding something (as a landed property, a position, or an office)" (*Webster's* 1993). It therefore incorporates, in the context of forest management, various combinations of use rights, stewardship, communal and individual ownership, state management, and so forth. This generality seems appropriate, given global variation.

12. According to Green (1986), *active participation* (in a health context) is "the conscious and intentional involvement of the individual or population in question, as distinct from the passive engagement of the individual or population in each of the activities or processes that follow": identifying their own goals or needs, setting their own priorities among goals or needs, controlling the implementation of programs or solutions, and evaluating or otherwise obtaining feedback on their own progress. The involvement of distinct stakeholder groups in forest management requires varying degrees of negotiation, perhaps at each of these steps.

13. See Palmer 1993 for a counter example from Newfoundland's fisheries.

14. Winthrop's (1991) first definition of *culture* (of many) is "that set of capacities which distinguishes *Homo sapiens* as a species and which is fundamental to its mode of adaptation."

15. Oppressive and unjust elements exist in all cultures; but conversely, human beings universally have difficulty thriving when their cultural systems have been disrupted.

16. Fuller discussion of this method was published by Colfer and others (1999a, 1999b). More detailed descriptions of the individual studies are available in the original reports (Tiani and others 1997 [Mbalmayo]; Tchikangwa and others 1998 [Dja Reserve]; Brocklesby and others 1997 [Mount Cameroon]; Sardjono and others 1997 [East Kalimantan]; Porro and Miyasaka Porro 1998 [Brazil]).

17. Traditionally, a Galileo study requires respondents to report their perceptions of the differences (often called "distances") among a set of concepts considered central to the definition of some topic, for example, "forests." The estimated dissimilarities are averaged across all respondents in any segment and projected onto orthogonal coordinate axes to produce a perceptual map, or space. Within this space, distances are predictive of attitudes, beliefs, and behaviors. Technically, 277 respondents in West Kalimantan estimated the pairwise dissimilarities among a set of terms including "forest" and 19 other concepts identified in previous analyses as pertinent to the perception of forests in Kalimantan villages. The resulting square-mean dissimilarities matrix then was analyzed in several ways, for example, in perceptual maps (multidimensional scaling [MDS]), charts, graphs, tables, and advanced artificial neural networks (ANNs). Perceptual maps were made using Galileo software, which produces

very precise representations of the dissimilarities in graphic form, and which allows transformations (rotations and translations) to common orientations for easy comparisons of data over time and across subsamples. Previous research has shown Galileo to be an appropriate model when holistic models of cognitive structure and processes are required, when precise results are desirable, or when a standard metric needs to be maintained across time or subsamples—for example, when time-ordered maps are needed, when maps are to be compared from sample to sample, and when the concepts to be mapped are known. Galileo modeling may be less appropriate when investigators are uncertain as to which concepts occur in the cognitive model or when reducing the time burden on respondents is crucial, an invariant metric over time and across samples is not needed, and precise results are not important (Woelfel and Barnett 1982, 1992; Woelfel and others 1986; Cary and others 1989; Woelfel and others 1989). When less is known about the concepts that need to be included, as is the case in preliminary studies, similar results can be obtained from CATPAC, a self-organizing neural network that reads text and uncovers the main underlying concepts. CATPAC makes it possible to work from in-depth interviews rather than quantitative scales, yet derive similar results (Cary 1995).

18. CATPAC (described in note 17) is a computer program that can perform this function quickly and easily from text (also tested in CIFOR's methods tests and included in Colfer and others 1999b).

19. We consistently use boldface to differentiate the locally defined concepts used in the Galileo studies from our own, presumably more general, meanings of the same terms.

20. We have tried this with and without columns differentiated by gender. Some researchers felt this differentiation on one form was unwieldy; others liked it. We do have gender-disaggregated data to be reported in the future.

21. This combines some of the approaches suggested by Pretty (1994), for example, persistent and critical observation, negative case analysis, and reflexive journals.

22. See Vayda 1996 for an interesting, philosophical discussion of methods of study and relationships among human actions and their environmental effects.

23. In the short run for environmental sustainability, this perceived level of security may suffice, because people will be motivated to take care of the resource if they think they will continue to have access to it. However, in the long run, human well-being will suffer when they do in fact lose such access.

24. Other images that come to mind include Herbert's (1984) descriptions of the spice worms moving beneath the sand across the landscape in *Dune;* moles appearing randomly in the "Whack-a-Mole" game popular in U.S. amusement parks; and Woelfel's image of lights switching on and off in a network of interconnected bulbs (paralleling the operation of the human brain) also captures some of this idea (presentation on neural networks by Joseph Woelfel to CIFOR, Bogor, Indonesia, in October 1996).

25. Prabhu has argued convincingly for recognizing a useful congruence between principles, criteria, indicators, and verifiers on one hand and wisdom, knowledge, information, and data on the other (Prabhu and others 1996, 1999).

26. This methods test was actually also carried out in several other Cameroonian sites by Chimere Diaw and associates, but the results were not available at the time of this analysis.

27. This research was led by Dr. Mustofa Agung Sardjono (an agroforester). Fieldwork was conducted by Edi Mangoppo Angie, Akhmad Wijaya, and Erna Rositah. The field test occurred in June and July 1997.

28. This research was led by Bertin Nkanje Tchikangwa (anthropologist), who was assisted by Sidonie Sikoua, Moise Metomo, and Marc Felix Adjudo. The research took place throughout most of 1997.

29. This research was led by Mary Ann Brocklesby (a social scientist), who was assisted by Priscilia Etuge, Grace Ntube, Joseph Alabi, Michael Anje, Victor Bau Bau, and John Molua. The field test occurred between July and September 1997.

30. The Brazil-based research was led by Roberto Porro, in partnership with Noemi Miyasaka Porro (both anthropologists), and took place in July and August 1998.

31. This research was led by Dr. Mustofa Agung Sardjono (an agroforester). Fieldwork was conducted by Edi Mangoppo Angie, Akhmad Wijaya, and Erna Rositah. The field test occurred in July and August 1997.

32. This research was led by Anne Marie Tiani (an ecologist). Field assistance was provided by Edouard Mvogo Balla, Annie Oyono, and Diesse Norbert Kenmegne. The test took place over a five-month period in mid-1997.

33. This research was led by Roberto Porro (an anthropologist), who was assisted by Noemi Miyasaka Porro. The test took place in July and August 1998.

Gender and Diversity in Forest Management

The question of identifying stakeholders for a given forest emerged as a critical issue in our research, from various perspectives. One of the first questions was how to identify those stakeholders whose well-being was most critical in sustainable forest management. In our research context, we concluded that seven dimensions could help us in making such an identification: proximity, preexisting rights, dependence, poverty, local environmental knowledge, a close link between the forest and the people's culture, and a power deficit (Colfer and others 1999c). We suggested a "Who Counts Matrix" to help in identifying the most important stakeholders in the field.

However, several thorny issues on this subject remained. Intracommunity differences such as gender, age, and wealth are addressed in this section. The Who Counts Matrix can help us identify broad categories of people; however, in particular forests, we often find considerable variation in local patterns of resource use, control, and benefits. We also have found that accessing marginalized groups can be difficult (for example, see Burford de Oliveria 1997; Colfer and others 1997c; McDougall 1998). In our attempts to identify the relevant stakeholders, we have supplemented our efforts with focal group analysis (Colfer and others 1999a), stakeholder sampling, neural network analyses such as CATPAC (Colfer and others 1999b), and cognitive mapping software such as the Galileo program (see Chapters 5 and 6).

Although we do not attempt to present a comprehensive survey of the literature relating to gender and forests, we want to share some of the material that has influenced our thinking. Analyses tend to fall into two main categories: those that are oriented toward demonstrating or proving the important roles of women in forests, and those that document methods and frameworks for addressing gender issues in sustainable management. Both are relevant for the subsequent analyses within this book.

Some of the earliest studies of this topic come from the field of anthropology. Margaret Mead's many works on gender-related issues are well-known. Building on growing theory about women's roles in general during the 1960s and 1970s (for example, Dubisch 1971; Sanday 1974; Rogers 1978), several studies began to look at women's roles in various contexts. Borneo— the island made up of East, West Central, and South Kalimantan, Indonesia; Sabah and Sarawak, Malaysia; and the Sultanate of Brunei—where women hold comparatively high status, has spawned a particularly large number of such studies (Dove 1980; Colfer 1981, 1982, 1983a, 1985a, 1985b, 1991; Sutlive and Appell 1991; Tsing 1993). Other notable Asian studies addressing this area include collections by Goodman (1985) and Atkinson and Errington (1990); the works of Weiner (1976), Wickramasinghe (1994), and Elmhirst (1997); and any publications by Madhu Sarin.

In Africa, more has been written on women and agriculture than on women and forestry, but there are some notable exceptions. Leach (1994) offers one of the most interesting and comprehensive studies of gender. Any works by Louise Fortmann, John Bruce, or Diane Rocheleau—though more focused on agroforestry systems—are likely to provide important gender and forest-related insights (see also Stamp 1989; Sigot and others 1995).

In Latin America, Murphy and Murphy (1974) produced an early and excellent longitudinal study of men's and women's roles as they relate to the forests. Siskind (1973) is another pioneer in this field. Townsend and others (1995) use the words of women colonists directly in describing life on the frontier in Mexico and Columbia. Schmink (1999) is a guiding light in Latin American studies related to gender and forests.

Besides these largely ethnographic approaches, the interface between population and forest quantity and quality has been recognized and discussed (for example, Colfer 2000a). Sen (1994) provides a nice discussion of some of the issues. Indeed, many of the issues that Sen addresses are relevant for gender and human well-being more generally. Some authors have also written pertinent works on gender and poverty (for example, Jackson 1996; Razavi 1999).

Two particularly interesting "how to" manuals are the Food and Agriculture Organization's *Gender Analysis and Forestry Training Package* (Wilde and Vainio-Mattila 1995) and the World Bank's *Participatory Development Tool Kit* (Narayan and Srinivasan 1994; see also ISNAR 1997). Increasingly, literature relating to participation tends to include a chapter or two on gender and diversity issues.

However, we have concluded that a need remains for analyses specifically focused on the concerns of women and men and, indeed, people categorized by various means as they relate to formal forest management. Chapters 1–3 of this book examine this topic in more depth, in Indonesia and in Cameroon. As one might expect in interrelated systems, the topic reappears from time to time throughout this book.

In Chapter 1, McDougall discusses some recurrent problems that plague assessors when dealing with intragroup variation (for example, gender, ethnicity, age, and caste) from a theoretical standpoint. She then describes her own experience testing several methods in a forest-rich Bornean context in East Kalimantan, Indonesia (Long Loreh, Bulungan Research Forest), focusing on differential success with men and women and with different ethnic groups. Her points are critical to remember when considering the question of representation in assessment or in ongoing management or monitoring.

In Chapter 2, Tiani writes about the experiences of Cameroonian women, including ideal roles, gender-related perceptions, women's particular uses of various forest products, and the different effects of change on men and women. She covers issues pertinent to most of the topics we address (stakeholder identification, security of intergenerational access to resources, and rights and responsibilities in management). Her contribution nicely demonstrates the futility of seeking single, direct causes in the interconnected systems that we are dealing with.

In Chapter 3, Brown and Lapuyade present the relationship between changing gender relations and forest use in central Cameroon. Their study site, Komassi, is in an area that has been logged consecutively for more than 50 years, but forest remains the dominant land cover. They primarily used rapid appraisal techniques and look at the differing perceptions of men and women about the changes that have occurred.

CHAPTER ONE

Gender and Diversity in Assessing Sustainable Forest Management and Human Well-Being

Reflections on Assessment Methods Tests Conducted in Bulungan, East Kalimantan, Indonesia

Cynthia L. McDougall

Gender and human diversity are critical dimensions of sustainable forest management (SFM) and human well-being (HWB). They are at the core of management decisions, tensions, and opportunities in tropical forestry. They also are complex and raise uncomfortable questions about the status quo and, as such, often appear to fit poorly with ambitious management or research agendas. For these reasons, despite their pivotal place in progress toward sustainable forestry, gender and diversity are often underplayed or even avoided in assessments of research and forest management.

In this chapter, I briefly explore the concept of gender and diversity analysis and its utility for assessing SFM and HWB (for management or research). My intent is to demystify the concepts through a theoretical discussion of gender and diversity analysis in HWB assessment that is supplemented by fieldwork examples. These examples are based primarily on my experiences with the HWB assessment and social criteria and indicators (C&I) methods test completed by the Center for International Forestry Research (CIFOR) in Long Loreh, Bulungan, East Kalimantan, Indonesia, with additional insights from the experiences of other CIFOR C&I methods researchers. The importance of recognizing the theoretical grounding of a research team vis-à-vis gender and diversity is highlighted.[1]

Then, I briefly address some gender and diversity issues relating specifically to the C&I used in the East Kalimantan assessment. I explore some

obstacles to the participation of women and other marginalized or less dominant groups of people in the assessment (or other research and forest management) processes and present some strategies for overcoming such barriers.
I conclude with reflections on the HWB assessment process from a perspective of gender and human diversity analysis.

My overall goal is to contribute to the larger pool of wisdom on these
issues so that more researchers, forest managers, and assessors can use gender
and diversity approaches in their work and thereby effect the kind of long-
term changes needed to maintain the world's peoples and its forests.

Concepts, Rationale, and Analysis

Gender and human diversity are overlapping concepts that constitute analytical
approaches for understanding the world around us. These approaches are
dynamic and relevant to HWB and SFM for both ethical and practical reasons.

What Are Gender and Human Diversity?

The term *gender* is probably one of the most used and least understood or
agreed upon terms in the field of SFM. It is commonly built into donor
requirements and strategic plans for SFM and other development research
and practices, yet it frequently is assigned different literal as well as implied
meanings.

The primary confusion relates to the difference between the meanings of
sex and *gender*. Whereas *sex* refers to the biological differences between
women and men, *gender* is a social distinction that is culture-specific and
changes across time (ISNAR 1997). In other words, individuals are born
male or female[2] (a biological distinction) and, via gender (or engenderiza-
tion), become men and women (a social distinction) by acquiring the culturally defined attributes of masculinity and femininity. Individuals then take on
the appropriate roles and responsibilities for these categories (Sarin
1997).The second main area of confusion arises because, even within this
framework, gender has been assigned different meanings by different users in
various contexts. Hawkesworth (1997) identifies interpretations that range
from "attributes of individuals" to "interpersonal relations" and from
"modes of social organization" to "questions of difference and domination."
The confusion created by such various meanings clearly runs some risk of
diluting the concept's utility. Fortunately, this confusion has been resolved
somewhat by a highlighting of the common thread that ties issues of the
"psyche to social organization, social roles to cultural symbols, normative
beliefs to the experience of the body and sexuality" (Hawkesworth 1997).[3]
This shared thread, *relationships,* can be understood as two interrelated
aspects: gender as "a constitutive[4] element of social relationships based on

perceived differences between the sexes,[5] and gender as a primary way of signifying relationships of power" (Scott 1986).[6] In other words, gender is both a key force or process that shapes or "socially constructs" relationships between men and women and a conceptual (or analytical) "torch" that points to or illuminates the distribution of power in those relationships.

Human diversity, more neglected than gender in discussions about SFM and related research, is an equally important concept. By *human diversity*, we refer not only to different ethnic groups but also to all the other significant dimensions of social and biological difference that crosscut gender and ethnicity, such as wealth, age, status, class, and caste.[7] As J. Scott (cited in Hawkesworth 1997) suggests for gender, human diversity is a force that influences (constructs) social relationships on the basis of perceived differences, as a way of defining and understanding power relationships.

The Relationship between Gender and Diversity

Over the past decade, gender has received some much needed and long overdue attention in forestry and research circles. This attention, however, has also generated concern that it may be "competing" with other elements of human diversity. A feminist critique has emerged that the "multiple jeopardy characteristics of many women's lives" (for example, race, economic status, and sexual orientation) raise serious concerns about generalizing gender as the primary analytical concept. Hawkesworth (1997) points out that if gender in reality "is always mediated by other factors such as race, class, ethnicity, and sexual orientation, then an analytical framework that isolates gender … is seriously flawed." This kind of oversimplification would only mask the complex and overlapping identities (and, in fact, the challenges or oppression) of numerous women and men. No individual can belong solely to a single group or category[8] of people—everyone belongs to many different groups, including those based on ethnicity, wealth, or status. Various dimensions of diversity (or identity) can reinforce positions of relative empowerment or disempowerment. In other words, gender combines with other forms of diversity to establish roles, relationships, and power structures.

This kind of critique is vital to the discussion on gender and diversity. It does not diminish the validity of gender as an analytical tool but offers a clear warning that gender analysis could become a victim of its own success. Considering gender in isolation—thereby drawing attention away from other dimensions of diversity—would run counter to the very foundations of gender analysis: the awareness of power relationships. Such an imbalance in a research or forestry initiative, for example, would risk reducing equity as well as effectiveness in outcomes, because it would fail to seek and take into account the other critical differences among stakeholders.

This potential pitfall may be minimized by considering a twofold role for gender analysis. At the macro (or generic) level, gender is a concept that raises awareness of human differences and prevents assumptions of homogeneity. It serves this role well because it is universally present as a key dimension. As Sandra Harding (cited in Hawkesworth 1997) explains, "The fact that there are class, race, and cultural differences between women and men is not, as some have thought, a reason to find gender difference either theoretically unimportant or politically irrelevant. In virtually every culture, gender difference is a pivotal way in which humans identify themselves as persons, organize social relations, and symbolize meaningful natural and social events and processes." Its very pervasiveness "opens the door to other social dimensions" (Schmink 1999). By sensitizing researchers and other actors in SFM, gender offers opportunities to introduce dimensions of diversity that may have otherwise been overlooked or ignored.

At the micro (or site-specific) level, gender plays a practical role as one of many forms of diversity. Depending on the individual context, it may be more or less important than other dimensions of diversity. The relative importance (and indeed the relationship between the different forms of diversity) has to be assessed individually. This kind of evaluation requires willingness on the part of researchers, forest managers, and assessors to be flexible and to avoid preconceived notions about the key dimensions of human difference in particular contexts.

This critique of gender (or of gender analysis approaches in research or forestry) contributes a valid and useful concern. Taking the above twofold approach to the concept of gender enables researchers or forest managers to address the concern by viewing it within a broader context and thereby maintaining consistency with its roots in issues of equity and analysis for change.

What Is Gender and Diversity Analysis?

Gender and diversity analysis is more than analyzing data by gender or ethnic group; it is an approach or methodology in which "gender" and "diversity" become analytical variables throughout the research or program, from planning to implementation, analysis, and evaluation. Furthermore, it addresses both the structure (that is, the roles) and dynamics (that is, the relationships) of human systems. It involves an ongoing process of exploring a spectrum of interrelated questions from role-based to relationship-based.[9] Questions include the following:

- What patterns of assigning meaning (beliefs) to events and things become evident between different groups? How do different groups value resources differently?
- What are the roles and responsibilities of the different groups regarding forest management?
- Who controls access to resources? Who makes decisions about them?

- Who benefits from each activity or enterprise? Who bears any associated costs?
- What are the relationships among those roles? What are the power dynamics?
- How do relationships and roles influence the decisionmaking of the group regarding resources?

Many practitioners and theorists increasingly emphasize that the analysis should be focused on the relationships rather than the roles, except at a very descriptive level, because roles offer a "static" perspective on issues that are based on power relationships and thus are, by nature, dynamic (Young 1988).

This important insight is linked to the change over the last two decades in the development field from the women-in-development approach to the gender and diversity approach.[10] This transition has improved understanding of the significant difference between changes in women's *condition* versus changes in women's *position*. Although women's material situations may improve, their relative social positions (and thus, long-term well-being) do not necessarily improve. This explains why major gains in well-being cannot be brought about by working with only women or with any other individual group alone. Because well-being involves the social relationships between women and men, both women and men must be involved (Sarin 1997).

A related and particularly relevant issue is that equity or inequity in relationships between genders or groups of people is often deeply engrained in the institutions and sociocultural fabric of a society. Because inequities are legitimized in traditions, customary laws, and religious symbols, they may be powerfully rooted in the society as well as in the minds of the very people who are marginalized[11] (personal communication with Barun Gurung, 1999). As Porro and Miyasaka Porro (1998) express, because we are all "born and raised in a gendered system of domination, both local men and women reproduce in their thoughts and practices the very system which imprisons them. Either resisting or being subordinated to it, women and men will in general express to the researcher the social experience they live."

This phenomenon was often and diversely present in our Kalimantan experience. Some of the most poignant examples were expressed by women, concerning equity in decisionmaking. During an activity that reflected the distribution of rights of the different forest managers or stakeholders, the women were unanimous in their allocation of far greater "power" in decisions to the men,[12] across all responsibilities. They also all expressed that this distribution of power was inequitable. Yet the women all agreed that, even though they did not like the current situation, it was impossible to imagine or describe a more equitable situation. Because they all were familiar with and accepted inequitable distribution in their community, they could not— at least initially—envision any other arrangement. Eventually, one woman offered an alternative vision that she saw as more equitable (her ideal situation), in which the women were far more involved in decisionmaking. Interestingly, however, even in the ideal, she allocated fewer rights to the women. She explained (and other women agreed) that women already have so many responsibilities—from tending rice fields and fruit gardens to household chores and childcare—and that increased decisionmaking power or rights would mean more tasks and responsibilities (such as attending meetings). Because she could not envision any of the traditional roles (and thus relationships) changing, any decisionmaking role would translate to additional burdens for women.

A final and important question is raised here about the nature of gender and diversity analysis. Because it is rooted in a movement for social transformation and is essentially political in nature, is gender and diversity analysis therefore also transformative about other relationships? Specifically, does it redefine concepts of subject–researcher relationships? Does it challenge traditional notions of objectivity? Certainly, gender and diversity analysis at least demands a certain degree of critical self-reflection by researchers in terms of their roles and relationships with the participants in (or subjects of) the research. This caveat—and the potential reflection on researchers' own identities, influences, and biases—poses a challenge (welcome, or sometimes unwelcome) for research teams.

SFM, Research, and HWB Assessments

Natural resource use is based on decisions, actions, and behaviors rooted in overlapping social and natural systems (Schmink 1999). Within this framework, gender and diversity are relevant to SFM because stakeholders' opportunities and constraints in decisionmaking (as well as resulting actions and behaviors) are determined by their different gender and diversity identities, which include their roles, knowledge, and responsibilities.

Because gender and diversity are critical in determining the choices that stakeholders make regarding forest management, understanding how these variables shape constraints, opportunities, interests, and needs can offer great insight for SFM research and management.[13] In this way, gender and diversity analysis contributes to creating a more accurate and complete picture of a complex social landscape. Unless explicitly addressed, gender and diversity differences are often forgotten (for example, the historical tendency to base information and explanation on the views of dominant or most accessible groups, or assumptions of homogeneity regarding values, access, or benefits) (Edmunds 1998; Schmink 1999). In the CIFOR team's HWB assessment experience in East Kalimantan, the most expedient assessment—one which drew on those individuals who were most accessible and willing to speak with us—would have indicated relative harmony and equity within the local community. The gender and diversity approach used in the stakeholder analysis revealed considerable perceptions of inequity in resources, benefits, and access between the ethnic groups as well as some inequity between men and women—all of which have implications for forest management.

Research and forest management that better reflect the experiences of nondominant groups is more likely to lead to policies or programs that respect those different experiences and are targeted to more sustainable and equitable impacts. Accurate and complete assessments also lead to more effective and efficient outcomes of research, policy development, or forestry and development programs. Wilde and Vainio-Mattila (1995) illustrate this point with a negative example of a community forest project in which a group of local men was invited to a joint planning meeting with foresters. The men requested and received 3,000 hardwood seedlings to plant for future use in making furniture and in carving. The seedlings were planted, and all subsequently died. The reason for this lack of success was that, although the women were responsible for caring for seedlings, they had never been consulted or properly informed. At a subsequent meeting between the foresters and both men and women of the village, the women expressed their interest in fast-growing softwood species for fuel and fodder. When the foresters provided a mix of these and the hardwood species, the women successfully planted and tended the trees.

Sarin (1997) relates a parallel case in Gujarat, India, in which a group of local men formed a forest protection group that banned extractive uses. The group assumed that women could meet their responsibilities for providing family fuel and fodder from private lands. In reality, the rules forced local women to walk much farther than before and to steal firewood from other communities' forests. This decision increased demands on labor and time, and as a result, many women were labeled as thieves. Only when the group's strategies were renegotiated, taking into account the gender and diversity needs, did it begin to succeed. Unfortunately, such examples—where important differences are (at least initially) ignored and unsustainable outcomes emerge—abound in forestry.[14] Especially given the increasingly scarce resources available for forestry research and management, the ability to increase effectiveness the first time around should be a powerful incentive to include this kind of analysis.

Within this discussion, a specific point must be made about the implications for equity, HWB, and SFM. HWB (a recognized pillar of SFM) is unevenly distributed across the globe. The current trend of the feminization of poverty offers a clear example. The majority of poor people in both the developed and the developing world are women, and poverty continues to grow faster among women (especially rural women) than among men (Ang-Lopez and others 1996). A gender and diversity approach tends to expose these "invisible" poor sectors of society and highlights the fact that well-being is not neutral or random. Recognition that well-being reflects patterns of gender and diversity creates better understanding and thus greater opportunity to address this inequity. Furthermore, it relates to the management aspect of diversity, whereby gender and diversity affect use patterns and options; frequently, more disenfranchised groups are linked to the depletion of forest resources because of a cycle of poverty and degradation: "The lack of progress in the eradication of poverty and the increasing proportion of women among the poor is the single most important threat to the progress of development and sustainability" (Ang-Lopez and others 1996).

Finally, gender and diversity analysis in SFM research and HWB assessments is also relevant from an ethical and practical perspective because of the potential empowering effect of participation. Although it is not without its risks, the positive involvement of marginalized or nondominant stakeholders in research and assessment processes can have powerful effects of raising awareness of their own situations ("conscientization"), which is one of the cornerstones of empowerment.

HWB in SFM Assessments: Reflections on Methodologies, Issues, and Experiences

From the outset, researchers involved in CIFOR's C&I methods tests and HWB assessments (at all sites) were requested to build gender and diversity analysis (or sensitivity) into their approaches, methods, and analyses. However, implementing this request in a responsive and adaptive way requires more than an add-on approach. "Building in" sensitivity to gender and diversity issues requires integration at the theoretical level, creating and trying methodologies that reflect those foundations, continually improving those methods in the field, and finally incorporating those methods into the analysis. In this section I explore these aspects, mainly based on my experiences in East Kalimantan,[15] and highlight some key issues and approaches of broader relevance that emerge from that example.

Awareness and Approaches: Grounding Oneself

Theoretical Frameworks. The approach developed for methods tests, or assessments, depends on the *theoretical grounding* of the test team. (This term refers to the theoretical focus consciously adopted by the research team as well as team members' unconscious assumptions or biases.) As already noted, gender and diversity analysis ranges across an entire spectrum. Whereas there are strong cases for the gender and diversity and power-based analyses, it also can be argued that other points on that spectrum can offer more or less appropriate approaches in different contexts or in different stages in the analysis. It may be useful for one research team to "choose to understand 'gender awareness' to mean simply making women more visible; others may focus on the differences between men and women's roles, even considering power" (Porro and Miyasaka Porro 1998).

A critical foundation of this process is that the team members develop initial and evolving clarity on where they see themselves in this spectrum, and why. Porro and Miyasaka Porro (1998), for example, established that for their assessment, it was

> more than simply necessary to understand the relations of power not only between loggers and peasants, but also between men and women, to address the inequalities involved in forest interventions ... we realized that neither looking at women only or using gender just as one more variable was enough.... There was a need to understand how development and conservationist policies were influencing the relations of power between men and women, how such relations were affecting their roles in dealing with forests, within their communities, and with other forest actors.

In fact, the team set out in advance a series of criteria that the methods would have to meet to be considered gender-sensitive. This planning step undoubtedly contributed to the high quality of the assessment. The ambiguities that could have arisen from multiple or undefined theoretical bases were thus avoided. An ongoing critical exploration of this question by an assessment team during all phases of research not only improves the research outcome but also creates a continuous learning process for the researchers.

Avoiding Pitfalls. Two interrelated issues emerged as needing attention in the East Kalimantan assessment: (a) "embeddedness" and assumptions and (b) shortcuts and habits. As researchers, we carry with us perspectives and beliefs adopted during our own socialization process (which in most cases took place in predominantly male-dominated, hierarchical systems). Because we are "embedded" in our own context of socialization, we may make assumptions about assessment participants, stakeholders, or contexts that may ultimately limit our understanding. During stakeholder identification, for example, preconceived notions about women's roles may prevent outside actors from perceiving the significance of women to certain aspects of forest management (for example, as bearers of significant knowledge about diverse species). As discussed earlier, some more marginalized stakeholders themselves may not easily realize their own role as important stakeholders in a forestry issue. Some women consider a role that sounds publicly important as beyond their understood domain—despite the fact that they may be the primary gatherers of fuelwood or important nontimber forest products, or keepers of other significant forest-related knowledge.

The related issue refers in part to the tendency to take shortcuts in problem-solving or make judgements by drawing on past experience or knowledge (heuristics). This tendency is not only natural but even can be expedient; yet at the same time (especially in combination with the other factors), it can lead to a lack of accuracy and dangerous generalizations (Bazerman 1998). It also refers to the natural move toward "easier" or more ingrained methods, habits, or skills, even when they may not be in line with assessment or research goals or may not be consciously selected approaches. Porro and Miyasaka Porro (1998) expressed the view that, even though they established the strategy of using stakeholder separation at the level of focus groups and then conceptual stakeholder integration at the level of analysis, "there was a tendency to isolate women's and men's roles, expectations, and practices and only look for similarities and differences between their relations with the forest, without making the connection between them…. We needed to constantly remind ourselves that building gender sensitivity in a project aiming toward SFM demanded more than data about a sum of individuals disaggregated by sex." They faced the common challenge that field realities (such as time constraints, complexity, and other pressures) are inclined to push gender and diversity analysis back to its more rudimentary forms in the absence of a conscious effort to maintain its integrity.

In East Kalimantan, even though our team had the specific goal of learning from participants, we became increasingly aware of our own embeddedness and assumptions as well as opportunities to fall back on shortcuts and habits. Great care was needed to avoid hearing most clearly only what we already anticipated to be true (or worse—even subtly conveying to local participants the notion of a "right" answer in the discussion). Within the team, a mixture of backgrounds (experience, ethnicity, and education) and the openness for some critical analysis and self-reflection offered opportunities to counter these tendencies. Having a researcher with training and experience in facilitation was useful in keeping the focus group discussions and other activities "directed" but not "led." During stakeholder identification and other activities, triangulating responses with different groups of stakeholders was important in circumventing assumptions, as was discussing specifics and examples with community members (to the greatest degree comfort would allow). It also was very useful to use the facilitation technique of "feeding back" statements and responses as clarifying questions during the course of discussions or activities to allow participants to correct the researcher's interpretations.[16]

Seeking Diversity. The ongoing process of stakeholder identification and understanding the complexities of stakeholder diversity emerged as an important issue in East Kalimantan. Tempting as it was (for reasons of time and simplicity) to compartmentalize stakeholders early on (that is, categorizing them as representing or belonging to a certain category), seeking to continually understand the complexities of stakeholders' diversity was crucial.[17] Each stakeholder belongs to multiple groups or categories simultaneously and over time. These groups and categories can either compete with or reinforce each other (in terms of power relationships). Stakeholders' identities, in other words, can be understood to exist as overlapping circles of identity. This perspective was important in East Kalimantan to the extent that it forced us to confront the illusion of uniformity in our "homogeneous" groups. The "homogeneous" women's group (all female, Christian, subsistence swidden agriculturists) was in fact far more complex than initially thought because of other factors, such as ethnicity and social status. These other identities were so divisive that they created considerable barriers to the participation of the Punan (the ethnic group of lower status) women. In an all-male series of gatherings, members of an ethnically mixed group appeared to be far less inhibited in their ability to exchange ideas; these men were accustomed to meetings because of their political or other leadership roles as village heads. Thus, among women, ethnically based status differences competed with the binding force of their common identity as agriculturists. However, among men, the identities of political figures reinforced the homogeneity of the group[18] (even though even more distinct differences existed among the men in terms of occupation).

Recognizing that stakeholders' power exists in relationship to each other and is historically based was critical for our analysis. The distribution of power between and within stakeholder groups referred to many things, from information flow to decisionmaking ability to the distribution of benefits. Tiani and others (1997) emphasize this point from their C&I methods testing experience in Cameroon. Anne Marie Tiani (personal communication, 1998) also says,

> the analysis should not stop at the level of interactions among different stakeholders concerned by the management of the forest.... Rather it should penetrate into the internal structure of the group so as to determine the practices that could determine sustainability for management. For example, it should question, do people have an equitable distribution of the resources put at their disposal? Are some social groups marginalized regarding the resources? What is the reaction of these groups regarding the resources? Looking at this level of interaction ... understanding power relations ... can help clarify actions and behaviors.

Impacts and Social Change. Gender and diversity analysis plays a critical role in shedding light on complex situations and highlighting issues or dynamics that may need resolution. It is also, at its core, oriented toward social change. (This concept is linked to the earlier discussion of benefits of a gendered approach—specifically, the empowerment aspect.) Rigorous gender and diversity analysis based on participation almost inevitably brings about some increased level of conscientization (which is a crucial element of change) for both local participants and researchers. This result may give pause to research teams who do not see themselves as active catalysts in the movement for social change.

Regardless of goals and comfort levels about catalyzing social change, once again, one priority is at least awareness of the implications of gender and diversity analysis—that is, understanding that the assessment process will very likely have some impact regardless of intentions. People are affected by even apparently small actions or decisions by the research team; intentional or not, activities may reinforce or undermine the status quo. For example, the decision (conscious or by oversight) to not invite a marginalized group of stakeholders to a meeting reinforces the public belief that they are less important. Asking men to invite women to sessions, or inviting groups to settings where they realistically cannot participate (either physically or socially) produces the same result. In East Kalimantan, some local men were very surprised that we sought groups of women (as well as groups of men) to undertake the activity and even more surprised that women had anything to contribute to mapmaking.

Acknowledgement from outsiders, proof of knowledge or ability, and consciousness and confidence-raising within a group can initiate or add momentum to a process of social change. Groups of women discussing power relationships or common concerns across ethnic boundaries (a relatively new phenomenon in our East Kalimantan site) increases awareness and alliance-building. A discussion about local rights may, in the long run, fuel local confidence to act. The potential downside of a gender and diversity approach is that, whereas it may offer long-term change and benefit, it also can create difficulties and/or conflict within a community because of its disruption of the status quo. Although this disruption may ultimately be a necessary part of making long-term changes regarding equity and other issues, it may trigger backlash against those involved and accidentally reinforce marginalization.

Assessment Tools: CIFOR's Generic C&I Template

Tools used in assessments, as with any intervention, are never entirely neutral. They may unintentionally favor some users over others or lead discussions and ideas in a particular direction. As such, it is always valuable to give the tools the same kind of careful consideration that the team undertakes regarding its theory and approach.[19] Our field experience focused on methods that rely on or feed into CIFOR's *Generic C&I Template* for SFM (1999).[20] CIFOR scientists and partners tried to build gender and diversity sensitivity into the template by using gender-balanced interdisciplinary teams in its development and by expressly encouraging the teams to consider these issues. However, because gender and diversity are context dependent and because the C&I set is intended to be adapted to each site, the template cannot refer to specifics of gender and diversity. Thus, the template encourages[21] and enables rather than "demands" the application of a gender and diversity approach.

As such, it is up to those who use the C&I template (and thus adapt it to local conditions) to recognize the openings and invitations set to incorporate gender and diversity sensitivity. Where the template reads "local communities," it is the responsibility of the assessor to consider the local stakeholder categories and groups as dynamic and complex institutions and to consider the issues of equity and power within and between the groups. In East Kalimantan, we tried to take the terms local people,[22] community, and so forth, as symbols to represent the many groups and categories and the relationships between them as well as a cue for us to consider which of the many overlapping groups and categories were relevant to each case. For Indicator 3.2.1, "Mechanisms for sharing benefits are seen as fair by local communities," we had to understand local communities as a generic term that needed substitutions by the multiple stakeholder groups and categories that we had assessed—Punan women; Punan men; Bugis traders; Kenyah Lepo' Ke women; Kenyah Lepo' Ke men; and within each of these categories, the subcategories of more and less politically connected, informed, and powerful.

Examining the C&I from this perspective, we were exposed to the dynamic between not only the timber company and the village at large but also between the subvillages, ethnic groups, religions, and genders. In fact, these dynamics turned out to be very relevant; access to benefits and communication with the company varied considerably between these overlapping groups. They appeared ultimately to be dominated not only by one ethnicity but also by a subvillage, one of the religious groups within it, and particularly by the men in that group. As a result, one specific subgroup had different opportunities and constraints in its decisionmaking framework regarding resources because it had far more access to information and assistance from the local timber company in the form of seeds and other agricultural inputs. This finding reinforces the crucial need for a stakeholder identification process that is thorough and inclusive and that acknowledges the tensions and complexities of forest situations.

One potential challenge to this kind of process is that most "assessors" (that is, researchers or formal forest managers), like most people, have been socialized to live in a male-centered, hierarchical world; they are embedded in it.[23] Combined with the other barriers to participation in C&I adaptation and assessment by less visible groups, this socialization often leads to assessors' focusing on the more dominant (often male) groups. Either homogeneity is assumed, or the dominant groups' features are projected onto the less dominant groups. To avoid or minimize this situation, the assessors can undertake awareness-raising activities for themselves as a part of designing their activities, such as reading background materials, participating in team discussions (for example, based on a few questions about their own professional and personal experiences with gender and diversity, their own potential preconceptions, and how the team will recognize and address biases), and involving a gender and diversity resource person early on. A good C&I adaptation and assessment process offers opportunities for consciousness-raising among all participants.[24]

Assessment Approaches: Challenges and Strategies for Accessing Local People's Participation

The assessment process, which strives for accuracy and completeness, relies on gender and diversity analysis to guide it in its early stages of stakeholder identification as well as in its actual assessment phases. One major challenge in East Kalimantan was that the very people it was most important to access or communicate with during these phases were often the most difficult to reach. By recognizing such challenges in advance, they can be better anticipated and addressed.

Challenges. Many factors can limit the effective participation of less dominant subsections of a community in research or assessment activities. They include many of the basic research biases (see Sutherland 1994), such as

- a *likeness bias,* whereby researchers focus on those groups that most resemble themselves (for example, in gender, culture, or level of education);
- an *off-the-beaten-track bias,* whereby researchers are reluctant to go far afield to meet with people; and
- a *language bias,* whereby researchers fail to communicate in the native language of the group.

Social or cultural norms of certain groups may also limit their speaking out in the presence of certain other groups, the most common example of which is women in the presence of men. In Long Loreh, East Kalimantan, only men would speak in a mixed-gender discussion group (whether ethnically homogeneous or heterogeneous). In an all-woman group, the women of the dominant ethnic groups spoke, and the Punan women spoke extremely rarely. In a group of all Punan women, however, women who did not speak in a mixed women's group spoke so energetically that the researchers barely had a chance to ask questions. In circumstances such as these, mixed-gender or mixed-ethnicity discussions, stakeholders may feel that they do not have the knowledge, ability, or right to contribute. Their participation may be actively or subtly discouraged or undervalued by others; they may risk violating social or cultural principles and expose themselves to risk of retribution by the community through participation.

Cultural or social factors, in the form of responsibilities and workload, also can limit women's and other groups' abilities to participate. In Long Loreh, the Punan worked in upland rice fields (*ladang*) that were farther from the village than those of other ethnic groups, and they worked them later in the season; both of these factors limited their accessibility for any village-based research. Commonly, local women have more responsibilities and less free time than men and thus have less time to be involved in research. Even when they are involved, they may still be at least partly involved with their children, either caregiving or having children drop in and hang on them (literally), so they are required to divide their attention in a way that is not necessary for male participants.[25]

The experience in Long Loreh also highlighted the extent to which some groups of people simultaneously have multiple characteristics (or sometimes identities) that overlap in a manner that reinforces distance from the research process.[26] The elderly, for example, often spoke only local languages, and little or no Bahasa Indonesia (Indonesian language); many had no formal education, and thus were illiterate; they tended to be shyer with outsiders; and were generally less mobile than the other individuals. Whereas a single identity or characteristic that distances a stakeholder from the research process may be addressed fairly readily, multiple "reinforcing identities" make it much more difficult. In fact, the elderly attended far fewer discussions than others and participated far less when they did. Our research team tried to include as many elderly people as possible by, for example, ensuring that elderly participants were always accompanied by a trusted relative or friend for support and translation, focusing on verbal or drawing methods and activities, and trying to keep activities informal and relaxed. Time and resources permitting, we could have visited elderly people in their homes and used more verbal activities, such as storytelling. As our experience revealed, unless such challenges are noted and efforts are made to proactively address those challenges early in the study, the risk that a group of people—in this case, a generation—with unique knowledge and experience will be overlooked is significant.

Researchers carrying out the methods tests in other sites encountered several challenges to accessing certain groups of people. For example, Anne Marie Tiani notes that in the Cameroonian C&I tests, stakeholders outside the local village were reticent to participate because much of the wood they used was cut illegally (personal communication, 1998). She also observed a more "structural" problem there in terms of accessing diversity. Although the researchers attempted to group equal numbers of women and men within the different stakeholder groups (as was recommended by the methods), those stakeholder groups were (sensibly) often organized by occupation, and the occupations tended to be heavily weighted by gender. The *ouvriers de scieries* (sawmill workers) group consisted of 24 men and only 4 women; the carpenters group was made up of 30 men, and no women; the sellers group numbered 30 women, with no men. Thus, some stakeholder groups, almost by definition, preclude representativeness of gender *within* the groups; in such cases, representativeness has to be sought *across* groups.

Strategies. In terms of accessing and "hearing" women, men, and people of diverse identities, one important approach is to seek *equality of outcome* (that is, the outcome of participation), not *identical opportunity to participate*.[27] In other words, it is the ultimate quality of the participation that should be equitable; the opportunity to participate (that is, the form and forum for participation) need not be identical or shared among all stakeholders simultaneously. In Long Loreh, for example, because of certain existing differences or barriers among the local peoples, it would not have been effective or wise to always invite men and women or people of different ethnic groups in equal numbers to the same meetings. Instead, we attempted to design and conduct the research and meetings in such a way that all stakeholders could participate as much as they wanted and in a safe environment, allowing them not only to contribute to the research but also, ideally, to gain something from the process. In many cases, we held discussions in single-gender groups or within (as opposed to across) ethnic groups.

Depending on the context and goals of the research or assessment, in some cases, it would be very appropriate to start with such a "homogeneous groups" approach and then move toward or iterate more processes with multiple stakeholders. Such would be the case, for example, in action research or management processes in which one or more groups lack the confidence or capacity to express themselves to other groups. The homogeneous group phase can be used to build capacity and confidence within those groups (and to build openness to listening in the others) before the different groups come together. Even in less action-oriented assessments or research, such a process of allowing the smaller homogeneous groups to gain confidence or clarity on the issues internally can be useful in laying the foundation for a meaningful, shared discussion or critique when a synthesis of the assessment or research is "returned" to the community.

Several well-documented strategies are used to increase the ability of less dominant groups to participate effectively in research such as HWB assessments. In terms of characteristics for success, it seems important that research teams should

- be gender balanced, have local language skills, and know local protocol and
- include a skilled facilitator to conduct the participatory sessions[28] (and have someone else take notes or listen).

In terms of approaches and actions, the assessors should

- invite women and other marginalized groups directly, not by proxy;
- go to the participant (not vice versa);
- avoid rushing through explanations of why the research may be relevant to different groups;
- understand that all groups will not act or interact in a similar way or at a similar pace;
- recognize that accessing the input of some groups may require more time than others, and plan for longer travel times, more cancellations, and slower meetings;
- be flexible and responsive with time, place, and pace of the meetings as well as with distractions, such as the presence of children or the participant's need to simultaneously carry out other tasks;

- be aware of group dynamics, and adjust processes as needed;
- understand the risks of breaching the status quo to the participants, and respect their right to be silent;
- use pictures, drawings, or other creative means as often as possible when literacy may be an issue for any of the participants;[29] and
- acknowledge the value of people's contributions both before and after discussions.

In our experience (and, it seems, in that of most other assessors), an underlying principle emerged that links all of the above strategies: flexibility and creativity in everything. In Long Loreh, when the Punan and Merap villagers (understandably) had no desire to return to their fields on the weekend, we tried to rework the transect walk method so it could be carried out indoors from memory. Even though an ideal participatory mapping scenario involves participants doing all their own drawing, when the women lacked the confidence to draw, I agreed to do a good part of the drawing while they directed me. (The men, who were more confident, did it themselves.) An earlier Indonesian C&I methods test team noted the same approach in their trials: "We realized that there was no standard strategy or approach even for the implementation of the same method" (Sardjono and others 1997; personal communication from Mustofa Agung Sardjono, 1998). They describe slightly different situations that required researchers to switch from house meetings to going with villagers to the forest or rice field. For example, when the team realized that it was going to be very difficult to get women to attend a group meeting because meetings and decisionmaking were considered the responsibility of the men, the team approached meetings another way. They instead invited the local women's church organization to prepare food and drinks for a men's meeting. This gathering gave the female researcher an opportunity to join the women, interact about the research, and initiate research topic discussions. This experience raised the women's comfort levels and confidence in the research process, and they became more willing to participate in meetings.

Conclusions

From a personal perspective, the most challenging and valuable part of the methods testing experience was facilitating the participatory research sessions. Taking on the role of facilitator means putting the theory of gender and diversity analysis into action. Discussion groups, focus groups, and mapping groups are microcosms that reflect all the tensions and complexities of gender and diversity in forest management writ large. In a group discussion with local peoples, especially when subject to outside pressure, it

An Excerpt from the Long Loreh Field Report

In retrospect, having taken the approach outlined in this chapter, the team felt generally satisfied with gender and diversity in terms of *accessing participants*—that is, people being present at sessions—and the balance of representation of different groups. The main weakness was the relatively low participation of members of the Punan communities, because we were never able to entirely overcome some of the overlapping barriers to participation. Some of these barriers may never be satisfactorily overcome without a much longer-term relationship and time to build trust. An additional note is that over the course of the human well-being assessment, we saw many of the same people participating again and again. This fact was positive in the sense that those people were interested in the issues and the process, but less than ideal in the sense that a small percentage of the total local population participated.

In terms of *accessing input*—that is, the degree to which all those involved in the research actually engaged themselves in discussions and were heard—the ideal would have been to have much more balanced participation and more people engaged from the less dominant groups, such as the elderly. Although the group-based methods both raised consciousness and allowed us to interact with many people in a short time, they also had some significant limitations. In the context of East Kalimantan culture, for example, it seemed especially difficult to create the kind of "space" needed in group discussions to allow some people to speak. Ideally, more one-on-one follow-up with people would be necessary to balance that limitation.

In summary, any success we experienced was most likely based on several linked elements. The effort to be very aware of the theoretical grounding was important, as was a continuous check regarding assumptions, biases, and shortcuts. Considering stakeholder identification to be an ongoing process allowed an acceptable level of complexity and accuracy in our understanding of local relations and dynamics. Trying to be flexible, creative, perceptive, and willing to abandon set plans enabled us to respond to situations and ultimately involve far more people than otherwise would have been the case. These strategies, as basic as they are, consistently seemed to be relevant in the field and, in retrospect, appear to have been appropriate for resolving the challenges we faced. The major limitation was the lack of time. A tight time schedule restricted our ability to pursue strategies more thoroughly, and to continue to learn and adapt. Most important, it also limited our ability to build trust in the research process and to offer local peoples the opportunity to take more legitimate ownership of the process.

would be relatively easy for a facilitator to "hear" only what he or she wants to hear about values or practices. It would be easy in difficult moments of a session to overlook people or to brush aside their opinions when they "make the process difficult" (for the facilitator, researcher, or assessor), for example, by voicing dissenting views or by having different needs for ensuring their participation. It would be easy to neglect the impact of a research process on local peoples, because it would make for simpler planning and implementation.

Extending this notion from research to the level of forest management, it would be easy (at least in the short term) to exclude local peoples from formal forest management planning in those many cases where they might have conflicting interests with other stakeholders. In fact, it would be easy to ignore gender and diversity issues in SFM, because to include them means challenging firmly held perspectives and perceptions and opening up planning processes.

Yet despite the perception of "ease" associated (at least temporarily) with ignoring gender and diversity at any level, such an approach also fundamentally minimizes learning, understanding, and potentially equity and positive change—especially for marginalized stakeholders. Researchers, facilitators, and formal forest managers are part of the matrix of gender and human diversity; choosing to ignore rather than acknowledge and address our places (as well as those of other stakeholders) in that matrix heavily influences the outcomes of our research, assessment, and actions.

Opening research and forest assessment and management to incorporate gender and diversity approaches is not easy, but it is possible, and it is desirable. The benefits and opportunities that emerge from taking on this challenge are necessary for maintaining multiple and diverse peoples and forests over the long term. If we acknowledge and address power issues through gender and diversity approaches, we may face multiple hurdles—but we also enable rigorous and insightful analysis of complex situations, as well as challenges to and opportunities for SFM. And in doing so, we may create the beginnings of processes that can positively shift an inequitable and unsustainable status quo.

Acknowledgements

The author was part of a field research team made up of Asung Uluk as a Center for International Forestry Research (CIFOR) research assistant and Ibu Terpina Ipo (Unjung), Ibu Lidia Unyat, Ibu Nyraisyah, Ibu Korlen Njau, Ibu Main Usat, Pak Jusuf Anye, Pak Usuf Lawai, Pak Hermann Songgo, and Pak Yunus Romains as community research assistants. The author acknowledges this team's work and commitment; the assessment (and thus also this chapter) would have been impossible without their efforts.

Endnotes

1. Several research papers strongly influenced my thinking and formulation of ideas for this text: Townsend 1995, Ibo and Leonard 1997, Selener 1997, Shearer 1997, Sarin 1998, and Ingles and others 1999.

2. In some instances, individuals are in fact born into a third biological category that combines both male and female characteristics: hermaphrodites. Although awareness is increasing about this group of individuals—which is quite diverse in terms of ultimate social identities and offers many important insights into this issue area—for the sake of simplicity, we focus on only male and female categories.

3. Harding offers a useful perspective that "feminists must theorise gender, conceiving of it as: an analytic category with which humans think about and organise their social activity rather than as a natural consequence of sex difference, or even merely as a social variable assigned to individual people in different ways from culture to culture" (cited in Hawkesworth 1997). The stress on the need to theorize is key, as is the emphatically made point that gender should not be conceived of as a causal force.

4. By *constitutive,* Scott means that gender "operates in multiple fields including institutions and organizations, as well as culturally available symbols and subjective identity" (cited in Hawkesworth 1997).

5. The question of the differences between the sexes is clearly a highly debated topic. In this chapter, I reject the notion of "biological determinism," or "essentialism." In other words, I do "not espouse [the idea] that all women share characteristics of nurturance, or lack of aggression; nor assume homogeneity in aims, etc., as a matter of essence.... It starts from the premise that very slight biological differences have been reinforced socially to create the appearance of considerable (almost unbridgeable) differences between men and women" (Young 1988).

6. Kanchan Mathur (cited in Sarin 1997) makes the poignant remark that "now, with the extensive use of amniocentesis to abort female foetuses, these socially constructed differences have started affecting females even before birth."

7. As one form of difference, *gender* can be understood to be subsumed within *human diversity,* as outlined in the following section. However, its ubiquity as a key (sometimes *the* key) form of difference in relations and roles among people creates significant value in also addressing it independently.

8. I refer to both *groups* of people (which can denote either a sense of membership to the group by its members or some political potential) and broader *categories* (which denote some shared, but not necessarily internally recognized, characteristic or position). Although the distinction can be important, because both concepts were equally valid within the context of the assessment, I refer to them alternately, without differentiation.

9. This analysis should be undertaken at many levels. As Mathur points out, gender roles "are embedded in four major social institutions: the household, the community, the market, and the state." These institutions define who gets "included or excluded, who has access to decision-making and resources, what gets done or is not done and who sets the rules ... as institutions are the sites for the production, distribution, and exchange of resources, it is important to know how they structure women's subordination and resourcelessness" (cited in Sarin 1997).

10. This transition also can be understood to be continuing in the current human diversity or stakeholder differentiation approaches, as espoused in this chapter.

11. One such example is an area in the eastern Himalayas where the women, despite having extensive knowledge of economically valuable crop genetic resources, have very little influence in decisionmaking about crop production (personal communication from Barun Gurung, 1999). Customary laws—such as those that deny property rights to women—reinforce those inequities as part of the cultural routine, not only for those men who benefit but also for the women who are marginalized by them.

12. Incidentally, the men also allocated greater power to the men than the women, but they perceived the gap to be far less than that expressed by the women.

13. Schmink (1999) rightly highlights a point that social research must constantly address: that gender and diversity analysis can offer opportunities for innovative alliance-building for resource management. She relates that "often women's responsibility for family subsistence and health means they focus more on livelihood systems and on the environment as opposed to the market-oriented perspective of men." From a conservation perspective, local women may be seen as key allies in conservation.

14. Sarin (1997) describes gender and diversity issues being ignored at the expense of sustainability at the household level in development: "Experience has disproved the projections of development pundits who believed in the 'trickle-down theory.' The assumption that targeting development interventions at male heads of households would equally benefit other household members, particularly women, has not been validated in practice."

15. See Chapter 4 for additional contextual information on a similar system in East Kalimantan and Chapters 5, 8, 12, and 16 for analyses from West Kalimantan.

16. For example, the researcher can follow up discussion points with a question such as, "Just so I am sure I understand what you are saying, I think I heard you say [feedback summary of statement/ideas]; what parts of that did I get right? What parts can you correct me on?" or "Am I right in understanding you to be saying [feedback summary of statement/ideas]?" (Note that the latter example, because it is a yes-or-no question, may be less effective because it invites more instant agreement and less thought than the former).

17. To meet this challenge, I tried always to be aware of the many concurrent identities of the people I was working with and talking to, as well as my own. When talking to a key informant, for example, if I approach her because she is a swidden agriculturist, I also seek as much information as possible about her other identities so that I can better understand who else she might "represent." Is she old or young? Rich or poor? Politically connected or politically marginalized? Is she typical of the group to which I have tentatively assigned her or is she, for example, a "poor" agriculturist who has far greater access to resources because of links with outside actors who donate seeds and tools?

18. This case highlights that *identity* is also intertwined with skills and opportunities to reinforce or develop those skills, not only because the men have the identities of similar roles ("political" figures) but because they had had the opportunity to develop similar skills and confidence in discussing their opinions in meetings through their political roles and tasks.

19. As described in the Introduction, the assessment was intended to generate insights into well-being and to assess the tools and methods themselves.

20. The *Generic C&I Template* (CIFOR 1999) from the Center for International Forestry Research (CIFOR) is designed as a monitoring tool to assist in assessing sustainability—including human well-being—at the (commercial) forest management

unit level (for more information, see Annex 1 in the Introduction and relevant discussion in the Introduction). CIFOR has also undertaken work on criteria and indicators for sustainable community-managed forest landscapes (see also Burford de Oliveira 1999; Burford de Oliveira and others 2000; Ritchie and others 2000).

21. References to "local people" and "communities" throughout the C&I set are contextualized by text on the significance of stakeholder diversity and analysis (Colfer and others 1999c).

22. In my opinion, using *local peoples* (plural) rather than *local people* would be more encouraging to assessors to consider the human diversity present in each site.

23. In this case we refer to *assessors* because, in accordance with CIFOR's *Generic C&I Template,* we are considering "outside" (non-community-based) parties undertaking (participatory) human well-being assessment for a Forest Management Unit. In co-management (or community management) assessment, community members would themselves take on the role of (self-) assessors by using criteria and indicators adapted to their context.

24. This case has a Catch-22, however. Where awareness of gender and diversity sensitivity is low—and thus where the potential consciousness-raising aspects of the C&I adaptation process would be most needed—the fundamental gender and diversity sensitivity processes that would allow for such insights are unlikely to be recognized or in place.

25. At least two of the CIFOR test teams (Porro and Miyasaka Porro, McDougall) specifically reported on this phenomenon as a factor in their CIFOR social C&I methods assessments.

26. Other methods testers in East Kalimantan (Sardjono and others 1997) divided stakeholders by age but apparently drew the line between youth and adult and did not specify the elderly (thereby creating an opportunity for this group to be overlooked in the process).

27. Although this approach may sound counterintuitive—especially in the current climate of appreciation for equal opportunity in other arenas (such as gender and organizational development)—it is an important perspective. *Equal opportunity* is most often interpreted to mean "the same opportunity," that is, in which identical opportunities are given to all players. This interpretation assumes that all players face the same constraints to acting on that opportunity. Kanchan Mathur (cited in Sarin 1997) alludes to the weakness of this assumption with an analogy of a fox and a crane to which food is offered on the same plate. The different access of the two creatures, despite apparently equal opportunity, is obvious. An approach of *equality of outcome* tries to overcome access differences by emphasizing the ultimate quality of the participation based on its outcome. In the analogy, the fox and the crane may be offered different foods on separate and different plates, thus addressing their individual needs and constraints.

28. This strategy, as well as sensitivity to the context and goals, is critical because of its major role in influencing the empowering and disempowering effects of discussions on local peoples.

29. For example, to improve the ownership of the discussion in a focus group with low literacy, our discussion notes ended up being a shared wall full of stick figures.

CHAPTER TWO

The Place of Rural Women in the Management of Forest Resources

The Case of Mbalmayo and Neighboring Areas in Cameroon

Anne Marie Tiani

Many studies have been carried out during the last decade on the concept of sustainable forest management (SFM), and criteria and indicators (C&I) have been developed to evaluate the sustainability of forest management (OIBT 1992; FSC 1994b; Burford de Oliveira and others 1998; Colfer and others 1998a, 1998b). A consensus has been reached that this management requires

- knowledge of different stakeholders, their interactions, and the impact of their activities on forest cover;
- access to resources by each of the stakeholders and also the equitable sharing of the benefits from the forest; and
- recognition that rural populations of forest zones are inevitably actors in the management of forest resources because of their proximity to, dependence on, and investment (time, energy, and resources) in the forest.

In general, when identifying stakeholders, a rural population is considered as a whole. Detailed studies reveal that this group is made up of subgroups who have different and sometimes contradictory interests in forest resource management. The origin of these differences may be historical, with roots in the occupation of the territory, in cases where the rural population is composed of many communities of different ethnic groups or with different activities (Brocklesby and others 1997; Tchikangwa and others 1998). Rarely have women been considered as a separate group in SFM. Field studies have shown that women have their own problems, perceptions related to their history, difficulties, and adaptive strategies (Weber 1977; Berry 1988).

In this chapter, I examine the specific situation of rural Cameroonian women in light of fieldwork data and in relation to Principle 3: "Forest management maintains or enhances fair intergenerational access to resources and economic benefits" (see Chapters 3 and 10 for analyses in similar circumstances in Cameroon and Chapter 9 for another example from Central Africa). This principle has been identified as fundamental to SFM. I seek to determine whether Cameroonian forest women believe that their access to resources is fair and secure, for them and their children, now and in the future.

Women represent more than 51% of the Cameroonian population, and 80% of them live in rural areas.[1] Rural women living in the forest areas of Cameroon are important, not only because of their numbers but also because they are the actors most closely related to the forest. They cannot be excluded from forest management that claims to be truly sustainable (CIFOR 1998).

I outline the conditions for women in this area in an effort to portray the social restrictions imposed on their access to resources, attempting to understand the survival strategies developed by these women to deal with social and economic difficulties and the consequences of these activities, as well as their impacts on forest cover. Finally, I propose some solutions for ameliorating living conditions of rural women that will permit them to be fully involved in the management of forest resources. These moves also will strengthen their roles as individual partners in SFM.

The data used in this analysis were collected during two field trips undertaken in cooperation with the Center for International Forestry Research (CIFOR). In March and April 1997, a study was conducted to test the ecological C&I for SFM in the area of SOLIDAM (Solidarité pour le developpement des villages d'Akak à Melan), near Akonolinga, about 200 kilometers southeast of Yaoundé. From April to December 1997, CIFOR carried out social science methods tests of social C&I (see Introduction for details). The main source of information for this research on the status of women was in the city of Mbalmayo and the neighboring villages of Akono (30 kilometers west) and Ekekam (60 kilometers east).

Location and Background

Mbalmayo—about 50 kilometers south of Yaoundé—was chosen for several reasons. Six wood-processing industries and many small sawmills are located in the city, and forest administrators—who are privileged partners in the management of Cameroon's forests—are also based there. Population density is high; Mbalmayo has about 50,000 inhabitants and is surrounded by the communities of Akono, Metet, Mengueme, and Ngomedzap. Vegetation (semideciduous equatorial forest of the central-southern Cameroonian pla-

teau, with a four-season climate) has been highly degraded by human disturbance. It is characterized by

- old fallow systems (*ekotok*) evident from the presence of oil palm and the
  density of vegetative regrowth,
- secondary forests (*essema*),
- swampy forests (*engas*), and
- blocks of residual primary forests (*afan adam* or *gobo afan*)—found mostly
  where saleable plants are scarce or where access is relatively difficult.

All local rural populations belong to the large Beti tribal group, but clans
differ from one village to the next. The Bënë live in Metet and Nkout, the
Mvog Manze in Yop, the Yanda in Akono, and the Etenga in Mendong.
Ewondo is the common language spoken in the region. In general, marriage
between members of the same clan is forbidden. Most adult women join
their husbands on marriage, so they are nonnatives to their village of residence. The Etanga are among the few clans who do not follow this rule.

Several ethnic groups live in Mbalmayo; the Beti and the Bamileke are the
most common. The Beti tend to work in the government service or for logging companies, whereas the Bamileke, who migrated from the west of the
country, work as carpenters, bricklayers, and mechanics and in other trades.

In the villages, the main activity is farming. Men grow cocoa, and women
cultivate food crops in a swidden system. In addition to farming, women
harvest nontimber forest products (NTFPs) and fish seasonally. Men also
hunt or tap palm wine for sale or home consumption. A group of illegal sawyers is particularly active in this area despite repressive measures by forestry
officials to curtail those activities. This illegal activity results because of the
area's proximity to Yaoundé as well as the high cost of sawn wood since the
devaluation of the CFA (Communauté Financière Africaine) franc; see
Chapters 3 and 10 for further discussion of this traumatic event.[2]

Village communities have settled on lands declared the National Domain
by forest administrators. The government confers rights of use of forest products on the communities. The local system of land appropriation, however, is
complex; plots of land are acquired through inheritance, by cutting trees
from communal virgin forest, by gift, or by legacy.[3]

Mbalmayo is essentially an industrial town with wood-processing industries of varying sizes. It is also a collection center for cash crops.

The Female Life Cycle in Cameroon's Forest Areas

Children of both genders belonging to the same family will, from early in life,
have different fates. Whereas the newborn female baby is called *ngon*, which
means "girl," the male child is called *nkudu*, meaning "the one who punches."
From birth, the destiny of each is defined: for one, to bear children; and for the
other, to fight and conquer (Tsala Vincent 1973 cited by Laburthe-Tolra 1981).

The girl is psychologically prepared for her role as wife and mother, and the boy is prepared for his tasks as a future household head by his father.

The life of a woman progresses in three stages: childhood, adolescence, and marriage.

- If there is enough money to cover school expenses, a girl child will go to primary school. If not, her brother will be favored. A girl will have difficulties continuing beyond primary school because high schools are rare in villages, and parents typically refuse to let their daughters go alone to distant places. In many cases, girls who leave the village for further study return later, pregnant.
- Once the girl reaches adolescence, she helps her mother with the daily tasks. She works to meet the needs of family members and to send money to her brothers who have gone to town to work, train, or pursue further studies. She may live in a kind of symbiosis with nature, as an element of nature. Her time and effort are centered on her environment, helping her parents to create and care for cash crop plantations, cultivate food crops, harvest fruits, and fish.
- Marriage is a radical change in the life of a girl. At this point, she loses all usufruct rights to the goods and resources she spent her life gathering or developing. They automatically belong to her brothers, who may remain in town or return to the village and who take them over when their father dies. She joins her husband's family—in general, the boy remains in his clan, even after marriage—where she will use the goods left by her married sisters-in-law. Sometimes the bridegroom, seeking emancipation or after a conflict with his family, migrates deeper into the forest to create his own *mvog* (village), in which case his wife will need to begin from scratch. As a young girl, she may feel that she is an economic instrument for her family; married, she remains as such to acquire, ensure, or maintain social rank for her husband.

This situation gives rise to many questions. Women do not have the right to own certain resources, such as trees (the tree is a sign of appropriation of the land). What property do they own or control, and what are the limits?

It could be said that the local population considers this form of management—which excludes a category of the population (women) from the benefits of certain resources—as equitable (Principle 1, Criterion 2; see Annex 1 in the Introduction). But who defines this equity? Men? Women? What are the resources?

The Concept of Resources

Resource is defined as wealth, natural products, goods, or natural means available in a country or a community. The farmers of Kan (Tiki and others 1996; Burford de Oliveira and others 1998) define resources as any elements of the environment, including people, that are not created by the people and that are used for survival.

In decreasing order of importance, the populations of Bitsok-Adzap classified forest resources as

- land, as the habitat for vegetation, wildlife, and an essential element in farming;
- wood from trees;
- wildlife; and
- aquatic animals.

(The resource "people" is excluded from this classification because it cannot be compared with the others.)

For farmers, the forest is seen as a complete ecosystem; it is an ecological ensemble made up of the environment and living beings. People, soil, plantations, and fallow lands are all elements of one whole. This definition, given by farmers for *forest,* differs from that of many current African regulations, in which *forest* means noncultivated, natural areas (Bahuchet 1993).

For each category of stakeholder identified at our study sites, the concept of **forest** is closely linked to that stakeholder's interests vis-à-vis resources. A CATPAC[4] survey (Table 2-1), carried out near Mbalmayo, shows that *forest* has several meanings (Tiani and others 1997).

The forest is primarily regarded as cultivable lands by the farmers of the region. Women retailers link **forest** to **foodstuffs** and **trees** because they buy foodstuffs harvested from the fields by farmers and because fruits, seeds, barks, wines, and firewood are taken from the trees growing in the forest. For carpenters and logging company workers, the forest is primarily a source of timber. It should be noted that these various stakeholders have all linked their existence to that of the forest. A perception common to most of these stakeholders is that **forest** is closely linked with **I** or **we** (Table 2-1).

The forest is linked to different concepts by the various actors, based on their respective interests: **forest–arable lands, forest–wood, forest–game.** Some have argued that linking resources to land is a relatively recent phenomenon. During the precolonial period, populations migrated periodically. Groups of extended families moved in search of hunting areas and fertile lands, which they recognized by means of their own fertility indices, such as the presence of some species[5] or the depth of humus. They would settle for about ten years, cultivate the land, then leave for another place. The land was regarded as useful insofar as it was rich in game and was fertile. It was just a

Table 2-1. *Mbalmayo Stakeholders and Important Concepts Related to Forests*

Stakeholders	Forest
Peasant farmers	I–live–fields
Retailers	foodstuffs–trees
Carpenters	wood–we
Sawmill workers	men–we–wood

substratum they used and then left to move on to new lands. Laburthe-Tolra (1981) confirms this migration: "Lineages go as far as possible toward virgin lands which give more valuable crops, toward new peoples to align with or to conquer, depending on the demographic expansion…." But he notes that Beti expansion during the precolonial period also corresponds to a process of segmentation of the lineage that was designed to emancipate the sons who would become heads of families. This last argument confirms the sociocultural aspect of the migration. In contrast, Laburthe-Tolra also writes that "there was no sharing of land during a traditional succession: ownership of land is a colonial invention." The land itself was never regarded as wealth.

The introduction of cocoa farming during the colonial period put an end to these migrations. Because of the life expectancy of the cocoa tree (40 years), its introduction brought about the private appropriation of land: ownership of trees led to ownership of the land on which they grew. By extension, fallow lands became private property. The status of cultivable lands changed suddenly. The land acquired a market value and became wealth liable to be passed on through inheritance (Weber 1977). In localities where the density of the population has reached a certain threshold, land belonging to the community (accessible to each of its members through land clearing) has become scarce. Most land has become private property; one can have access through inheritance, a complex system of loans (whose duration is equal to the life expectancy of the borrower), marriage, or legacy. Thus, the land to which access becomes more and more difficult gains greater value to be regarded, nowadays, as the most important wealth after human beings themselves.

We have discussed a hierarchy of forest resources that varies according to the interests of various forest stakeholders. The farmers of Kan classified forest resources as important for land, wood, game, and fish (Tiani and others 1997). We found similar classification systems for men, women, and youth at all the sites where CATPAC was tested. It should be noted that NTFPs were not mentioned in the classification process, during the study of Burford de Oliveira, or even during the implementation of the CATPAC method. Brown and Lapuyade (1999), who worked in the South Province of Cameroon, note that "generally NTFPs do not seem to be considered as a main source of income…" (see also Chapter 3).

The Place of Women in Resource Management

Cultivable Lands

In precolonial Beti society, the land did not constitute wealth but a resource to be worked. Men's sole farm-related activity was clearing land. They hunted, gathered useful forest fruits, waged war, and fished (Weber 1977). Only women carried out farming activities. Harvests belonged to the family head who shared them among the family members according to need. Food-stuffs were regarded as wealth. Women, young boys, and girls could be seen as a free workforce to be mobilized by the men to produce the foods necessary to preserve their social status.

Viewing land as wealth began with the introduction of cocoa farming. Colonialists presented cocoa as a source of significant income, which immediately increased interest in the tree. Men then began planting young cocoa plants after clearing land. They required women's help to establish the young trees, care for them, and later harvest the cocoa beans. Thus, women's work in cash crop plantations was seen by men as the continuation of the former female duty to participate in her husband's activities. In return, women expected their husbands to meet their needs with the revenues accruing from their labor. Unfortunately, consensus on this issue was and still is rarely attained.

The question arises of why women can't create their own plantations. Berry (1988) responds, "That they have apparently not done so is the result of several factors which have combined to restrict women's ability to commute jurally recognized claims to tree crop farms into effective control of the trees or their fruits." Some people have argued that constraints are created by men, anxious to preserve their access to free labor, prevent women from creating plantations, and consequently, from acquiring lands. Over time, these constraints have become social rules. In some cases, women are not allowed to plant perennial trees or any species with a high traditional value, such as plantains. The creation of a plantation requires an abundant labor force and a relatively long time before production. It is difficult for women to mobilize such a labor force while men continue to rely on their wives, children, and relatives for that purpose. The customary organization of labor excludes the possibility of a woman creating her own plantation; she follows her husband and helps him. Women are also too busy with food crops and multiple household chores to be able to devote time to plantation development. Despite the fact that modern law, which theoretically prevails over customary law, advocates gender equality, women have found it difficult to defend their interests against a man's when conflicts arise over resources.

Cocoa farming brought by colonialists could have been a unique opportunity for women to acquire their own lands, but this opportunity was not embraced. Therefore, "women are never on lands that belong to them but on those belonging to their fathers, to a brother, to their husbands, or to a son" (Weber 1977).

Women, unable to own perennial tree plantations, put a lot of effort into subsistence food crop farming, either on or in association with their husbands' plantations. This situation has strengthened a gender division of labor that already existed during the colonial period, based on the natural physical abilities of men and women. What was, initially, only a division of labor according to specific abilities has, some argue, become a system of exploitation supported by gender-specific interests.

Timber Resources

Wood is the second most important resource provided by the forest. It can be used in various ways, for example, timber for export, local or regional timber, and firewood. Each kind of wood can have one or many users. Logging companies export logs and retail wood. Illegal sawyers, most of whom are farmers, provide timber to the carpenters and the shipwrights of the region. Firewood collectors supply the local and neighboring markets with fuelwood for households.

During our fieldwork, almost no women were working in the wood sector. The logging companies in our research sites employ a few women, mainly for secretarial work; only one woman works as the manager of a local carpentry business that belongs to her husband. In general, illegal sawyers and firewood collectors and sellers are men; women only collect the amount of wood needed for their own household use. So, logging activities relate only to men.

However, it would be a mistake to assert that wood, as a source of cash income, does not interest women. The CATPAC results show that both men and women think that the forest is a place for farming, logging, and hunting (Table 2-1). Very few mentioned NTFPs, despite their importance in the daily lives of women, in towns as well as in villages. Even for women, NTFPs are regarded as secondary products of the forest.

Women do not carry out more activities in the wood sector because of several constraints not easily overcome in the region at this stage. They include physical, material, and financial requirements as well as cultural norms (some would say prejudices) that prevent women from doing wood-processing work, which is traditionally regarded as men's work. Thus, what remains for women is what no longer[6] interests men: NTFPs. "This reflects the cultural preferences of precolonial times where food eaten raw, such as fruit, was associated with women and children" (Guyer 1984, cited in Brown and Lapuyade 1999).

Nontimber Forest Products

Formerly considered unimportant, NTFPs have attracted the attention of researchers in recent years. Various aspects of their management have been studied, especially the economic perspective (Tsagué 1995; Ndoye 1998; Ndoye and Kaimowitz 1998; Bikie and others 1999; Eyebe and others 1999). C&I have been developed to allow the certification of sustainably managed NTFPs (FSC 1994b; Pierce 1996). Ruiz Pérez and others (1999) identified an important relationship between NTFPs and women.

The forest in Central Africa contains a wide range of NTFPs with multiple uses, in modern and traditional pharmacopoeia, in nutrition, as a pecuniary source of revenue for households, and as a source of employment (Ndoye and others 1998; Eyebe and others 1999). The main NTFPs harvested and commercialized in the dense humid forest of Cameroon are *Cola acuminata, Dacryodes edulis, Elaeis guineensis, Garcinia kola, Garcinia lucida, Gnetum* spp., *Irvingia* spp., and *Ricinodendron heudelotii*. In terms of quantity and the margin of profit from sales, *Dacryodes edulis* appears to be most important. This product is harvested and sold by men (Ndoye and others 1997/98, 1998).

The study site (Mbalmayo and its surroundings) has specific characteristics as a part of the dense humid forest of southern Cameroon. *Gnetum* spp., whose natural territory is limited to the northern shore of the Nyong River, is almost totally absent. Only a small quantity of *Dacryodes edulis* is collected from the forests of the area because most of this product that is sold in the markets of Mbalmayo is imported from either Yaoundé or Makenene (about 250 kilometers northwest of Mbalmayo) by Bamileke traders. In addition, only small quantities of *Cola acuminata* and *Garcinia cola* are found.

Ndoye and others (1997/98) found that, in 1996, *Irvingia* spp. harvested in the Mbalmayo area was sold in the markets in Mbalmayo as well as in Mfoundi and Mokolo (Yaoundé). Part of the *Cola acuminata* sold in Mbalmayo is harvested in the area, and part is received from markets in Yaoundé. This finding suggests that the quantity of *Cola acuminata* produced locally is very limited. *Garcinia lucida* sold in this same market comes from Mvog-Mbi (Yaoundé), and *Garcinia kola* seems to be unknown in the area. Moreover, these authors note that in Central Africa, 78% of people involved in the management of NTFPs are women. In the Mbalmayo area, this number is likely to be higher because of the scarcity of some products, such as *Dacryodes edulis,* generally sold by men.

From the above discussion, it may appear that Mbalmayo and its surroundings are poor in certain NTFPs. Such is not the case for other products that are plentiful and provide for the needs of markets locally and in Yaoundé. Various NTFPs are available year-round (marantacea leaves, raffia, palm nuts, and bark) or seasonally (mangoes, honey, caterpillars, and many kinds of fruits). They are abundant from July to October, and the period when fruits mature coincides with a slowdown in agricultural activities, so women have enough time to collect NTFPs.

Many NTFPs are used by the population, including honey, lianas (woody vines), mushrooms, snails, various insects, fish, game, wrapping leaves, fruits, and palm nuts. Some of these products have a high market value. The pits of dried or roasted wild mangoes (*Irvingia gabonensis*), locally called *andok*, are used to thicken soups. Rural women can sell a five-liter bucket of these nuts for 5,000 CFA at the city market, whereas the same quantity of shelled groundnuts (an agricultural product) garners less than half that price. *Ezezang* (seeds of the *Ricinodendron heudelotii*) also are used to thicken soups, and their price is two or three times higher than that of groundnuts. Palm nuts, grown in fallow lands, contain a fatty juice used across the region almost every day for cooking vegetables and leaves. Marantacea leaves, which are collected in low-lying areas, are used to wrap foods for steaming.

Except for products such as raffia, palm wine, and cane, the management of NTFPs is controlled by women, from collection to commercialization through transformation or processing. Nowadays, these activities constitute an important source of income for the household and give a certain level of financial autonomy to women.

Wine tapping is done exclusively by men. They fell palm trees to tap the wine that they sell to women, who then use local techniques to make a popular drink with a high alcohol content called *odontol* or *arki*. This product, which is highly prized, is sold in the village or in the markets of nearby towns.

Moabi oil, extracted from the fruits of *Baillonnella toxisperma,* is eaten and also used in cosmetics to make hair-conditioning products. This oil is now scarce because continued logging has made the source tree rare near villages. Competition between people and wildlife is strong, because both rush to collect the tasty fruits.

From their harvest in villages to their distribution in towns, NTFPs pass through five stages:

- *Harvest*—Women farmers or collectors collect or gather the NTFPs. Whereas the main activity of these women is farming food crops, they collect NTFPs or fish seasonally.
- *Processing*—The collectors or sometimes village women specialize in the processing techniques.
- *Collection and transport*—Typically, farmers or traders who live there go from house to house in the village to buy household surpluses at cheap prices. They accumulate and carry these goods to the markets of neighboring cities. These women traders also carry on farming activities part of the time.
- *Wholesale-to-retail distribution*—Women retailers who live in towns buy products in bulk and then resell them in the same market or elsewhere.
- *Retail sale:* Consumers in urban areas buy the products.

Along this chain, NTFPs pass from collectors living in rural areas to consumers living in towns. Each stage in the process is dominated by women.

During the season of scarcity or periods of intense farming activities (when women are busy with their farming duties), retailers leave urban areas to buy directly from the producers.

Social Changes and Their Consequences for Rural Women

Significant changes have been occurring in traditional societies, prompted by various factors.

- Efforts by the government to expand primary and secondary education into all villages during the past ten years have been successful, and the number of girls in secondary schools is growing comparable to that of boys in some areas.

- The media is accessible to almost all rural populations and is continuously introducing and fostering new and different lifestyles.
- The influence of the media is supplemented by increased mobility and communication between people, bringing about the intermixing of peoples, ideas, and religions.
- An economic crisis led to price reductions for the main cash crops beginning in 1986, and then the CFA franc was devalued in 1994.
- Dualism as a result of adopting customary law and modern law has paved the way for a third more uncertain trend, jurisprudence.

The very foundation of traditional social life has become unstable. Divorce is common. Children born out of wedlock are more numerous, as well as women of child-bearing age who have never married. Polygamy is declining due to the influence of churches. Pokam and Sunderlin (1999) found in their research of 38 villages of the Center, South, and East provinces that in 1997, 20.8% of heads of households were women—single, widowed, or divorced. Moreover, the number of women (>15 years old) absent from their village increased from 41% in 1974–1975 (Franqueville 1987, cited in Pokam and Sunderlin 1999) to 60.9% in 1997. Matrimonial migration has had an economic imperative, where women left villages to go to towns in search of jobs.

Security of Intergenerational Access to Resources

Traditional rules of the Cameroonian forest area deprive young women of their rights to family goods when they marry, ceding the benefits to their brothers. This loss is more or less compensated by usufruct rights acquired by brides in their new family. Young women have the right to use their husbands' lands to cultivate food crops. This right is valid even after their husbands' deaths, as long as they stay in the clan.

However, the status of women who have never been married or whose marital links have been interrupted is important when considering the concept of security of intergenerational access to resources. Three cases of this status (divorced women, never-married women with children, and married women whose only children are girls) are analyzed below.

Divorced Women. After marriage, women lose their rights (of property or of usufruct) to the goods of their natal family. On divorce, they lose the usufruct rights they had gained in their marital families. When a woman returns to her own village, she has no legitimate claim to even the smallest share of resources that she contributed to gathering before the marriage. She and her children have to beg her brothers to lend them, temporarily, a plot of land.

The argument put forward to support this state of affairs is that these rules, apparently unfair, contribute to protecting the community against social evils such as divorce. Thus, the argument goes, no married woman will consider leaving her husband's household. But this argument no longer holds true, given the ongoing changes in values. In fact, the number of divorces is increasing, particularly among young people.

Never-Married Women. On the death of her father, a never-married woman becomes the property of one of her brothers as part of the inheritance. She has the value of a good, like land or livestock. She continues to work in the plots of land that belonged to the family and that have become her brother's. However, the person with legitimate use rights is now the brother's wife. Therefore, the situation of the never-married woman is marginal, comparable to that of a stranger who may leave at any time.

The situation of her children is even more difficult, because all that is granted to them is seen as alms. In the past, children "with no father" were called "sons of the village" and they were treated as *tsali* (slaves). Their mothers remained in the village as servants of their brothers and worked as such. They could be excluded from a group at any time, and they did not share the inheritance; they were almost condemned to remaining as slaves and, by that ruling, single (Tsala Vincent 1973, cited in Laburthe-Tolra 1981). As the number of children without fathers increased, they were frequently (and the tendency persists today) attached to the lineage of their maternal grandfather, especially when he did not have other children or only girls.

Whatever the case, such children are still marginalized. What security of access to resources do such women and their children have?

Married Women Whose Only Children Are Girls. When a married woman has only female children, the resources (particularly land and plantations) that she and her daughters spend their time working will belong to others after the death of her husband. In the case of a monogamous union, her brothers or her husband's nephews will inherit; in cases of polygamy, her husband's male children.

What incentive is there for a woman to manage resources in a sustainable manner when it is clear that only others will receive the benefits?

Consequences of Women's Insecure Access to Resources

The Gender Division of Labor. Surveys carried out near Mbalmayo (Yonta 1995; Tiani and others 1997) show that farmers prefer to plant exotic fruit trees (for example, *Mangifera indica* and *Persea americana*) over "wild" fruits (for example, *Baillonnella toxisperma* and *Ricinodendron heudelotii*). The former will always be considered their property—exotics cannot be confused with forest trees—and planted trees may be inherited by their offspring. This is not the case for local species that will, sooner or later, be considered forest trees (technically, farmers should seek authorization from the government before cutting a wild tree). Thus, local species will be left for loggers. Yet these wild fruits have a useful role in the lives of local people living in forest areas, for food and as one of the main sources of household income. This preference for planting exotic trees and neglecting native trees has important consequences for SFM and is closely linked to the lack of secure intergenerational access to resources.

What is the attitude of women toward such practices that exclude them from inheritance? We have argued that women are not motivated to create perennial plantations even in places where they have the opportunity. Most women prefer short-term investments. They reject cash crop farming such as cocoa, coffee, and palm oil in favor of annual and biennial crops such as corn, groundnut, and cassava. This preference reflects an established gender division of labor: cash crop farming, a profitable but fairly long-term investment, is left to men; food crops that are essentially annual or biennial are left to women.

Increased Deforestation Caused by Food Crop Farming. Originally meant for subsistence, food crops have been revalorized because of increased demand in towns with the growth of urban populations. Urban expansion has been stimulated by national population growth and high levels of rural-to-urban migration since the decline in prices for cash crops. Importation of foodstuffs was reduced with the drastic drop in cash flow during the economic crisis of 1986 and currency devaluation in 1994. Increased mobility at the international level due to improved roads, greater mobility of the population, and better access to foreign markets has generated massive exports to neighboring countries, particularly Gabon (see Chapter 9). All these factors have led to an increase in the value of foodstuffs.

Pokam and Sunderlin (1999) show that the revalorization of food crops and the increasing demand for them prompted farmers to increase their areas of farmed land. New farms have been created by clearing old-growth forest, thereby reducing the total forested area that remains.

Every year a family clears between 0.18 and 3 hectares of forest lands to create farm land for food crops (Bahuchet 1993). At the same time, they abandon the plots of land that they had been cultivating two or three years earlier after the eventual harvest of all that was sown there: corn, groundnuts, and many secondary products; followed by yams, sugar cane, pineapple, and pepper; then finally cassava and bananas, which continue to produce for many months. People strive for fallows of 10–15 years (reduced to 3–5 years in areas with high population density). However, a fallow is never completely abandoned, because the clearing of a plot is traditionally the proof of appropriation of the land by a family. So, the wild trees left on the plot during its clearing now become the property of the family and are important sources of and habitats for NTFPs.

In general, the area reserved for family farming is part of a rotational system. Fields are often found on old fallow lands or in secondary forests. However, the young are willing to clear the virgin forest, either to get their own plots or because its soil is more fertile and favorable to banana and cucumber cultivation.

If women were to invest in perennial plantations, with their labor and time requirements, they probably would devote a great part of their efforts to taking care of existing plantations, and therefore, the clearing of old-growth forest could be stabilized.

Conflicts over Acquisition of Resources. When the interests of the various stakeholders diverge, there is competition and conflict for resources. Trees such as *Pycnanthus angolensis, Irvingia gabonensis,* and *Baillonnella toxisperma* have multiple uses.

Villagers cut young trees that surround the village for poles (to make huts) and bark (for medicinal use). Illegal loggers cut them when they are a bit older to sell to carpenters as timber. In general, loggers cut old and senescent trees, but very often the authorized minimum exploitable diameter for many species corresponds to the flowering and fruiting age (personal communication from Nsangou Mama, 1997). Loggers thus cut trees that should be reproducing, which explains the observed rarity of young plants of some species. Women collect caterpillars and fruits from the trees without felling them.

Whatever the basis of these divergent interests in the same resource, the activities of some stakeholders affect others. Those who are most penalized are the women, who are obliged to go farther from the village, walking long distances to collect NTFPs that they formerly collected behind their homes.

The Change of Perspective in Favor of Rural Women

To identify the rights of each stakeholder to manage the forest around Mbalmayo, a pebble-sorting method[7] was used (see Chapters 12, 13, and 14 for comparable analyses in other locales). Overall, the state is perceived to hold 51% of the rights to manage the forest (Table 2-2), followed by local populations (27%), logging companies (13%), and finally local craftspeople (10%). All stakeholders agreed that the main manager of the forest was the state and that local populations played a very important role, if only because of their proximity to the forest. Men were seen to hold 71% of the rights and women 29%.

Women, on average, felt that rural women had 10% of the rights to manage the forest, whereas men considered that right to be only 8%. These results can be interpreted as the manifestation of consciousness by women themselves about their rights and roles in resource management.

For this study, the most significant result focuses on the perspectives of different age groups on the roles of rural women (Table 2-3). These results suggest that the perception of the sharing of rights to manage forest resources from one generation to another that is favorable to rural women may be

Table 2-2. *Rights to Manage the Forest, by Sex of the Respondent*

	Respondents		
Managers	Women	Men	Mean
State (male)	36.25	37.32	37.05
State (female)	13.15	13.58	13.47
Male local population	21.89	17.75	18.77
Female local population	10.00	7.67	8.25
Local craftsmen	5.79	7.12	6.79
Local craftswomen	2.57	3.28	3.10
Male loggers	7.55	9.01	8.65
Female loggers	2.80	4.27	3.91

Table 2-3. *Rights to Manage the Forest, by Age Group of Respondent*

	Respondents			
Managers	Adults (36–64 years)	Youths (15–35 years)	Aged (65 years+)	Mean
State (male)	39.45	35.22	31.17	37.05
State (female)	13.52	13.93	9.83	13.47
Male local population	18.52	17.43	30.50	18.77
Female local population	8.06	8.65	6.83	8.25
Local craftsmen	6.24	7.25	7.88	6.79
Local craftswomen	2.22	4.33	1.29	3.10
Male loggers	8.20	8.90	10.50	8.65
Female loggers	3.80	4.29	2.00	3.91

changing (see also Ruiz Pérez and others 1999). That change is in line with the overall direction presently recognized in changing societies. The ongoing feminist movement may be changing the ways of thinking in favor of women, a trend to which youth seem to be more sensitive. Women have fought efficiently against the economic crisis since 1986. When men were incapable of ensuring their families' livelihoods, women reacted immediately and positively, putting in place strategies to increase their revenue and to take care of their entire families (Pokam and Sunderlin 1999; see also Chapter 3).

Prospects

Given the increase in the share of management rights assumed by rural women from one generation to another, it is possible that this positive change may carry over into the future. It is even probable that the pace of change may increase because

- the feminist trend is growing more intense;
- rural women are increasingly aware of their position of privileged resource managers;
- women are increasingly acting as full partners and no longer simply as complements to men;
- women are organizing themselves into networks of production, aided by national and international nongovernmental organizations;
- women are creating their own plantations and are increasingly managing their own incomes; and
- women produce foods not only for their own consumption but also for trade (for example, rural women living near Mbalmayo produce corn— not a traditional food—to sell to breweries and to fodder producers).

Such changes may represent harbingers of important improvements in women's conditions in rural Cameroon, an important element in improving human well-being more generally.

The drop in the price of cocoa (from 420 to 250 CFA) in the 1980s and other commodities in 1986 had dramatic consequences for rural areas where the sale of those products was the principal source of income. At the same time, the improved market for foodstuffs in urban areas and foreign demand combined with macroeconomic factors, prompting men to turn to food crops for trade (Bopda 1993). Pokam and Sunderlin (1999) note that between 1974–1975 and 1987, the proportion of male cash crop farmers declined from 83.7% to 27.1%, whereas that of male food crop farmers grew from 6.5% to 32%. The activities and sources of men's revenue diversified during this period of economic crisis. Brown and Lapuyade (1999; see also Chapter 3) confirm this tendency and add that women have revalorized NTFPs and increased the production and commercialization of food crops such that family needs formerly attributed to men are more ensured by women.

The production and marketing channels for food crops are still controlled by women. However, if men continue their tendency to switch to food crop production, problems may increase between men and women relating to access to arable lands. In the near future, we can expect an unavoidable intensification of farming activities in areas with high population density.

Conclusions and Recommendations

Important aspects of traditional social structure did not acknowledge separate roles for women in the management of resources but rather as components in the service of men. Excluded from the management of land and trees, women developed strategies that allow them to enjoy the benefit of some resources at their level of access. They occupied the spaces vacated by men.

When men hunted, fought wars, and collected, women farmed the crops. While spending time on cash crop plantations, women also continued to plant crops in association with NTFPs for domestic usage. Even when men and women seemed to carry out the same activities, like fishing, the nature of the activities varied with the sex. Men fished with nets or lines in big rivers, whereas women fished in dams or streams.

With the advent of the economic crisis, women increased their investments in crop farming. At the same time, they developed the area of NTFPs and systematized their trade. The economic crisis also brought about the massive involvement of men in crop farming, an area that they had previously neglected. What other adaptive strategies are women likely and able to develop if this domain becomes too crowded?

Brown and Lapuyade (1999) observed that the "spheres of action" of men and women are comparatively separate (see also Chapter 3). It is therefore important to consider women as managers in their own right and not as assistants to men.

Future studies are needed to better understand the constraints that hinder the success of women in a sector as important as the management of wood. Measures need to be taken to facilitate the work of women by

- improving access to sources of investment and financing,
- providing extension advice on advanced farming methods to augment agricultural yields,
- offering subsidies for entry into agricultural activities,
- improving the transport of food crops to towns by the stabilizing of transport prices, and
- strengthening the appropriate markets.

To reduce anthropogenic pressures on the forest, women should be encouraged to undertake activities not linked to forest exploitation. At the village level, resource management has to be organized to protect trees with multiple uses near the village and to prevent the disappearance of NTFPs in the vicinity. Research and extension programs on the techniques of conservation and processing of food crops and NTFPs to reduce postharvest losses are necessary. Species with multiple uses could be domesticated and planted.

Several questions are still to be answered. Does the local population share benefits equitably? Are some social classes marginalized vis-à-vis security of intergenerational access to resources? What are the attitudes of those marginalized groups toward resources? We need to be able to respond to such questions when evaluating intergenerational access to resources. We hope that the answers will lead to an understanding of the reasons for some behaviors. Important aspects could remain hidden if only the broad picture of the interactions between various "stakeholders" is taken into consideration.

Important issues that need attention in the assessment of human well-being in Cameroon include a better understanding of

- recent social phenomena, such as increases in divorce and single lifestyles, and the multidimensional consequences of these situations on children and on resource management;
- the causes of the minimal presence of women in the wood sector and its impact on sustainable resource management; and
- security of intergenerational access to resources for all social groups during the establishment of any assessment process—at times, a few individuals claim to represent a larger group of people, when in fact they act only in their own interests, at the expense of other members of the community they pretend to represent.

Endnotes

1. National Statistical Data from Cameroon's 1987 Population Census.

2. Between 1980 and 1986, US$1 was worth about 250 Communauté Financière Africaine francs (CFA). In 1986, the price of cocoa, for instance, dropped from US$1.70 to US$1 per kilogram. With the 1994 devaluation, the value of the U.S. dollar doubled and reached 550 CFA. In 1997, the price of a kilogram of cocoa varied between US$1 and US$1.30, though the price in CFA increased from 250 to 700 CFA.

3. *Inheritance* here refers to the formal inheritance patterns and *legacy* to more informal, individual bequeathing of goods.

4. The CATPAC method is described in the Introduction. Our sample included men and women of different ages with different jobs all linked to the forest: logging company workers, district workers, forestry officials, farmers, carpenters, and female retailers. Only one question was asked to each interviewed person. In town, the question asked was, "What is the importance of the forest in your life?" In villages, the local language was used: "*Ebang Afan labele ndiane fé a feg dzoe?*" ("What do you think the word forest means?") The answers were either written or tape recorded and then computer analyzed.

5. These species include *Ceiba pentandra, Triplochiton scleroxylon,* and *Erythrophloeum gabonensis.*

6. Before the introduction of cocoa farming, fruits were harvested from useful multiple-use forest trees by men (Weber 1977).

7. The pebble-sorting method used to measure rights and means to manage is described in the Introduction. The sample groups chosen for this method included men and women of different ages with various jobs related to the forest: workers in logging companies, forestry officials, local populations, craftspeople, and town council workers. During the survey, individuals 15–35 years old were regarded as *youths;* individuals 36–64 years old were *adults;* and individuals more than 65 years old were *aged.*

CHAPTER THREE

Changing Gender Relationships and Forest Use

A Case Study from Komassi, Cameroon

Katrina Brown and Sandrine Lapuyade

Economic and environmental pressures affect access to and use of forest resources, and these dynamics affect men and women quite differently over time. Women are especially dependent on nontimber forest products (NTFPs), but the role of these products has changed markedly. All forest products harvested are now commercially traded in much of Cameroon, compared with only a decade ago, when few products had commercial value. Whereas men have been able to diversify their livelihood strategies, women have less room to maneuver and increasingly rely on diminishing forest resources. This situation has profound impacts on the way women and men perceive change as well as on the current and future management of forest resources.

Not all of the characteristics that influence social change are fully investigated in this study; our focus is on the dynamics of change, and identifying and separating those factors that are experienced differently by men and women within the same community. The analysis is therefore informed by the work of social anthropologists such as Guyer (1984), Leach (1994), and Goheen (1996). We espouse a political ecology approach that considers how local livelihood systems are shaped by wide political or economic factors occurring on a broad scale. Our approach also acknowledges the interaction of gender with social, political, cultural, and ecological factors (for example, see Rocheleau and others [1996] on feminist political ecology).

The primary focus of this chapter is intragenerational aspects of access to resources and how they change over time with respect to changing gender relationships. Intragenerational issues concern the differences between people of the same generation at one point in time, whereas intergenerational analysis examines differences between past, present, and future access. Our analysis centers on the changes or dynamics of intragenerational access, particularly in terms of gender differences, but captures some aspects of intergenerational dimensions as we discuss how access has changed over time.

In attempting to reflect the dynamics of social relationships alongside environmental and economic changes, we focus on the gender relationships affecting change in forest cover and livelihoods. We first discuss the context of the study in southern Cameroon: the characteristics of the region and the research site. Next, we examine how men and women perceive change differently, the implications for changing gender relationships and livelihoods, and access to forest and other natural resources. Then, we explore the use of forest products and their contributions to livelihoods and welfare, how this is differentiated by gender and other social characteristics, and how access has changed over time. In the conclusion, we discuss the implications of the findings for the well-being of men and women and for forest management.

The Research Context

The research was undertaken in the village of Komassi in southern Cameroon (see Figure A2 in the Introduction). In collecting data on livelihood strategies, group meetings and in-depth conversations with individuals were emphasized. Participatory rural appraisal techniques (such as wealth ranking, village transects, mapping, and matrix ranking) were used together with direct observation. The transects used followed the main foresters' track, which is used daily by men and women to access some of their cultivated land and the forest.

Because the village is quite small (32 households), a village census identified the head of each household and his or her spouse or siblings. This census gave us an opportunity to explain the research aims clearly and to become familiar with the people and their household structures. Two large group meetings were held with men and women separately to identify major issues regarding their changing circumstances and to discuss natural resources management issues. Subsequent meetings were held with focus groups to discuss specific issues, such as the changing use of forest products or income opportunities. Semistructured interviews were conducted with 26 individuals (men and women) chosen through judgement sampling. We also held in-depth conversations with informants selected according to their place in the household structure (wife, co-wife, widow, widower, unmarried sister) or their social status. By using several methods to elicit information, we hoped to cross-check the data and overcome potential biases of participatory rural appraisal methods, such as may result from group discussions in which socially constructed testimonies are put forward.

The humid forest zone covers approximately 18 million hectares of southern Cameroon, forming the western margins of the large tract of forest of Central Africa. The zone is home to more than 4.7 million people, of whom at least 2.2 million live in rural areas and are primarily dependent on agriculture and forest-related activities. (This figure is cited by Ndoye and Kaimowitz [1998] as the most recent reliable figure, but it is likely that the rural population has increased considerably since 1987, when the estimate was made.) The deforestation rate in Cameroon is believed to be twice the regional average, with an estimated annual average of 1% over the past two decades (Laporte 1999). The major causes cited are population growth and shifting cultivation. Three factors have been identified as contributing to recent high rates of forest conversion: return migration from urban to rural areas, a resulting expansion in areas of cropping, and increasing logging intensity. These factors are related to the economic crisis and downturn in oil revenue in Cameroon. However, these generalizations mask very important processes and dynamics in changing land use patterns. For example, increases in areas of cultivated land reflect not only population increases in rural areas but also a shift in farming systems that now grow food crops for cash. This phenomenon has been documented at a macro level (Sunderlin and Pokam 1998) as well as in microscale analysis (Brown and Ekoko 2000).

In 1984, Cameroon's humid forest zone contained an estimated 425,000 small farms, which cultivated approximately 800,000 hectares. Major crops were cocoa, coffee, cassava, groundnut, maize, cocoyams, and plantains. Much of the area farmed is under a system of fallow farming that integrates forest and cropping. The system is distinguished by the patterns of fallow use and the continuum between forest, fallow, and cropped areas (Brocklesby and Ambrose-Oji 1997) and by the gender division of labor in agriculture (Guyer 1984; Brown and Lapuyade 1999).

Forests in southern Cameroon have been commercially logged since colonial times, and until the 1960s, a steady though "modest" (Van Dorp 1995, 1) trade with Europe was maintained. Since the country's independence in 1960, the domestic timber industry has been encouraged, but large-scale commercial logging did not really expand until the 1980s. Logging operations focus on a few species, and harvesting is very selective—to the extent that some companies almost specialize in one species. However, the low rate of offtake does not translate into sustainable practices; on the contrary, it is associated with large amounts of waste, unskilled felling, and poor road construction (Gartlan 1992). In some parts of the Centre and South Provinces, as around Komassi, forests may have been logged successively several times by different logging companies over the years. The forests around Komassi have been logged for more than 50 years; the first loggers were recorded in 1944 (Brown and Ekoko 2000).

Penetration by logging companies has undoubtedly affected human–forest relationships around Komassi. However, the most significant changes have been brought about as a result of external shocks caused by the so-called economic crisis. This crisis has had a profound impact on rural and urban livelihoods as well as on the operations of large-scale logging companies. After a decade of economic boom based on the petroleum industry, in the late 1980s, Cameroon implemented a Structural Adjustment Program following a sharp economic downturn. The decline in the price of export crops, the withdrawal of state services, and the devaluation of the local currency (the CFA [Communauté Financière Africaine] franc) had severe repercussions in both the rural and urban populations of the country that affected rural livelihoods as well as forest cover and resource management (Tchoungi and others 1995; Reed 1996; Kaimowitz and others 1998). Three key effects of the economic crisis are return migration to rural areas, increased cultivation of food crops for cash, and an apparent increase in forest clearance for agriculture (Eba'a-Atyi 1998; Ndoye and Kaimowitz 1998; Sunderlin and Pokam 1998). It is against this backdrop that we analyze gendered perceptions of change.

Gendered Perceptions of Change

Men's and women's accounts of the major events that have affected their lives over the past 10–15 years reveal interesting differences in their perceptions of change. Life is perceived by women to be far more difficult because of the greater cash demands and poor terms of trade, whereas men think their standard of living is steadily improving (Table 3-1). These changes are described relative to several different aspects of livelihoods and farming systems. The information was compiled from responses to the question "What are the major events that have changed life in Komassi over the past 20 years?" in group interviews with mixed-aged groups of men and women.

The ways in which men and women experience change reveal how the patriarchal system existing in Komassi defines separate spheres of action for men and women. The findings initially suggest a strict division of gender roles along the lines of the classic (and now discredited) public/private dichotomy (Stamp 1989, 24), with men responsible for the "public" domain and women limited to the "private," domestic domain. Our study shows how the respective spheres of influence and power are far from simple or static. In fact, women's "public" activities have always been and are increasingly important, especially when considered in light of other factors, such as the impact of the local political economy and power relationships within the household.

Table 3-1. *Perceptions of Events that Have Changed the Lives of People in Komassi, Cameroon*

Events mentioned by both men and women	
Men's comments	*Women's comments*

1973: Opening of the village health center

1997: The health center expanded to Integrated Health Center

	"*The health center is a good thing but medicine used to be free, and now we have to pay.*"

1980: Building of the road to Dzeng

"*The road makes it easier to go to Yaoundé to sell and buy food crops and medicine. It also makes it easier for the governor to come and visit the village.*" "*The negative effects of the road are negligible.*"	"*They say that the road has brought development but the crisis came with the road. We were better off without.*" "*The road is dangerous for women when they go to sell their crop in Yaoundé. There are many accidents.*"

1987: Building of the water tower to provide village with communal tap water

"*Wells were dug in the village well before. Families who had enough money would have a well dug and make water available for free to other villagers.*"	"*Since the system was put in place by Scan, water has broken down, people from neighboring villages complain they do not have access to water anymore, but in Komassi we have easy access to water because wells were dug in the village.*"

continued on next page

Although men express interest in events that have affected domestic life, such as the opening of the health center and the provision of water, they put more emphasis on the ways in which those changes have had an impact on the status and prestige of the village (Table 3-1). The presence of more services and institutions, such as the training center, the market, and the "House of the Party" (local representation of the political party currently in power), has turned Komassi from a little hamlet into an important village with a central role for many people in the district. These aspects of change were not highlighted by women.

Men did not discuss agricultural issues, which were central to women's concerns. Although men have turned to the cultivation of food crops in recent years, their livelihood strategies are not yet so closely dependent on food crops as are those of women. Women are still predominantly responsible for the cultivation of food crops, and their livelihood strategies are mainly based on the commercialization and consumption of these crops. More important, women's crops are used to feed their families, and the surplus that is sold is devoted to the basic welfare of all household members, including

Table 3-1. *Continued*

Events mentioned by men only
1965: Building of corrugated roofs
1970s: Opening of the agricultural extension service
1980: Building of a bigger church and subsequent development of the market *"The market is our pride."*
1995: Opening of the training center for brick laying and joinery *"This improved the atmosphere in the village as more people came to live here. Moreover, students make furniture for people in the village who can buy them without having to pay for transport."*
1996: Electricity is brought to the village *"People can have radios and TVs."* *"When women come back home late they can still work."* *"We can drink fresh beer."* *"Electricity is less expensive than petrol and does not have to be paid as often."*
Opening of the logging tracks *"In making the fields more accessible, the tracks have increased the enthusiasm of cultivators."*
Opening of the "House of the Party" *"If you are a member, they can find a place for you."*

Events mentioned by women only
The price of food crops has decreased.
Cocoa plantations had to be abandoned.
Cassava suffers from an unknown disease.
Crops are eaten by animals (mainly hedgehogs).
The price of consumer goods has increased.
The price of transport has increased dramatically.
The clearing of fields is more difficult because hiring a chainsaw is more expensive.

their husbands. On the other hand, when they are growing food crops, men have discretionary power to decide how to use their production. Men's crops therefore do not necessarily contribute to household welfare.

Men's opinions that life is easier now than a few decades ago are supported by apparent trends in their incomes. Men claim that they earn more now than in 1980 or in 1988 (Table 3-2). Although the information does not reveal exact figures, it is possible to conclude that their incomes have increased by about one-third since 1980. On the other hand, women per-

Table 3-2. *Changes in Men's Sources of Income, 1980–1998*

Source of income	1980	1988	1998
Cash crops			
Cocoa	•••••	•	••
Coffee		•••••	••
Crafts industry		••	••••
Food crops	•••	•••••••	••••••••••••
NTFPs			
Cane	•••	•••	••••
Game meat	•••••	••••	••••
Fishing	••••••	••••	•••
Palm oil	•••	••••	•••••
Palm wine*a*	••••••	••••••	•••••••••
Waged jobs			
Civil administration	••••••••	••••••••	•
Pension			••••••••

Note: The matrix was produced by men using stones to express the relative importance of their different sources of cash income.

a Includes *odontol* (distilled wine) and *nkouet* (fermented wine).

ceive their incomes as having declined since 1988 (Table 3-3). Income data has not been corrected for inflation or valued by purchasing power in a systematic way; it is solely based on perceived gross values and trends.

The data in Tables 3-2 and 3-3 were derived from ranking exercises carried out with separate groups of men and women of mixed ages. Participants were asked to identify their main source of cash income and to attribute several stones for this source of income.[1] Each matrix was built up slowly and systematically by making comparisons between one source of income for the three years (1980, 1988, 1998) and then between different sources of income within each year. Thus, the number of stones does not represent absolute levels of income, only proportions between years and between activities.

Women's incomes today rely nearly as heavily on food crops as two decades ago. However, some important changes have occurred as a result of the declining price of food crops. Since 1988, the relative importance of the various food crops as cash earners for women seem to have very clearly shifted. Whereas cocoyam, cassava, plantain, and groundnuts were the main cash earners in 1982 and 1988, crops such as cucumbers, sweet bananas, and maize have become equally important. Similarly, in the past ten years, women have developed a strategy of creating added value to their crops by selling more processed food, such as cassava rolls and doughnuts. Cassava rolls are perceived as bringing at least as much income as some major crops such as cocoyam. However, their production is very time-consuming, and some of

Table 3-3. *Changes in Women's Sources of Income, 1982–1998*

Source of income	1982	1988	1998
Food crops			
Cassava	• • • • • • • • • •	• • • • • • • • • •	• • • • •
Cocoyam	• • • • • • • • • •	• • • • • • • • • •	• • • • •
Cucumbers	• • •	• • •	• • • • • •
Groundnuts	• • • • • • • • • •	• • • • • • • • • •	• • • • •
Maize	• • •	• • • •	• • • • • •
Plantain	• • • • • • • • • • •	• • • • • • • • • • •	• • • • •
Sweet bananas	• • •	• • • •	• • • • • •
Tomatoes	• • •	• • •	• • •
Processed food crops			
Cassava rolls	• • •	• • • •	• • • • • •
Cooked yam	• • •	• • •	• • •
Processed NTFPs			
Cooked game meat	• • •	• • • •	• • • • • •
Odontol[a]	• • •	• • • •	• • • • • •
Palm oil	• • •	• • • •	• • • • • •
Trade sales			
Beer and sugar	• • • • • • • • • •	• • • • • • • • • •	• • •
Herring	• • • • • • • • • •	• • • • • • • • • •	• • • • •
Rice	• • • • • • • • • •	• • • • • • • • • •	• • • • •

Note: The matrix was produced by women using stones to express the relative importance of their different sources of cash income.

[a] Distilled palm wine.

the poorer women find it difficult to find enough time to produce them. This is especially true of women who live alone and do not have the financial means to buy men's labor to clear their fields and help with other tasks. Their workload is therefore heavier, and time available for the production of cassava rolls is scarce.

The importance of trade in beer, sugar, herring, and other processed goods has also decreased considerably over the past ten years. Before 1988, trade in these products represented the second source of income for women after food crops, but now fewer and fewer women are able to generate enough initial capital to buy those items for resale. This situation is caused by the combined effect of the increase in price of those products and the reduced income generated through other sources such as food crops.

Finally, it is important to note the increasing importance of two main kinds of NTFPs as sources of income for women; palm trees and game meat now represent their second source of income.

Over the past ten years, food crops have become the main source of income for men (Table 3-2). This trend also has been observed at the national level (Sunderlin and Pokam 1998; see also Chapter 2). The shift from cocoa to food crops as the main cash earner in Cameroon has not occurred as a direct result of any single factor and was predicted by some observers well before the economic crisis based on the assumption that increased urbanization would lead to a higher demand for processed and staple foods: "In the mid- and late 1970s N'Sangou argued that a high proportion of Yaoundé's food still came from the immediate hinterland .... In fact he predicted that by 1990 Eton peasants would get most of their income from traditional food for the market rather than cocoa" (Guyer 1984, 62).

In addition to the shift toward food crops, men's sources of income seem to have diversified. A few households have started coffee plantations to replace cocoa. The craft industry is another source of income that men have developed since the late 1980s. Services such as repairing shoes, tires, or electrical devices have become a central element of some men's livelihoods. (Komassi benefits from the presence of a high proportion of civil servants who work in the school and the training center. The so-called external elite, who have comparatively high purchasing power, have encouraged the development of these service activities.) The increasing number of cars that drive between Komassi and Yaoundé provide some business for the mechanics and the *docteur des roues* who repairs tires. Men with good connections or relatives in urban areas also earn income from renting houses they own in the village or in Yaoundé.

It is not so much diversification but a shift in the relative importance of the products as cash earners that characterizes changes in women's sources of income. Although they were not listed as main sources of income, other activities have been developed by women over the past couple of years. One of them is collecting sand from the Komo riverbed to sell to building entrepreneurs. However, women's sources of income are still mainly from agriculture. Opportunities outside the food crop sector are limited and increasingly difficult to access. The few women who have nonagricultural sources of income are struggling to maintain them. This is the case for women trading beer or sugar as well as for the village dressmaker.

Our findings suggest that men have diversified more successfully than women; they have sources of income such as pensions and crafts. Almost no new sources of income for women have developed (Table 3-3), and individual interviews revealed that women who used to sell beer, sugar, rice, smoked fish, and other goods can no longer afford to engage in this trade because of prohibitive costs and capital requirements.

Increasing demands for cash in a context where some men have lost their financial power leads to shifts in responsibilities; the roles of men and women are being redefined as circumstances change. Women are increasingly responsible for paying school fees and for medicine, whereas men increasingly use the hoe to cultivate food crops. Although men still explain that using the hoe and leaning over the earth is bad for them and that women are naturally better suited for the job, some also admit that it is necessary to make a living. The division of labor and responsibilities varies between households, depending on other income opportunities, the internal structure of the household, and the availability of labor. Interviews with women often suggested that things had changed to their disadvantage:

- "Before when a man asked his wife what she did with the money, she could answer it did not come from his work. Many men did not even look at their wives' fields; they did not put a foot in them. They had their cacao plantations that earned them good money."
- "Young men are getting worse. Women have two fields and men have only one. They want the money."

Our data suggest that, although an increasing number of men are turning to food crops, a minority of them actually use the hoe, as the majority of men plant plantain rather than groundnut or cassava. Furthermore, when men decide to grow groundnut and cassava, women often provide most of the labor required on those fields, whereas "men work when they want; they are not obliged." Women's workload has increased over the past two decades as a result of this shift as well as increased cash demands, worsening terms of trade for their major crops, decreasing fertility of the soil, and, for some, reduced male earning ability. Although women's income may be crucial as the last buffer against dwindling financial resources, their power to renegotiate their roles and responsibilities seems to be limited. For example, one woman said, "When you get married, you know that your husband has to look after the children. Before we had children he gave me money, but now that I have children he knows I am not going to leave; he does not need to give me anything. Even school fees have been split and sometimes I have to pay for them on my own. I often think I have become the man of the family. I have to pay for everything."

Importance of NTFPs in Livelihoods

An important feature of the livelihood systems of rural dwellers in the humid forest zone is their reliance on forest products. Ndoye and others (1997/98) document the extensive use of NTFPs, highlighting the interdependence of farm and forest in rural livelihoods. A wide range of products is used, and NTFPs have become increasingly commercialized in recent years. There are important differences between households and also within households. Cash income from the sale of forest products (honey, medicinal plants, vegetables, and fruits) is especially vital for women (Watts and Akogo 1994; Ndoye and others 1997/98). In addition, findings from several sources indicate that

NTFPs may be particularly important as a source of both subsistence and marketed products for poorer households (Ruitenbeek 1996).

The forest not only is at the center of farming systems in this part of Cameroon but also provides men and women with a range of products that are both culturally and economically valuable. The role of the forest is evolving as the social and ecological environment changes, and it is evolving differently for men and women.

Intragenerational Access to NTFPs

Men refer to hunting as the most important activity related to NTFPs, even though game meat does not represent one of the major sources of income for them (Table 3-2). Fishing is second most important, whereas gathering fruit, nuts, and mushrooms is considered to be strictly the women's responsibility. When a man finds a tree bearing fruit or nuts, he may eat some on the spot but he does not get involved in gathering them. He returns home to tell the women and children in his household where they can find the tree and leaves them to deal with the collection. This behavior reflects the cultural preferences of precolonial times, when "food eaten raw, such as fruit, was associated with women and children" (Guyer 1984, 28).

When asked about forest products, men speak more spontaneously of wild game than of palm trees. This response is surprising, because palm wine has been a more important source of income than game for at least the past two decades. Three main factors explain their response. First, palm trees can be grown and may not be seen strictly as a forest product that you can find only in the wild. Second, NTFPs generally are not considered a main source of income. Because palm wine is a main source of income, it falls into a different category of product. Finally, hunting is seen as a "noble" activity strongly related to male status. In the same way that men are/were planters (not cultivators),[2] they also are hunters (not gatherers). Hunting is part of their social identity, which partly explains their willingness to speak about it first.

Men rank boas high on the list of animals that they hunt (Table 3-4). Boas are sought for their flesh as well as their skin and are said to sell for a good price. However, killing a boa is a rare occurrence, so although the revenue from a boa is high, it is not a reliable source of income. Even so, men insist that a boa is the most important prey when hunting. Traditionally, only the old and spiritually strong are allowed to eat the meat of certain snakes (Guyer 1984, 29); this special feature adds to the importance of hunting as a status activity for men. Indeed, social values appear to be important in motivating hunting activities. In addition to highly valued but rare species, such as boas, more common animal species also provide men with cash and food. The next most important species after boa are deer, porcupines, hares, wildcats, a few others, and then pangolin, which is sought for the flavor of its meat.

Table 3-4. *Animals Hunted by Men and Their Relative Importance*

Name	To eat	To sell	To make other products	To cure	Rank
Boa	•••••	•••••	•••••	•••••	I
Deer	•••••••	••••••••	•		2
Porcupine	•••••••	••••••			3
Hare	••••••	•••••	•		4
Wildcat	••••••	•••••			5
Wild boar	•••••	•••••			6
Tortoise	••••	••••		••	6
Monkey	•••••	•••••			6
Viper	•••••		••••		7
Antelope	••••	••••	•		7
Pangolin	••••	••••			8
Salamander	••••	••••			8
Xerus (*rat palmiste*)	•••••	••			9
Caiman	••••	•••			9
Crocodile	••••	••			10
Bird	•••	••			11
Hedgehog	•••	•			12
Grass snake	••••				12
Chimpanzee[a]					
Gorilla[a]					

Note: The matrix was produced by men using stones to express the relative importance of the different species hunted.

[a] Although these species were clearly hunted, men decided not to rank them as part of the matrix because hunting these species is illegal.

Women use a wide range of NTFPs. During group meetings, women insisted on listing no fewer than 18 products, arguing that they were all very important. Because of the many products identified (Table 3-5), it was difficult to compare products effectively. During individual conversations, wild mango (*Irvingia gabonensis*) and *njansang* (*Ricinodendron heudelotii*) were the two products mentioned most often by women as sources of food and income. Over the past 20 years, processed NTFPs have become twice as important a source of income for women. However, none of the NTFPs identified as important (Table 3-5) was spontaneously identified as a major source of income (Table 3-3). This discrepancy highlights the significance of nonfinancial values of NTFPs for women.

Although NTFPs are increasingly used as a source of income, their value may be defined by two other main criteria: versatility and substitution (for

Table 3-5. *Women's Use of Nontimber Forest Products*

Product name			Uses				Rank
Ewondo	English	Scientific name	To eat	To sell	To cure	To build	
	Papaya	Carica papaya	••••••••••	••••••	•••••		1
Ofum be	Lemon		•••••••	•••••	••••••		2
Adjap	Moabi	Baillonella toxisperma	•••••	•••••	••••••	••••••	3
Ibouma	Guava	Psidium guajava	•••••••	•••••	•••••		4
Tom		Pachypodenthium staudtii	••••••	•••••	•••••		5
Mvut		Trichoscypha acuminata	•••••	•••••	•••••		6
Njansang		Ricinodendron heudelotii	•••••	•••••	•••••		6
Ekong		Trichoscypha arborea	•••••	•••••	•••••		6
Biton	Palm nut	Elaeis guineensis	•••••	••••••			6
Abel	Cola	Cola spp.	•••••	•••••	•••		7
Essok		Garcinia lucida		••••••	•••••		8
Assa	Plums-safoutier	Dacryodes edulis	••••••	•••••			8
Ofumbi	Oranges		•••••	•••••			8

Local name	Common name	Scientific name	
	Liana		9
Fia	Avocado	Persea americana	9
	Grasshopper		10
	Larvae		11
	Fish		11
Ekouam		Cola pachycarpa (Ekom)	11
Angongi		Antrocaryon klaineanum	11
Ndo'o	Wild mango	Irvingia spp.	12
Evoué		Cola lepidota	12
Mvonde	Coconut	Cocos nucifera	12
	Caterpillar		13
	Snail	Achatin sp.	13
Ezeng		Leea guineensis (essong)	13
Evoula		Vitex sp.	14
	Straw		15

Note: The matrix was produced by women using stones to express the relative importance of the different forest products they gather according to different uses.

groundnuts). First, it is important for a product to have multiple uses. For example, the products used for medicinal purposes as well as for food and income are ranked higher than those that have no medicinal use (Table 3-5). Second, if a product can replace groundnut as a key ingredient for food, it is considered to be more useful. Groundnut plays a central role in the diet, and in times of shortage, products such as *njansang* (*Ricinodendron heudeleotii*) and wild mango (*Irvingia gabonensis*) can replace groundnut in cooking.

Although NTFPs may constitute an important source of supplementary income, their economic value is reduced by the seasonality of their production. An elderly woman noted, "all these products help a lot, but you don't find them all the time." The importance of NTFPs in children's diets was often emphasized by women, but their value as cash earners seems secondary. Women are very enthusiastic about explaining the different uses of forest products; they seem eager to share their knowledge and to list as many products as possible. However, knowledge regarding the properties of these products for medicine is kept secret and is not shared between individuals. The various medicinal uses of NTFPs were not explored as part of this study.

Changes in the Use of NTFPs over Time

Two major interrelated changes have affected the use of NTFPs by the people of Komassi. First, the increasing need for cash has led to the marketing of a greater proportion of these products. Second, many of the products mentioned by the people of Komassi are increasingly difficult to find. NTFPs are sold more often, but whereas women's incomes depend increasingly on processed NTFPs, men generally rely less heavily on NTFPs than 20 years ago. Many men and women stated that fishing is no longer worth the effort and that hunting is increasingly difficult.

This trend is clearly attributed to the economic situation of Komassi:

- "The crisis has fallen on the rivers and on the trees."
- "As soon as we realize that we can make money with some forest products, they stop growing."
- "Because life is difficult, people cultivate anywhere. We burn the forest and the trees disappear."
- "Since people are now eager to sell *tom* (*Pachypodenthium staudtii*) because it earns cash, they cut the tree to access the fruit. You do not find much of it anymore."

Although hunting is thought to be a more lucrative activity than producing palm wine, the relative importance of hunting as a cash earner (Table 3-2) has decreased markedly over the past 20 years. The most significant change is the shift from selling raw meat to selling it cooked. Since the early 1990s, women have started selling more meat dishes in an attempt to compensate for the loss of income from their food crops. Men confirmed that the number of hunted species has decreased over the past ten years. They see the presence of too many hunters as one of the major causes for this decline. The noise made by loggers and the felling of trees are two other main reasons given. Loggers are thought to be damaging the forest. In addition, logging tracks are said to have facilitated access to the forest by poachers (see also Chapter 9). The proximity of Yaoundé and improved access to Komassi are also believed to contribute to increased poaching.

Since the 1970s, the availability, collection, and use of NTFPs has changed considerably (Table 3-6). Men and women cite three reasons for the growing difficulty of finding some NTFPs: the need to cultivate larger areas of land (and consequent reduction in forest area); increased harvest due to commercialization, which results from an increased need for cash; and logging activities. Access to and use of forest for cultivation also has changed. We find that although the land closest to houses, *nkoa'nfo*, is still being cultivated by some, a growing number of people now have fields deeper in the forest, *nko'o*. Fields were open in *nko'o* for the first time around 1978. Reasons given for this recent cultivation of forest land farther into *nko'o* are the construction of logging roads, the decreased fertility of land closer to the village, and increased damage from domestic animals close to the village.

The uncultivated area of *nko'o* is thought to be larger than the area already opened up for agriculture. Therefore, the local people do not perceive any scarcity of land and believe that in ten years time it will be possible for the Komassi residents to cultivate land closer to the village again. Meanwhile, the expansion into the forest is encouraged by the need to cultivate larger areas as a result of (not ranked)

- increased need for cash,
- widespread pest damage to crops,
- population growth in Komassi,
- declining fertility of the land (except on newly cleared forest land), and
- a greater number of men cultivating food crops.

Although no evidence links diseases that attack fruit and weeds that invade the forest to forest clearing, they are identified as important causes of the disappearance of NTFPs used by women.

The decreasing availability of NTFPs from the Komassi forest as a result of these various factors and their concomitant growing value as sources of income have brought about some changes in the management of tree resources. Women stressed the increasing need to ask for authorization to gather fruit such as papaya[3] and wild mango on other people's fields. For example, in interviews women observed, "Before you could pick papaya without asking. Now one has to ask for permission because we know it can be sold." The same is true of other forest products, such as cane. The village chief explained, "In some villages they have cane furniture workshops, but

Table 3-6. *Long-Term Changes in the Use of Forest Products by Women*

Name	1970s Gathered	1970s Sold	Changes and comments	1998 Gathered	1998 Sold
Cola nut	• • •		"Started decreasing about ten years ago. You can sell a lot if you know how to keep it in the ground. Sells very well during the dry season."	• •	• •
Ekoam	• • • • • •		"Seldom sold because it is very hard. It is only because of the crisis that we sell it but we should keep it for children."	• • • • •	•
Ekoang	• • •		"If you do not fell the tree you cannot pick the fruit unless you are very brave. This is why there are fewer now."	• •	• •
Esok	•		"Does not grow here naturally. Children go and get it elsewhere so that we can use it in making palm wine. Otherwise many people come to sell it here because they know we do not have any."	•	
Evoué			"Does not grow in this area. It is like *okok* (*eru*)."		
Guava	• • • • •		"Although it does not suffer from a disease, it is disappearing—but we plant it."	• •	• •
Komen	• • • • • • •		"Now we do not give them away anymore. Only if you are a fool."	• • • •	• • • •
Lemon	• • • • •		"There were a lot in the old days; now they are very rare."	• •	• •

continued on next page

not here in Komassi. However, cane comes from Komassi's forest. Before they could come and take it, now they have to ask for authorisation."

The growing trade in NTFPs is partially seen as a negative change that affects the lives of women and children. Some women complain that NTFPs should not be sold but kept for feeding children. Forty years ago, papaya fruit was eaten only by children, completely disregarded by adults. When school attendance became more widespread, papaya became increasingly valuable as food for children who had to spend most of their day at school. Although it still plays an important role as food for children, today it is listed among the most important cash-earning fruits. An elderly Komassi resident reflected,

Table 3-6. *Continued*

Name	1970s		Changes and comments	1998	
	Gathered	Sold		Gathered	Sold
Mboal	••		"Prepared with smoked fish."	••	•
Moabi	•••••••		"We used to find many of them but they have been logged. With the fruit we make oil that is much more expensive than palm oil."	•	•
Mvut	•••••••		"Before we could pick so much we did not know what to do with it."	•••	•••
Njansang	•		"It is when men started to marry Eton women that we started using it. Here women do not know how to prepare it but when you know, it earns a lot of money."	•••••	•••••
Palm nut	•••••••		"This is the most important of all because we use oil for everything."	•••••••	•••••••
Papaya	•••		"We started selling them about five years ago."	•••••	•••••
Plums	••••••		"We used to share them with people; now there is a disease."	•••	•••
Tom	••		"If you have a tree in your field, you are rich."	•	••
Wild garlic	•••••		"Now we sell it because the Bamileke like it very much, but the weeds kill it."	••	••
Wild mango	•••••••		"It is about five years ago that we started selling it."	••••	••••

"Before we could bring back baskets full of fruit, and children could eat. Now, we sell everything. Money has spoiled everything."

Thus, monetization of NTFPs has impacts on several members of the community, but women in Komassi generally see these changes as having negative effects on their well-being and that of their children.

Implications for Well-Being and Forest Management

The analysis of changes in livelihoods and in the use of forest products highlights several different aspects of intra- and intergenerational well-being. The way that men and women perceive and articulate changes in their lives over the past 20 years is strikingly different. The differences can best be expressed in the concluding statements of the focus group meetings. Men believed that "our standard of living is improving constantly. Death is the only thing that makes life hard, but it is natural." Women, on the other hand, were unanimous that the quality of life had declined and agreed that "we sell less than we used to, but we have to work more because of the diseases and all the other problems. Maybe this is the end of the world." These perceptions point to intragenerational and, specifically, gender inequities in terms of well-being.

We have highlighted some significant changes and differences in livelihood patterns that relate to and strongly influence these disparities in well-being. Although many people have been adversely affected by the economic crisis and other economic and social changes, evidence indicates that women generally have borne the brunt of the costs of adjustment (Due and Gladwin 1991; Afshar and Dennis 1992; Sparr 1994; Elson 1995). In Komassi, men appear to have diversified their sources of livelihood. They earn a living from shifting patterns of crops grown[4] (particularly from commodity to food crops for cash), have access to nonfarm income, develop small trade and craft businesses, and have access to pensions and remittances. In contrast, women appear to have less room to maneuver and depend increasingly on selling food crops (with declining terms of trade) and on NTFPs, adding value where they can by processing food and other products. In short, men have the means to diversify in several ways, whereas women are forced to rely more on natural resources for their livelihoods.

Forest management also has been affected by change. The scarcity of NTFPs has been noted by women in the village, possibly caused by overexploitation, conflicts with competing uses (especially logging), and difficulties over access. Women are increasingly dependent on forest resources for their livelihoods, but every product they collect is now sold for cash, whereas 20 years ago, NTFPs were used for subsistence needs and given as gifts. Everything in the forest now has a commercial value, and monetization is beginning to change the way villagers view the forest. It is now a potential commercial resource that outsiders seek to exploit (Brown and Ekoko 2000). The commodification of these products is not seen as necessarily beneficial. Women feel that in general, life was easier when their livelihoods depended less on earning cash and their natural environment could provide them with most of their subsistence needs.

Conclusions

Our study seeks to capture and explain the dynamics of changing gender relationships and forest use over time. We used participatory rural appraisal methods to explore perceptions of change and the values of different products in the livelihoods of different actors. They provide qualitative insights into the changes and the differences, which can complement insights provided by other studies and more quantitative analyses. Importantly, they highlight the way in which change is significantly differentiated—in our case, the difference between men and women is striking.

The intergenerational dimension of gender power relationships also is a focus of this study. Women have been important cash earners in southern Cameroon for more than three decades (Guyer 1984), but their success has not eliminated unequal power relationships between men and women. In some households, the woman has become "the man of the family" in the sense that her wages must replace the man's declining income, so she has become the more important wage earner.

The power that women have gained in decisionmaking within the household is still limited. Even if some women feel that they are playing a man's role, they have not acquired the decisionmaking power of men. This study shows that power relationships, although constantly renegotiated, are still shaped by "women's weaker bargaining strength within the family" (Agarwal 1990, 228). Additional study of the intergenerational dimension of well-being should focus on analyzing the ability of younger women to renegotiate gender roles within the household and gain greater opportunities to respond positively to the pressures of rapid change and development. Investigation of these intergenerational differences would enhance our understanding of the rapidly shifting dynamics of social and forest relationships.

Acknowledgements

We thank Dr. Ousseynou Ndoye, Dr. François Ekoko, and Florence Munoh (Center for International Forestry Research [CIFOR], Cameroon) for their kind help in Cameroon. We are especially grateful for the time, patience, and support of all the villagers in Komassi who participated in our research efforts. Carol Colfer and Ousseynou Ndoye made extremely helpful comments on earlier drafts of this chapter.

This research was undertaken through a collaborative research project with CIFOR, funded by the U.K. Department for International Development (DFID). The views expressed here are those of the authors, and all errors remain ours. Our views do not necessarily reflect the views of DFID or CIFOR.

Endnotes

1. Our matrix ranking method is similar to CIFOR's pebble-sorting method described in the Introduction (see also Chapters 2, 7, and 11).

2. From the end of colonization until recently, men were viewed as planters (of cacao) and women as cultivators. Although the shift toward food crops has turned some men into cultivators, men are referred to as planters more often than as cultivators. This situation may have encouraged men to maintain some kind of plantation on their land and to choose to grow perennial crops such as plantain rather than groundnut when turning to food crops. Tree crop plantations are also a way to gain permanent customary rights.

3. Strictly speaking, papaya is not an NTFP because it is an introduced species that does not grow in the wild. However, it is often planted in fallow land, and during group meetings women listed it as a forest product.

4. Although the terms of trade have declined for food crops, those for cocoa and other commodities have fallen even more drastically. Combined with the increasing urban markets, the low prices for cash crops have provided incentive for men to shift to food crops.

A Conservation Ethic in Forest Management

The idea of a conservation ethic is a popular one, and one that has held serious attention within western environmental groups over the years (see Thoreau 1957; Leopold 1970). After reading numerous tracts relating to this topic, we settled on a dictionary definition for *conservation:* "a careful preservation and protection of something; esp: planned management of a natural resource to prevent exploitation, destruction, or neglect" (*Webster's* 1993). *Ethics* is defined as "a theory or system of moral values (the present-day materialistic ethic)." Together, these definitions imply that a *conservation ethic* is a theory or system of moral values pertaining to the planned management of a natural resource to prevent exploitation, destruction, or neglect. We see this definition as intimately related to people's worldviews, that is, their perceptions of how humans and nature are related.

A large and related body of literature—going back at least to Rousseau's "noble savage"—argues that indigenous or tribal people have a special link with nature, that they manifest beliefs and actions that serve the functions of conservation, and that this set of practices and beliefs constitutes a conservation ethic. Kemf (1993, xvi), for instance, states in the introduction to her collection of articles about people in protected areas that "indigenous people, who number around 300 million today, are the traditional guardians of the Law of Mother Earth, a code of conservation inspired by a universally held belief that the source of all life is the earth, the mother of all creation" (see also Banuri and Marglin 1993).

Barbier and others (1994, 85), who divide humanity into "ecosystem people" and "biosphere people," express a more moderate view. The *ecosystem people* (most of the people discussed in this book) "subsist ... largely on resources produced or gathered from their immediate vicinity … [they] often develop a strong social culture that reflects their close interaction and inter-dependence on the environment, both within and across generations."

Biosphere people—like most readers of this book—are characterized as being "more transient, less community based and … thus less well 'bonded' with the environment." Mangel and others (1996) provide seven principles for the conservation of wild living resources, along with numerous mecha-nisms for implementing them. Many of their principles and mechanisms echo (or, in fact, predate) conclusions in this book. Numerous books have appeared in recent years urging westerners to change their worldview and more meaningfully acknowledge human integration into the natural world.

Some observers have argued that women are more likely to hold to Earth-preserving worldviews. In their collection on ecofeminism, Diamond and Orenstein (1990, xi–xii) identify three philosophical strains related to current efforts to defend the Earth:

- "[that] the Earth is sacred unto itself, that her forests, rivers, and different creatures have intrinsic value";
- "because human life is dependent on the Earth, our fates are inter-twined"; and
- "from the perspective of indigenous peoples, whose connection to native lands is essential to their being and identity, it is both true that the Earth has intrinsic value and that we are also dependent on her. Thus we … walk a fine line between using the Earth as a natural resource for humans and respecting the Earth's own needs, cycles, energies, and ecosystems."

Roszak and others (1995) contribute to the writings on ecofeminism and take a more formal psychological view on our need to both restore the Earth and heal our minds. Sewell's (1999) psychological approach emphasizes the important role of sight in our relationships with nature; Head and Heinzman (1990) include a section called "Saving the Forests and Ourselves" in their collection. Pedersen (1996) talks about "the global ideology of nature" in a book that examines Asian perceptions of nature (Bruun and Kalland 1996). Environmental ethics, as Whitten and others (1996) use the phrase, is close to what we mean by a conservation ethic. They discuss philosophical and religious bases for such beliefs, as well as "deep ecology" and the Indonesian state philosophy, Pancasila. Terborgh's (1999) book, *Requiem for Nature,* pre-sents a pessimistic view of the future, ascribing fault largely to our inability or unwillingness to concern ourselves with the degradation of the world around us. AtKisson (1999) goes beyond this pessimism—though acknowledging similar negative trends most convincingly—to suggest some constructive steps forward.

At its core, the question of a conservation ethic is a cultural issue, and numerous anthropologists have written about the views and relationships of indigenous peoples to their environments (good representatives of anthropological analyses are Clay 1988; Croll and Parkin 1992; Redford and Padoch 1992; Redford and Mansour 1996). Fairhead and Leach (1994/95) take a still deeper look in an African context. They convincingly argue that the views of nature held by more powerful outsiders in Guinea—which they call "the degradation discourse"—have so successfully defined reality that a demonstrably false view of ecological history in the area has been accepted by almost everyone. They conclude their book with a damning accusation that brings home the importance of issues such as people's perceptions of nature:

> It is hard to underestimate the importance of the degradation discourse's instrumental effects on many aspects of Kissidougou's life. These have impoverished people through taxes and fines, reduced people's ability to benefit from their resources, and diverted funds from more pressing needs. They have accused people of wanton destruction, criminalised many of the everyday activities, denied the technical validity of their ecological knowledge and research into developing it, denied value and credibility to their cultural forms, expressions, and basis of morality, and at times even denied people's consciousness and intelligence. The discourse has been instrumental in accentuating a gulf in perspectives between urban and rural; in undermining the credibility of outside experts in villagers' eyes; in provoking mutual disdain between villagers and authority; and in imposing on the former images of social malaise and incapacity to respond to modernity.

Leach and Mearns (1996) have assembled comparable analyses in several African countries. In Indonesia, the dominant view for years was that the Dayak and other forest peoples were wanton destroyers of the forest, despite an ever-increasing number of studies and analyses (such as Chapter 4) that showed the sustainable nature of their traditional systems, particularly compared with the effects of "modern" interventions such as mechanized logging, large-scale plantations, and resettlement schemes (see also Langston 1995 for an ecological history of the Blue Mountain region of Oregon that sounds somehow similar, though phrased differently—Langston is an ecologist).

We began the research reported in this book with some ideas about the relevance of a conservation ethic for forest people. The forest peoples we knew did not have the romantic and idyllic views of nature represented by the stereotypical western environmentalist (either for themselves or for indigenous peoples). Indeed, many aspects of life in the forests of Kalimantan caused fear among the people: demonstrably dangerous animals; slippery surfaces, sometimes at great heights, that could result in frequent and harmful falls; poisonous plants; and frightening spirits, to name a few.

However, traditional systems of land use had conserving functions. Among the Dayak, for instance, their swidden system had functioned in such a way as to maintain a near-continual ground cover, dramatic levels of biodiversity, and clear rivers with abundant fish. Although not the primary intent, conservation was a result of their normal practices. The Dayak also had some practices that were overtly conservationist (such as taking only the amount of wildlife they could use, taking care with fire, not reusing a rice field when there were signs of declining fertility). Taboos and beliefs about animals and spirits also appear to have had a conserving function (see Colfer and others 1997a; Wadley and others 1997).

But it is easy to push the conserving elements of traditional cultures too far. Polunin (1983) analyzes traditional marine reserves in Indonesia and New Guinea and expresses a concern that we share: do these systems have the capacity to respond appropriately to the kinds of massive change to which they are being subjected? Zerner (1994) examines the changing definitions of traditional marine tenure systems, by different actors, most recently responding to international and national conservation discourses. Hames (1991) offers a reasoned critique of the idea that Amazonian indians purposely conserve wildlife, whereas Kaskija (1999) extends a similar critique for the Punan of East Kalimantan.

Perhaps the most likely conclusion is that forest peoples have elements of their systems that conserve and other elements that do not (particularly during periods of rapid change, as many are experiencing at the current time). Harms' (1987, 256) comment reflects our views nicely: "Because the relationship between nature and culture is mediated by human choice, it is both dynamic and unpredictable."

Yet the idea of a conservation ethic surfaced repeatedly as a possible contributor to more sustainable management of the environment in the studies that resulted in this book. Many have wanted to identify ways to assess the presence of a conservation ethic and, indeed, to measure it.

In our work, we initially examined the idea of a conservation ethic as a possible dimension in determining "who counts" in sustainable forest management. We tried several ways to "measure" the concept: participant observation (Chapter 4), the Galileo program (Chapters 5 and 6), and the CAT-PAC program (Chapter 2). In Chapter 4, Sardjono and Samsoedin provide the most contextualized analysis of this topic. They attempt to describe a conservation ethic among the Benuaq Dayak of East Kalimantan. They present an introduction to Benuaq traditional knowledge and resource use patterns, arguing that Benuaq traditional management warrants recognition and support from the government and other would-be forest managers. Besides identifying aspects of the traditional system that relate to a possible conservation ethic, Chapter 4 provides an ethnographic backdrop for other Kalimantan-based chapters. The area is intermediate on the forest-rich/forest-poor continuum and demonstrates important changes that are occurring over much of Kalimantan at this time.

Chapter 5 documents our use of a sophisticated computer program to measure local people's cognitive maps relating to natural resources in West Kalimantan, Indonesia (a forest-rich area). The analysis focuses on gender and ethnic differences in perceptions. We hoped that a pattern of cognitive distances between, for instance, the concepts **forest** and **me** might provide us with a straightforward and replicable method to (a) measure aspects of a conservation ethic and (b) clarify its significance in sustainable forest management, by comparing results in forests of varying qualities. Chapter 6 is our first example of the kind of cross-site comparison we had initially planned, using comparative Galileo data from Kalimantan, Cameroon, and Brazil.

Much of the relevant literature does not clearly differentiate what people think about their environment from what they do with it. In Chapter 4, Sardjono and Samsoedin discuss both beliefs and practices. In Chapters 5 and 6, the authors focus on people's views of their relationship with nature. The degree to which such views are reflected in practice—and thus have or do not have important direct effects on the sustainability of forest management—is the subject of much debate, and cannot be resolved here. Our conviction is that because people's views affect their behavior—though to varying and probably unpredictable degrees—it is important to understand those views.

CHAPTER FOUR

Traditional Knowledge and Practice of Biodiversity Conservation

The Benuaq Dayak Community of East Kalimantan, Indonesia

Mustofa Agung Sardjono and Ismayadi Samsoedin

We conducted a study using participant observation to examine the concept of a conservation ethic among the Benuaq Dayak community of East Kalimantan, Indonesia. In this chapter, we describe the institutional arrangements of the Benuaq Dayak and the relationships between the people and the land. An extensive discussion of Benuaq swidden cultivation and related knowledge is included.

We present problems associated with the maintenance of Benuaq traditional knowledge and traditional mechanisms for biodiversity conservation as we observed them and conclude with some policy recommendations that we developed based on our findings.

A Conservation Ethic in a Traditional Community

We have found the concept of a *conservation ethic* to be attractive and potentially useful, particularly for stakeholder identification. Sustainability would seem more likely to exist in contexts where local stakeholders consider it desirable. Yet we, like many others, have been frustrated about how to operationalize this concept. Traditional systems typically contain traits or features that serve to maintain or protect natural resources and biodiversity. We have argued that the existence of such traits suggests some thought processes on the part of local people that could be described as a conservation ethic.

However, we have succeeded in generating neither a clear definition of nor a set of steps that can be used in assessing the presence or absence of a conservation ethic. Even so, we believe that our results are of interest for people who want to assess sustainable forest management and who are interested in community-based forest management as part of a conservation strategy.

Although we do not yet have access to an in-depth study of traditional knowledge and know-how related to biodiversity maintenance, several findings indicate that such a body of knowledge and practice exists among local native communities in Kalimantan, Indonesia. They include customary legal regulations and relevant local terminology among several local groups, such as

- the naming of stages of forest regrowth in the agroforestry systems of the Kenyah Dayak communities at Long Segar, Muara Wahau, and Long Ampung (Colfer 1983a);
- the terms *munaant, simpukng,* and *lembo,* which are indicative of similar management practices in home and forest gardens among the Tunjung and Benuaq Dayak groups in Barong Tongkok, East Kalimantan (Sardjono 1990);
- the use and cultivation of land based on vegetation types among the Simpang Dayak community at Ketapang, West Kalimantan (Djuweng 1992); and
- the land management system of the Banuaka Dayak community in Kapuas Hulu, West Kalimantan (Frans 1992).

Several researchers (for example, Gonzalez 1999) have described Dayak agricultural systems as being characterized by a "mosaic" of stages of forest regrowth, used for foods, medicines, natural pesticides, and timber for building construction. Soedjito (1994) noted that the Kenyah Dayak at Long Sungai Barang, Apo Kayan, maintained at least 91 species of beneficial plants in their gardens and 25 species in their upland rice fields. Furthermore, the Dayak community at Hulu Sungai Bahau maintained and cultivated 58 local paddy varieties, whereas the Krayan Dayak at Long Bawan are known to have sustained 37 local paddy varieties.

The collection of data about traditional knowledge, innovation, and practices that contribute to biodiversity is a potentially important and underused mechanism for evaluating biodiversity conservation. Before we can build a clear, precise, objective picture of traditional knowledge in East Kalimantan, substantial gaps in the data need to be filled. Such information can make a significant contribution toward clarifying definitions of terms such as *community-based forest management* and *community forests.*

In addition to our interest in the abstract concept of a conservation ethic, a further objective of this research was to gather information about the traditional knowledge of the Benuaq Dayak, especially concerning biodiversity and genetic resource management. We expected that a careful examination of their community-based forest management system would highlight the importance of learning, recording and improving on traditional knowledge, current community practices and innovation. Finally, we hoped to identify factors that influence and encourage social change in contexts where traditional knowledge has been found to be "alive and well."

Methods

Several methods were used in this study, including a literature review of research on the Benuaq and other Dayak communities in Kalimantan and their natural resource management strategies and an exploratory description of their systems of natural resource use. We also used primary techniques for data collection, such as direct interviews with key and opportunity informants, group discussions (with participants categorized by age and gender), and field observation. Secondary data from various sources also was examined. Data analysis involved qualitative description based on key information.

The Study Area

Our research took place in Damai, a subdistrict in Kutai District (now West Kutai, according to the 1999 governmental reorganization), at the upper end of the Kedang Palu River, a tributary of the Mahakam River (see Figure A1 in the Introduction). Four villages (Temula, Dempar, Jontai, and Sembuan) were selected as study sites.

The research area is immediately adjacent to the North Barito District in the province of Central Kalimantan (115°30'E, 0°15'S). The Damai subdistrict (kecamatan) consists of 18 villages, most of which are eligible to be part of the Indonesian government's program to help the "villages left behind" (Instruksi Presiden Desa Tertinggal, or IDT)—the poorest of Indonesia's villages. Damai is adjacent to Central Kalimantan, to the west.

The area's remoteness is reflected in its distance from Samarinda, the provincial capital and nearest city. Traveling between the locations actually requires two trips: a 21-hour longboat ride between Samarinda and Melak (the subdistrict capital city), plus 2 to 3 hours' drive by car between Melak and Temula (the village nearest to Damai). Other sites involved additional travel on foot or by canoe driven by an outboard motor. In effect, this remoteness means that the area has been neglected (compared with more accessible communities) in the regional development process.

Natural Conditions

The climate in the research area is wet and tropical with moderate winds. The average temperature throughout the year is 25 °C. According to subdistrict office records, the annual rainfall is 1,500–3,500 millimeters, with 40–100 days of rainfall per year, on average.

The topography is variable; altitude ranges between 200 and 1,000 meters above sea level. Slopes vary from somewhat steep (15–20%) to very steep (>40%). Lowland valley areas, plains, and swamps also have important uses and potential for agriculture and rattan gardens.

Soils are variable but generally dominated by the infertile yellow-red podzolic, latosolic, and adsolic types with clay and sandy clay texture. The erosion potential (and reality) is very high, particularly in riverine areas, where flooding occurs regularly. Mineral deposits of gold, coal, and limestone are common. Inhabitants living upriver on the Nyuatan River have a gold-mining tradition. According to local people, several exploration companies are surveying the mining potential near the study area. Biodiversity in the study area does not differ greatly from the flora and fauna in primary and secondary humid tropical forests elsewhere in East Kalimantan. Numerous kinds of fauna (both terrestrial and aquatic) and endemic flora are regularly seen. Important dipterocarp forest, so valuable for its timber, remains within this region, with common trees including *Shorea laevis* (*bengkirai* or *jengan*, varying by community), *Dryobalanops* sp. (*kapur* or *ngoi*), *Dipterocarpus* sp. (*keruing* or *apuin*), and *Shorea* sp. (*meranti* or *lempukng*).

Sociocultural Context

The main ethnic group residing in the study area is Benuaq Dayak. Even though the history of the Dayak has been well documented (for example, Riwut 1979; Pemda Kaltim 1990), the precise details of the settlement process of the Benuaq Dayak remain unclear. The absence of written material about their history means that researchers must rely heavily on oral traditions, which may vary both by locality and the researcher's skill to elicit the stories. However, the Benuaq Dayak represent a subgroup of the larger Luangan group that surrounds them (Weinstock 1983) and probably came from the upper Barito River in Central Kalimantan (Widjono 1992).

In fact, the Benuaq Dayak living in the study area are made up of two subgroups: the "original Benuaq" people in Temula and Dempar villages (downstream), and the Tuayan people in Sembuan and Jontai villages (upstream). The group's hospitality to newcomers has resulted in new permanent community members through family or marital connections, jobs in the church, teaching, trading, and employment in nearby companies.

The level of education in the communities is low, perhaps because of a lack of facilities and personnel willing to teach in this remote region as well as a low interest in education within the communities. The low educational level plays a role in the people's high dependence on natural resources. Generally, people earn their daily living by farming and gardening (rice, rattan, vegetables, or fruit trees). In recent years, several families have experimented

119

with tree crops (such as rubber, coffee, and candlenuts) in their upland fields or their home gardens.

As is common in other East Kalimantan Dayak communities, the Benuaq maintain many communal aspects of their way of life. Their organizations and patterns of behavior are based on the strong and valued traditional legal rules that continue to regulate daily life. The existence of traditional values is reflected in the fact that the traditional leader is more highly respected than the village headman, as demonstrated in the handling of fights, theft, and marriage. For certain problems, particularly those related to village government, the authority of the village official remains greater. In several villages, during the time of our study, the two positions were held by one person so that both the community and the village government were directly managed by that individual.

Although it is less obvious, the Benuaq recognize a kind of traditional social stratification in daily life. There are two kinds of community groups: one considered noble (descended from spirits) and the other common. In recent years, this stratification has declined because the whole community has come to claim equal rights and responsibilities. Even so, on some occasions, the "nobles" quite clearly maintain some of their previous dominance, particularly in holding certain administrative posts.

During our field observations, we witnessed several of the surviving traditional ceremonial activities. The *kwangkai* (the reburial of the bones of someone dead for several years) and *beliatn* (traditional medical care by shamans) are two examples. As a result of the influence of governmentally recognized religions that have gained in prominence in recent years, several traditional ceremonies have disappeared. The role of the formal religious leader as priest or pastor in people's daily lives has increased.

Results and Discussion

An Overview of Benuaq Institutional Arrangements

One of the most essential elements in Benuaq community philosophy is embodied in the word *tempuutn,* which covers the basic concepts and beliefs that characterize Tunjung and Benuaq Dayak cultures. *Tempuutn* is also a ceremony that lasts from a few hours to several weeks or even months, which is believed to heal, create peace, increase plant fertility, or be a last rite to honor the spirit of someone who has died. *Tempuutn* is particularly important in the narration of the history or the origin of anything related to the spirit or soul, to special animals, plants, or to customs and traditions that must be obeyed by humans (Madrah and Karaakng 1997).

Tradition in the Benuaq language is synonymous with *regulation.* The five essential elements in Benuaq tradition are healing (*beliatn*), funerary traditions, marriage, upland rice farming, and conflict resolution traditions (personal communication from Ujung, 1998; see also Madrah 1997; Madrah and Karaakng 1997).

In this study related to human well-being, healing,[1] upland rice farming, and conflict resolution are most relevant traditions. *Beliatn* is a ceremony to heal the sick or to restore balance in a situation or condition. Both men and women can be ceremonial healers (called *pemeliatn* or *tukang beliatn*). Such healers get their powers by means of ascetism, dreams, or wealth or by birth. One type of related tradition is called *beliatn gugu tahun* or *nalitn tautn*. This ceremony is carried out over 24–105 days to obtain an abundant harvest. It is considered necessary when a sin such as incest has been committed that has resulted in illness, long dry seasons, or contagious diseases. Such beliefs represent the Benuaq people's conceptual link between their own behavior and the environment of which they are a part.

Traditions related to upland rice farming have a pervasive influence throughout the Benuaq Dayak subsistence system, including traditional forest management, ownership of upland fields, prohibitions and regulations about land and other resource use, and disputes over natural resources. One critical feature is the recognition of a community territory when using farming and forest areas. The traditional forest area is not available for the Benuaq Dayak to use for their fields. This regulation is generally accepted, with serious social and spiritual sanctions in cases of violation. Our findings suggest that farming use by the Benuaq of these forested areas is highly improbable.

Under inheritance traditions, the elder children of both sexes receive more than the younger ones, the argument being that they have provided more service to their parents. If the oldest child leaves for a significant period of time, younger siblings who continue to help their parents can become eligible for the larger inheritance. If parents die without dividing their land among their children, the eldest son will decide the distribution.

Local leaders (mantiq) typically leave their positions to their descendants, but the family connection is insufficient to ensure that the post will be inherited. The incumbent must be recognized as having knowledge (obtained from education [*mangaziig* or *bakajiig*]) and be acceptable to community members. In the past, the traditional leaders also were the official local government leaders.

Today, leaders are chosen by Government of Indonesia procedures. The *mantiq* may remain as the traditional leader, and the same person can become both an official and traditional leader, as has happened in Dempar. Traditional leaders are selected by the communities, who make a recommendation to the village headman. The proposal is then passed on through the subdistrict headman, to the district leader and the provincial governor, who makes the final decision. However, the decree will be signed by the district leader.[2]

Above the traditional leader with his subordinate group of elders, a "great traditional leader" represents four to seven lower-level headmen. The extent of authority of the great traditional leader follows a river. The Damai subdistrict has three main rivers and a branch (Kedang Pahu, Idatn, Nyuatan, and the Nyuatan branch) and three great traditional leaders, each with more than five common traditional leaders subordinate to them in 19 communities. Such a traditional network of relationships among communities could be a valuable resource in promoting conservation efforts.

Benuaq People and the Land

The forest is an essential part of the lives of the Benuaq, fulfilling many of their needs, particularly those who live in relatively isolated areas. Their dependence on the forest has generated regulations and arrangements to enable use of the forest in a reasonable and sustainable manner. These regulations and arrangements, inherited over time, have served an important function in the form of beliefs, customs, and traditions. Forest management has been controlled by these inherited systems of behavior, for example, when opening up forest area for upland rice cultivation or for harvesting wood and nontimber forest products (NTFPs). Such traditional customs and rules have served the same function as externally defined, formal natural resource management regimes.

Upland rice cultivation is the central subsistence activity of the Benuaq. Farmers routinely strive for a balance between cultivation and fallow times explicitly intending to maintain the fertility of the soil, particularly for rice cultivation. Complementary activities on these fields are the growing of horticultural crops and rattan. Hunting, fishing, honey gathering, and other NTFPs also supplement the Benuaq diet and income. NTFPs are collected during certain periods in the rice and rattan agricultural cycles and as a combination of leisure and productive activities (gathering forest fruits, for instance, can provide opportunities for courtship or enjoyable hunting expeditions). The Benuaq are unusual in their traditional practice of rattan cultivation, and regulations and management arrangements relate to that crop. Rattan gardens are grown in upland rice fields or in specially reserved sites.

Land holds a special place for the Benuaq in their worldview; agreed traditional legal norms are synchronized with the dynamics of everyday life. Norms regulating land ownership, resource use, and the community's natural wealth remain important. Among the Benuaq, as among many Dayak groups, land serves one of the most vital functions in their lives and has correspondingly important associated rules and regulations. Land relates to basic needs, serves as a symbol of material prosperity, and is closely related to social and cultural beliefs as well as to political values.

Scharer (1963, cited in Widjono 1992) reports that the Benuaq Dayak believe the universe is filled with magical powers. Whenever the rules of the universe are functioning properly, these powers are harmoniously balanced. But human behavior has the potential to disturb these rules of the universe, and destructive magical power can be stimulated. This belief, linking people to their environment, adds a note of compulsion to Benuaq Dayak attempts to maintain those rules of the universe.

The Benuaq recognize a mystical nature in both human and nonhuman living creatures in the universe. They recognize dwellings for these creatures in a land above the sky, another below the sky, and a land of the soul inhabited by their ancestors. God is believed to inhabit both the land above and below the sky. In the above-the-sky representation, God is a hornbill, which is also a symbol of courage and power. In the below-the-sky representation, God is described as a dragon, symbolizing fertility (Nanang 1990, cited in Widjono 1992).

The Benuaq Agricultural System

In this section, we provide an ethnographic description of the forest farming system of the Benuaq Dayak. We present issues of land tenure and use, discuss the conceptualization of farm–forest links, and briefly describe the agricultural cycle.

Land Tenure and Use. Land used for upland rice farming can be covered with underbrush, secondary forest, or old growth. In the case of secondary forest, it typically has been in fallow for 10–20 years. However, if rattan (*uwe* or *we'*) is planted and growing on that land (a rattan garden, or *kebotn we'*), the fallow period may be prolonged for more than 20 years.

A particular family may either inherit land from parents or clear old-growth forest. In the latter case, the family gains traditional ownership rights to that land. When landowners are not making use of their own land, other families may be allowed to use the land for seasonal crops (not usually for woody or other tree crops). The land within the traditional territory of the community—which, in some sense, can be considered community property—is available for use only with the express permission of the individual landowner (that is, the family that cleared it or inherited it). The differences between individual and community land ownership and use rights overlap and are not as clear as in western countries.

To avoid disputes over land, families discuss boundaries when they first consider clearing the land. Borders tend to be natural features, such as rivers, tributaries, ridges, valleys, and certain trees. They may be marked by long-lasting trees such as *Durio zibethinus* (*durian* or *ketungan*), *Artocarpus champeden* (*cempedak* or *nakaatn*), and *Shorea* spp. (*tengkawang* or *oraai*) to clarify borders and strengthen any claims. The rotation period of any field depends on the landholdings of the particular family. Families with claims on more land have longer rotations; those with less land, shorter. Budiono (1993) calculated the hectarage of families in the Benuaq community of Bentian Besar to be between 1 and 40 hectares; the average is 11.94 hectares per household. The area needed for subsistence is between 0.6 and 2.5 hectares, yielding 2–7.5 tons of field dried rice, or an average of 1.32 hectares with 3.95 tons.

Farm–Forest Links. The Benuaq spiritual link between farming and the environment is embodied in the decision to begin the yearly agricultural cycle on the basis of the heavenly cycle. Traditionally, the Benuaq Dayak community recognized the appearance and size of particular stars as an indicator to begin upland rice production. The stars in question were referred to as "seven stars" (probably the Pleiades) that indicated the beginning of site selection, usually in April or May, and *belentik* or *poti*, a star visible from the beginning of the clearing activity up to planting. To begin burning, *belentik* must be exactly overhead. If this rule is not followed and burning is carried out before or after this time, paddy production is believed to be lower.

Within this agroforestry management system, the Benuaq Dayak community recognizes forest successional stages based on the age and type of vegetation that dominate the area selected for a rice field. At Temula and Dempar villages (downstream sites),[3] the recognized successional phases are as follows:

- *Umaq* or *ume*: A newly opened site for upland rice cultivation (first year) is usually derived from young secondary forest (*kurai*), old secondary forest (*bengkar bengkaletn*), or old–growth forest (*bengkar*).
- *Baber* or *boaq*: The site continues to be used for upland rice (second year). The soil is still considered to be fertile, and the yield is still satisfactory. Physically, the vegetation begins to sustain herbaceous plants other than rice, such as shrubs and bushes with a height of less than one meter. Dominant varieties include *padekng* (*Imperata cylindrica*), *buncar* (*Paspalum conjugatum*), and *kanau* (*Pteridium aquilinum*).
- *Kelewako*: The site is still considered to be fertile and productive enough for continued upland rice farming (third year). The height of other vegetation types present reaches two meters, and the dominant types include *buncar* (*Paspalum conjugatum*), *jojot* (*Musa* sp.), *sempiring* (*Cyprus* spp.), *padekng* (*Imperata cylindrica*), *mukng* (*Blumea balsamifera*), and *bekakang* (*Melastoma malabraticum*).
- *Keleweo* or *baling batekng*: This phase is similar to *kelewako*, but the tree trunks remaining in the field have begun to deteriorate, hence the name *baling batekng*. The site is likely to be abandoned as a rice field because soil fertility and yields will have begun to decline. Pioneer vegetation types, mainly weeds such as *padekng* (*Imperata cylindrica*) and *buncar* (*Paspalum conjugatum*), may cover the whole area.
- *Kurat ure*: About five years after use as an upland rice field, vegetation present has reached pole size (perhaps 15–20 centimeters in diameter). Dominant types include *kleboto* (*Trema tomentosa*), *mawah* (*Mallotus paniculatus*), and several climbing types of vegetation (Piperaceae).

- *Kurat tuha* or *batekng*: In the secondary successional stage of 5–15 years after cultivation, the vegetation present has reached the tree stage; example species include *naga* (*Schima* sp.), *mahang* (*Macaranga* spp.), and the shoots of Dipterocarpaceae.
- *Bengkar bengkaletn*: Old secondary forest that resembles old-growth forest in both composition and structure.
- *Bengkar*: Old-growth forest that has never been cleared completely for upland rice farming but is used for hunting, rattan collection, honey gathering, and other activities because it is close to the village.
- *Bengkar mentutn*: Old-growth forest that has never been cleared for upland rice farming and is rarely or never visited by local inhabitants because of distance.

Figure 4-1 is a visual representation of the stages of succession recognized by the Benuaq Dayak communities. Within this overall view of the environment, various techniques are used to select sites for upland rice farming. The Benuaq assess soil fertility and the suitability of primary and secondary forests for upland rice farming based on the presence of the following characteristics:

- dominant cover crops such as *isaq-isiq* (*Stachyprinum jagorianum*), *tempereh* (*Pandanus* sp. 1), and *kajang* (*Pandanus* sp. 2);

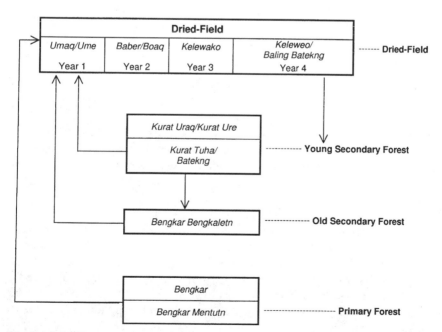

Figure 4-1. *Forest Successional Stages Known among the Benuaq Dayak (Benuaq and Tuayan)*

- soil color and texture (suitability is determined partly by thrusting a bush knife into the soil—soil sticking to the knife indicates probable high clay content, a good sign for farming);
- proximity to a river (silt-laden lands closer to rivers are believed to be of better quality); and
- absence of resin-producing trees (*meranti* or *lempukng* [*Shorea* sp.] and *kapur* or *ngoi* [*Dryobalanops* sp.]).

In addition to their concerns about soils, local people take into account some other important issues when selecting sites:

- protection of fruit trees such as *durian* or *ketungan* (*Durio zibethinus*), *lai* (*Durio kutejensis*), *mangga hutan* or *encapm* (*Mangifera* spp.), *cempedak* or *nakaatn* (*Artocarpus champeden*), and *rambutan* or *bua* (*Nephelium* spp.) and
- protection of trees used for nesting by honeybees, such as those known as *tanyut,* which include *benggeris* or *puti* (*Koompassia* sp.), *kenari* or *jelmu* (*Canarium decumanum*), *kapur* or *ngoi* (*Dryobalarops* sp.), *bengkirai* or *jengan* (*Shorea laevis*), and several species of *meranti* or *lempukng* (*Shorea* sp.).

Certain taboos are also respected; for example, if something *tuhing* (prohibited) or *nyaho* (sign that something is about to happen) is sighted in the upland rice field, clearing will not go ahead. Some *tuhing* and *nyaho* include

- groups of honeybees gathering at the cleared field and nesting in trees that are not normally used by bees for nesting,
- animals dying without apparent cause (not killed by human beings), and
- unusual bird songs heard when the field is to be cleared.

These farm–forest links reflect the ecological knowledge of the Benuaq Dayak and highlight various concerns relevant for natural resource management.

The Agricultural Cycle. The Benuaq agricultural cycle is focused on rice cultivation. Although the steps in a swidden cultivation system appear simple—clear, burn, plant crops—they make up quite a complex scheme of management.

Selecting Forest Type and Identifying Borders (Ngeraakng). New (*baber* or *boaq*) or slightly older (*kelewako* or *kuratn*) fields that have been planted with rice may be used again if the fertility remains sufficiently high (as indicated by tree size, leaf color, and soil color). Secondary forest (*batekng*) is considered the ideal location for upland rice farming, because soil fertility is considered to have recovered. Old-growth forest (*bengkar*) is also considered a good habitat from which to cut upland fields because it may provide the highest yields.[4] The size of trees, however, makes clearing difficult and drying a time-consuming process.

The boundaries of any new rice field typically are discussed in April or May, before any activities occur in the fields; this discussion is referred to as *sempekat enak umaq.* Families and community members whose fields will be adjacent place markers to identify boundaries. Borders may consist of fruit trees, bamboo (*Bambusa* spp.), *sungkai* (*Peronema canescens*), kapok or *kapoq* (*Ceiba pentandra*), or *dadap* (*Erythrina* sp.).

Location (distance from village and rivers, slope of land) and preexisting property rights are important considerations in choosing a site and boundaries. Traditionally, to determine whether land was of suitable quality to produce good yields, two methods were used.

- A sharpened internode of bamboo was driven into the ground. The more soil retained in the bamboo when it was withdrawn (indicating clayey texture, not sandy soil), the more fertile the soil and the better the expected yield.
- *Akar malakng*, a type of root, was split vertically into eight parts and bound together in the middle. The ends were then tied to each other systematically. The middle binding was untied, and the rootband dropped on the ground. If it made a circle shape, the land was fertile; if it formed an asymmetric or dispersed pattern, the land was infertile.

These methods are no longer used by the Benuaq. They believe that the soil in their area is more fertile (*tana' lingau*) than in neighboring areas. Currently used indicators include the composition of plants growing on the forest floor, soil color and type, slope, distance from rivers and streams, and the size of trees. The people also avoid clearing areas with fruit trees or trees that commonly harbor bees' nests—for both pragmatic and sacred reasons. The Tuayan do not allow the eating of greasy foods at the site to be cleared.

Cutting Shrubs (Nokap). This process is carried out alone or in exchange labor groups using bush knives between April and June. Its purpose is to clear bushes and shrubs to facilitate the cutting of the big trees more safely. Once dry, the shrubs provide fuel to ensure a good burn. This cutting process normally takes three to five weeks, depending on the tools, labor available, and the amount of vegetation on the site.

Cutting Trees (Temekng *or* Nowekng). Cutting, done by men with an axe and a chainsaw, begins with trees of pole size and above. The dead trees then dry out, contributing to a good burn that will clear the field well and produce abundant ash. The cutting is designed to allow sunshine to reach the ground. The Benuaq from Temula and Dempar refer to this activity as *temekng,* whereas the Tuayan Dayak in Sembuan and Jontai call it *nowekng.*

Pruning (Nutu). Individuals or exchange labor groups may prune in June and July. The process involves cutting stems and branches that have been felled into small pieces to facilitate the drying process. This step is only necessary in fields cleared from forest.

Drying (Oikngjoa *or* Ekayjoa). Cut trees are dried for three to five weeks in July and August before burning. The drying process depends on the number of days of sunshine and the size and number of trees cut. The Tuayan turn the wood and cut it into smaller pieces to speed up the drying.

Burning (**Nyuru**). Burning, which takes place in August or September, removes the slash and any remaining living plants. Ash from the burn is useful for increasing soil fertility. The removal of living plants also reduces weed problems for the rice crop. Several precautions are taken to reduce the danger. Burning begins at the bottom of a slope, taking into account the wind direction. If a person undertakes burning alone, a partition is made around the field to ensure the fire will not escape. Typically, people cooperate in groups of at least ten and with the knowledge of neighbors who guard their own fields.

Reburning (**Mongkekng** *or* **Menukng**). The remaining unburnt branches are piled and reburnt. The Benuaq from Temula and Dempar villages call this step *menggekng* or *mongkekng*, whereas the Tuayan (from Sembuan and Jontai) call it *menukng*.

Planting (**Ngaset** *or* **Ngula**). Rice is planted between August and October (after burning) in a rice field–cum–rattan garden. It is planted cooperatively (known as *pelo*—with both men and women alternately exchanging their labor with other families) or by individual families on their own fields. Men wield the dibble stick, and women drop five to ten seeds into the holes (*moyas*), around 30–50 centimeters apart, depending on land characteristics. Rattan, fruits, and other secondary crops (vegetables, tubers) are planted about three months later.

Weeding and Field Maintenance (**Ngejikut** *or* **Ngerikut**). The main crops are cleared of weeds that disturb and inhibit growth twice a month. The Benuaq call it *ngejikut;* the Tuayan, *ngerikut*.

Harvesting (**Ngotepn** *or* **Ngotow**). Every attempt is made to harvest mature rice between January and April, before bad weather or pests can destroy the crop. Individual families and work parties may be paid "wages" in paddy rice. Rice panicles are individually cut with a finger knife. The harvested rice is usually stored in a rice storage barn in the upland field before being removed to a similar barn in the village or home.

The Maintenance of Benuaq Knowledge and Traditional Mechanisms for Biodiversity Conservation

Rapid economic development in Damai and surrounding subdistricts has had negative consequences for the existence and potential value of traditional knowledge related to biodiversity among the Benuaq Dayak in the research area. Knowledge and practices based on local wisdom (see Table 4-1) have been implemented for generations but are no longer widely known among young generations of the Benuaq.

Land use patterns and traditional crafts on which the people depend function within a subsistence economy that is unable to compete with develop-

Table 4-1. *Some Traditional Knowledge and Practices Concerning Daily Activities of the Benuaq Community that Reflect a Conservation Ethic*

Activity	Traditional knowledge and practices
Dry-field cultivation/upland rice farming	• Giving priority to old secondary forest, to maximize harvest without disturbing succession process • Avoiding wildfire by using firebreaks, accounting for wind direction and slopes, keeping an eye on the fire • Working together with manual tools (without mechanization) and use of fertilizer and/or other chemicals • Using different local rice varieties of *pare* and *pulut* (*Oriza* spp.) • Following fallow period and rotation principles to let the soil recover and maintain fertility
Rattan harvest/ collection	• Growing rattan with other useful plants (sown or naturally regenerated) • Cutting rattan selectively to ensure sustainability
Fruit tree gardening and fruit collection (*simpukng*)	• Conserving forest fruit trees (in situ conservation) for food security • Gardening fruit trees (ex situ conservation) in different places—such as close to farmlands, settlements, and rivers—to result in ecological and socioeconomic benefits and balance
Collecting vegetables and condiments	• Collecting young leaves (for example, rattan), ferns, and condiments from the forest (usually on the way to or from farmland) for daily consumption
Hunting and fishing	• Periodic hunting and fishing, individually or in small groups (not more than five people) • Using traditional or simple tools (blowpipe, traditional spear, "local production" gun) that have a low impact on the environment • Targeting hunted species, particularly those that are also farm pests (deer [*Cervus unicolor*], wild pig [*Sus barbatus*]) • Hunting game mainly for daily consumption and sharing with neighbors before selling excess (if any) • Prohibiting the use of *tuaq* (*Derris elliptica*), a small plant used for fish poison

continued on next page

ments, such as new income sources and large-scale commercial enterprises. This situation has caused a decline in local natural resources on which residents depend. Another important change has been the introduction of an artificial ecosystem (through the development of plantation estates) that, many believe, promises much profit in a short time; however, long–term sustainability is seriously in question.

Table 4-1. *continued*

Activity	Traditional knowledge and practices
Collecting honeybees	• Prohibiting the cutting of different species of *tanyut* (honeybee trees)—*benggeris/puti* (*Koompasia* sp.), *kenari/ jelmu* (*Canarium decumanum*), and *durian/kalang* (*Durio* sp.)—and different species of dipterocarps • Collecting in groups, with a system of sharing
Collecting wood for construction and fuel	• Collecting construction wood for self-consumption only (not commercial), usually from traditional forests and fallow areas • Prioritizing fallen wood, dead wood, and unproductive trees for timber, usually saving them during clearing for dry-field cultivation • Using fallow areas and bushes as firewood sources • Choosing only a few selected plant species as firewood (based on burning and fire quality, ash residue, and toxicity)
Damar resins, saps/ *malau*, barks/ *gembor*, aloewood/ *gaharu*, other raw materials collection	• Making seasonal collection part of daily activities • Collecting mostly from fallen resins (for *damar;* from many species of dipterocarps), selected trees (that is, they do not cut or tap every tree in a stand; for *gaharu*, from *Aquilaria* sp., and for *malau*, from *Palaquium calophyllum*), and dried trees (for bark of *gembor/Alsiodaphne* sp.)
Medicinal plant use	• Not allowing forest activities to disturb medicinal plants • Cultivating some medicinal plants close to settlements and frequently using them for traditional cures (*beliatn*)

Source: Primary data (1997).

In addition to external problems, there are internal concerns related to the maintenance of traditional knowledge about and potential for biodiversity (Table 4-2). Partly a result of interaction with the outside world, young people are reluctant to respect and maintain the traditions of their parents. For example, upland rice farming is no longer considered an appropriate profession for a high school graduate, and the use of forest plants for medical treatment is seen as an inferior alternative to "modern" medicines from a local government clinic. Indiscriminate rejection of traditional practices and beliefs, even when they may have strong and lasting value, underlines the possibility that important and useful traditional knowledge and potential, proven over generations, may be lost in the future.

Conclusions and Recommendations

Our observations, analysis, and discussions have brought us to the following conclusions about traditional systems of belief and resource management:

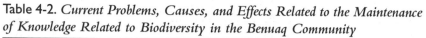

Table 4-2. *Current Problems, Causes, and Effects Related to the Maintenance of Knowledge Related to Biodiversity in the Benuaq Community*

Problem	Causes	Effects
Shortage of resin (*damar/* dipterocarps), sap (*malau/Palaquium calophyllum*), aloe-wood (*gaharu/Aquilaria* sp.), and bark (*gembor/Alsiodaphne* sp.).	• Resin-producing trees are hard to find (>10.0 km from the village). • Market for resin has collapsed, so nobody looks for it. • Sap from the bark of *gembor* wood, used for plywood glue, is rare and takes too much time to collect given priorities on upland fields. • The price of *gembor* bark is low (Rp 300/kg fresh, Rp 700/kg dry), further reducing interest in collection.[a] • The amount of aloewood (*gaharu*) has declined through excessive and indiscriminate cutting.	• Overall diversity of the subsistence system is reduced; dependence on upland rice and rattan is increased. • *Gembor* bark market has collapsed, reducing incomes.
Shortage of wildlife	• Reduction in habitat (forest area). • Hunting grounds are now about 10 km from the village.	• People have reduced their meat consumption, substituting fish. • People have started to raise chickens, ducks, and pigs.
Decreased rice yields, supplying barely enough for daily needs	• Soil fertility has declined due to shortened fallows in upland/dry fields. • Increased levels of pests (rat, monkey, deer) are seeking food in upland fields because of reduced forest area. • Farmers are uncertain about which new technologies are appropriate and helpful for increasing production.	• People's incomes from upland yields are lower, especially during the long dry period. • People are confused, concerned, and worried about their and their children's futures.
No buyers for hand-made products (mats, baskets, various rattan backpacks, for example, *lanjung* and *anjat*), with rare exceptions	• Community members lack knowledge about markets. • Community members lack innovation in designing new motifs. • Youths are not interested in learning to weave.	• Interest in producing crafts in the village has declined. • Crafts are made for subsistence use only.
Low selling prices of *sega* (*Calamus caesius*) and *jahab* (*Calamus trachycoleous*) rattans (price: Rp 250–350/kg undried/fresh)[a]	• *Jahab* and *sega* rattans are sold in large amounts because of widespread popularity in other countries. • Community members lack innovation or new and appropriate technology for rattan production. • Cultivation of other commodities is minimal. • Rattan is sold in undried (fresh) form. • Rattan and produce from upland fields are sold at low prices because of immediate need for money.	• People's options for income generation have become severely restricted, primarily to only rattan.

continued on next page

Table 4-2. *continued*

Problem	Causes	Effects
Reduction in availability of fish, particularly in dry season	• Fish poison or electric fishing kills all fish. • Strangers steal fish. • Fish populations decline in the dry season. • Fish farming has not been widely adopted.	• The community buys fresh and canned fish. • People feel hopeless about the situation.
Forest foods rarely used	• Knowledge of good forest foods has been lost to younger generations. • Existence of home gardens has resulted in people's increasing dependence on home-grown foods. • Distance to the forest reduces people's willingness to gather foods.	• Local knowledge is being lost. • People are moving toward dependence on home gardens.
Decreased honey gathering	• Although *tanyut* (for example, *Koompasia* sp.) trees, which typically house wild honeybees, exist in the area, the honeybees no longer come. • People increase their dependence on dry fields because honey is not available.	• No income is derived from honey production.
Limited use of medicinal forest plants except among the older generation	• The young are not interested in learning this lore because artificial medicine is provided and encouraged by the government. • Traditional medicine (*beliatn*) is recognized only by certain people.	• Local knowledge is becoming extinct.
Forest fruit prices quite low (durians sell for Rp 500–1,000 each)[a]	• It's a "buyer's market," so any price offered is accepted. • Overproduction of fruit results in low prices.	• Prices go down with seasonal overproduction. • Community attitude to fruit prices is "better something than nothing at all."

Source: Primary data (1997).

[a] US$1 = 2,500 rupiah (Rp; 1997), although at the beginning of 1998, devaluation had eroded the value of the rupiah to US$1 = Rp 15,000. The local people felt no significant effect of the devaluation, because local trade is conducted in rupiah.

- Among the local communities in the research area, traditional values remain strong, even under conditions of rapid socioeconomic and cultural change in recent years.
- Traditional regulations and values relate to the whole landscape; natural resource use patterns tend to emphasize equity in benefit sharing and sustained production.
- Local inhabitants maintain an interdependent relationship with the forest.
- What is called "socioeconomic progress" by outsiders has led to changes in people's motivations and ways of thinking about natural resources,

including forests. A logical consequence of these changes, in combination with the external influences that have shaped the landscape in the area, has been the degradation of the forest and serious reduction of its benefits for the surrounding people.

- Learning, documenting, and reinforcing local people's benign practices in natural resource management and biodiversity conservation represent practical alternatives to current, centralized land use planning that has led to the current deplorable situation; a solution that makes use of local people's knowledge and social capital can contribute to solving the problems of the region.

These conclusions lead us to propose that

- economic development of the area should be in line with the conservation of positive traditional values,
- efforts to increase the economic attractiveness of traditional management should be maintained and intensified, and
- it is very important to support every possible effort to implement appropriate policies taken by both the forestry department and the local government.

Our original aims of this study were to explore the concept of a conservation ethic and to examine the changes occurring in the Benuaq system of agricultural management from the standpoint of the conservation of biodiversity. We have found that the traditional system has much to recommend it in terms of traditional knowledge and lifestyles. However, it clearly is under threat. If the social capital represented in the Benuaq way of life is to be used effectively for the benefit of local forests and local people, its value must be explicitly recognized, consciously strengthened and revitalized, and encouraged and supported by other stakeholders. At this time in Indonesia's history, the likelihood that this may happen is greater than has been the case in the past.

Acknowledgements

Fieldwork was conducted during the extremely dry months of July to December 1997 under the framework of a broad community-based forest management research program of the Biodiversity Foundation, in cooperation with the MacArthur Foundation, Asia Forest Network, and Indonesia's Forestry Research and Development Agency.

To all the local communities in the four research villages who shared their experience and knowledge, we are most grateful. We also thank Wijaya and Edi (from The Social-Biosphere Foundation/Yayasan BIOMA) for support in data collection, Pak Paulus Matius (from the Faculty of Forestry, Center for Social Forestry, at Mulawarman University) for plant determination and some cultural explanations, and Dr. Carol J. Pierce Colfer (Center for International Forestry Research [CIFOR]) for critical comments and editing. Yvonne Byron's careful editing for language also has been a great boon.

Endnotes

1. *Editors' note:* This issue of human health has been highlighted as a principle in CIFOR's *Generic C&I Template:* "The health of the forest actors, cultures, and the forest is acceptable to all stakeholders." We have not yet investigated it in the systematic way we have dealt with the other two main principles discussed in this book.

2. With the issue of Act No. 22/1999 on Local Government, all administrative procedures, including the selection of traditional leader, will probably change.

3. The alternative Tuayan Dayak terms for each successional term, from the upstream Benuaq communities, are also provided.

4. Colfer with Dudley (1993) found the same perception among the Kenyah Dayak, but examination of long-term land use and harvest data showed that rice fields made from old secondary forest produced the highest yields over time. Important issues may include the advantages of gaining rights to the land cleared and a possible reduction in time spent weeding in old growth.

CHAPTER FIVE

Assessing People's Perceptions of Forests

Research in West Kalimantan, Indonesia

Carol J. Pierce Colfer, Joseph Woelfel, Reed L. Wadley, and Emily Harwell

To what degree do people who live in forests have a conservation ethic? How can we quickly and reliably assess how close people feel to the forest? How closely integrated are the lives of forest people with their environment? These are some of the value-laden but important questions that recurred throughout our research on human well-being. Previous experience with cognitive mapping (the Galileo method described in the Introduction) suggested that the approach might be helpful for dealing with these questions.

One issue we explicitly did not address in this research is the degree to which people's views of their environment and their place in it affect their behavior toward the environment. We view this issue as a variable link in both time and space. The Galileo method is designed to examine aspects of cognition only.

We conducted research in four villages, among two ethnic groups in and around the Danau Sentarum Wildlife Reserve (DSWR) in West Kalimantan, Indonesia[1]:

- *Wong Garai:* A 14-household longhouse lying along the northeastern periphery of the DSWR. In this fairly typical local Iban situation, rice is cultivated in long-fallowed hill swiddens and short-fallowed swamp swiddens, and households rely on nontimber forest product (NTFP) collec-

---

This chapter is adapted from Colfer and others 1996a.

tion, supplementary cash crops of rubber, and long-term circular labor migration (see Wadley 1997; Colfer and others 2000a). These Dayak people are a mixture of Christian and animist.

- *Kelayang:* An Iban longhouse situated to the south of Wong Garai, just within the eastern boundary of the reserve. The 28-household community (including a 14-door longhouse and several separate houses) is situated on the banks of a major river and is heavily involved in fishing for both consumption and sale, but not at the expense of rice agriculture, which is primarily on swamp swiddens. Its Christian and animist residents collect NTFPs for consumption and cultivate some rubber, collect rattan, and perform wage labor to generate cash; however, most need to supplement their agricultural income by selling fish products (see Harwell 2000a).

- *Nanga Kedebu':* A small Melayu fishing community of 108 people (Colfer's 1992 census) located in the heart of the DSWR (also discussed, like Danau Seluang, in Chapter 16). These fisherfolk are integrated into a money economy, selling their fish (both processed and fresh) to buy rice and other staple foods. The inhabitants are formally registered as residents of the larger community, Selimbau, on the Kapuas River, from whence comes a yearly inundation of additional fisherfolk during the dry season (the de facto population increased to 199 in October 1992). They are Muslim and share significant common cultural features with related peoples described by Firth (1966), Harrisson (1970), Scott (1985), and Furukawa (1994).

- *Danau Seluang:* A Muslim Melayu fishing community, located near the southeastern border of the DSWR, roughly the same size as Nanga Kedebu'. However, situated closer to more dry land than Nanga Kedebu', agriculture plays a slightly more significant role in the natural resource management system. It is formally affiliated with the larger village of Jongkong, on the Kapuas River.

After describing the necessary site-specific implementation of the method, we provide examples of the variety of analyses that can be carried out on the results. In the concluding section, we discuss questions that have arisen and problems that remain for future research.

Method

The conditions we wanted to assess in this study include the presence or absence of a conservation ethic, a feeling of closeness to the forest, and an intimate link between local culture and the forest. (For a technical description of the Galileo method as applied in the DSWR, see Annex 5-1.)

We began by selecting locally appropriate concepts pertaining to people–forest interactions, based on our familiarity with the area. The concepts included **fish, wood, rattan, honey, garden, animal, food, earth/soil,** water, I (the respondent), **man**, **woman**, **village/home**, and **money**. We also included several other concepts that we thought could help us with particular analytical questions: **price/value**, **good**, **future**, **spirit**, and **fire**. One of the strengths of this method is that local concepts are not defined by the researcher but rather reflect local usage and relationships among concepts. Research results simply show the positioning of locally important concepts on a *cognitive map*, a plot that results from averaging the cognitive measurements made by the local people.

At the DSWR, we selected four communities, two Iban and two Melayu, because we knew they followed different systems of resource use. We also tried to interview roughly equal numbers of male and female respondents to assess the magnitude of gender differences in perceptions of natural resources. Because the communities were small, our goal was to interview every teenager and adult; there were very few refusals.

Local assistants conducted the interviews under our supervision. Although both researchers and respondents found the task challenging, we were reasonably satisfied that the local people understood the process and its purpose. Besides explaining the measurement concept (which was easier for the Melayu to understand than for the Iban), we reassured people in all communities that there was no "right" answer, that their own perceptions were the important thing. Many actually enjoyed the intellectual challenge of the task.

The three kinds of conditions of interest to us (a conservation ethic, a feeling of proximity to the forest, and a forest–culture link) have been identified as related to sustainable forest management, though the exact causal links are not clear. Nor are the values (in terms of cognitive distances, in this case) that would indicate the degree to which these conditions apply. One of our goals, as we tried these methods in various contexts, has been to gain a better understanding of the range of variation (in people's feelings of closeness to the forest, for instance) and how this variation correlates with forest conditions and forest sustainability.

Results from West Kalimantan

The most fundamental output of the Galileo program is a *means matrix,* which is the mean response (from all the respondents) computed for every pair of concepts. Put another way, it reflects the mean distances perceived by the community in question between every concept and every other concept. The program provides extensive descriptive and inferential statistics, including standard deviations, standard errors, indices of skewness and kurtosis, sample size, maximum and minimum values, and other more global statistics; for our purposes, we were satisfied with fairly simple analyses. The statistical precision of the measures in this study was excellent, with mean distances ranging from about 5% for the full sample (277 cases)[2] to about 6–9% within gender segments (about 100 females and 170 males) and about 10–15% within each village (roughly 70 cases per village).

One strength of the Galileo approach is the multitude of ways in which one can examine the data produced. We have selected three ways of analyzing the data. First, we provide cognitive or perceptual maps (plots) for each community, through a gender lens. Then, we look at the ethnic differences in perception. We conclude with a discussion of three concepts that we consider closely related to sustainability: **good**, **future**, and **forest**.

Cognitive Maps in Four Communities and for DSWR as a Whole

The Galileo plots[3] show a cognitive map for the total data set (DSWR; Figure 5-1), two maps disaggregated by gender (Figures 5-2 and 5-3), and two maps disaggregated by ethnic group (Figures 5-4 and 5-5). Although the plots can represent only 3 of the 20 dimensions in this multidimensional space (much as a photograph presents us with only two of the three dimensions our eyes normally see),[4] they provide an appealing visual representation of the (approximate) thinking of a group of people about forests and other natural resources.

Reading the plots involves first noticing the two dimensions represented by a particular concept's placement on the horizontal grid. The third dimension can be read by following the "X" on the grid up or down to the "O" at the other end of each vertical line. The "O" is the placement of the concept in the three-dimensional space. In Figure 5-1, for instance, **spirit** is quite far from **money**, because **spirit** is below the grid to the right front and **money** is above the grid to the left center. These plots indicate that people in the DSWR typically do not see the concepts **money** and **spirit** to be particularly closely related—in marked distinction to **garden** and **water** (both central and below the grid) or **man** and **woman** (both left front horizontally and below the grid).

The concepts **spirit** and **fire** are displayed as rather peripheral (Figure 5-1). Our reason for including **spirit** in the list of concepts was not because it

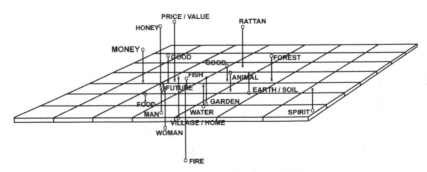

Figure 5-1. *Cognitive Map in Data Set for the DSWR*

came to our minds as a locally appropriate concept in this cognitive domain but because we wanted to examine the DSWR situation (a comparatively sustainable one) in light of the considerable literature on the close spiritual connection between forest peoples and their environment (see the collections by Banuri and Marglin 1993; Kemf 1993). Others suggest that women are particularly environmentally sensitive (Diamond and Orenstein 1990; Gomes and Kanner 1995; Roszak 1995). These data do not support such contentions for this West Kalimantan context. Among these peoples, **spirit** is the most distant from the other concepts and particularly distant from **woman** (see Figures 5-2 and 5-3).

We add that the terms used for spirit in West Kalimantan are not sufficient to represent all the connotations that go with the English term *spirituality*. Additionally, the Iban concept of *spirit* includes both a positive and a negative element; among the Melayu, *spirits* seem invariably bad. In fact, a wealth of qualitative data suggests a significant spiritual link, in the western sense, between Bornean peoples and their environment.[5]

Fire, also distant from other concepts, was likewise not a concept that emerged from our knowledge of people's views of the people–forest link in Kalimantan. However, fire is perceived as a significant threat to the DSWR's unique ecosystem. We wanted to know how people perceived fire in relation

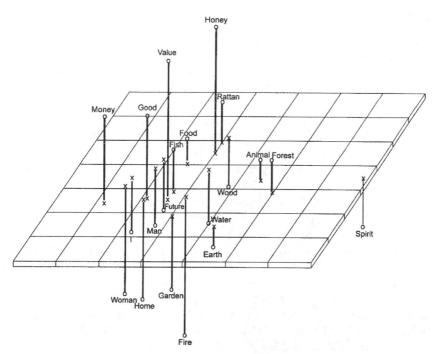

Figure 5-2. *Cognitive Map of Women in Data Set for the DSWR*

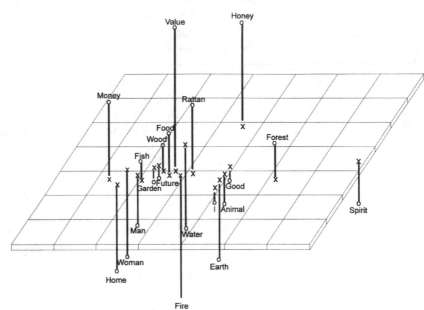

Figure 5-3. *Cognitive Map of Men in Data Set for the DSWR*

to the other concepts we were examining. The 1997–98 El Niño also spurred added interest in the implications of fire in Borneo.

Rattan and **honey** were selected as relevant for both ethnic groups. For the Iban, these products are only two of a vast repertoire of regularly gathered NTFPs; for the Melayu, they are the two most important. In the Melayu context, these products also have been the focus of the local Conservation Project activities to increase production and local incomes (Figure 5-4). The Kelayang Iban have also long been involved in rattan collection and sale to timber companies and, more recently, processing of handicrafts for sale to the Conservation Project (Figure 5-5). Note the smaller distance between **honey** and **money** among the Melayu, for whom honey collection can constitute a major income source at certain times of the year.

Another interesting feature of these cognitive maps is the fairly uniform closeness with which **man** and **woman** are perceived. They are slightly closer among the Iban than among the Melayu, which again fits with our expectations. Men and women seem to be perceived as much closer than they typically are in American studies of this kind (Newton 1977; Newton and others 1984). The comparative lack of gender stereotyping in some Bornean groups has been documented in other studies (see Dove 1981; Davison and Sutlive 1991; Drake 1991; Mashman 1991, for gender studies among Iban groups).[6] Our data seem to support these previous conclusions.

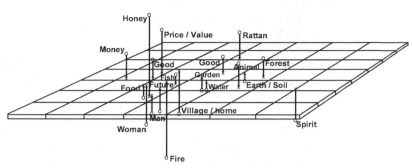

Figure 5-4. *Cognitive Map of Iban in Data Set for the DSWR*

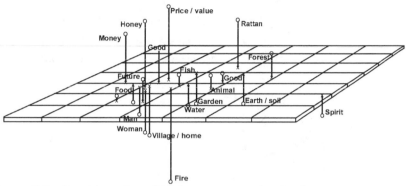

Figure 5-5. *Cognitive Map of Melayu in Data Set for the DSWR*

In including **price/value**, we were trying to get at the idea of *value* as something abstract and related to values (in a noneconomic sense), but the local terms (*rega* in Melayu and *berega'* in Iban) always implied both. This usage is consistent with previous analyses of other Bornean groups (see also Colfer and others 1997a). Only the Iban analysis (Figure 5-4) shows this concept to be relatively integrated with the other concepts. This result may reflect the Iban's higher sociocultural or philosophical valuation of the natural environment than the Melayu, which would be consistent with our impressions as well. The Iban and other Dayak, though scorned at the national level as "backward," exhibit a marked self-confidence and commitment to their own way of life which does not characterize the Melayu. Although cash is much more important in Melayu daily life, it may be that these Iban—like some Kenyah Dayak—consider money to be of less value than locally produced goods (because money quickly disappears, but rice, for instance, can be stored and more easily kept for emergency needs).

Whereas the plots provide the best holistic view of these data, the most accurate reflections of people's estimates of distance are the actual means. Tables 5-1 and 5-2 show the distances men and women perceive various concepts to be from one of our central concepts, forest. In the eyes of the average DSWR woman, the cognitive distance between forest and fish is 3.5, whereas the average DSWR man's perception is that forest and fish are 3.81 units apart (farther). These means are accurate to about ±5% for the total sample, ±6–8% for the gender segments, and ±10–15% within each village. Obviously, a great number of tables and figures could be created, depending on which concepts were of interest for any particular research problem. Table 5-1 shows results from the total data set (which we call "DSWR"), reflecting the distances estimated by all the respondents in all four communities. Table 5-2 deals with each of the four villages individually.

For the DSWR as a whole, the concepts perceived as closest to forest are earth/soil, wood (predictably), and animal. Rattan, water, and honey are also fairly close. Women express the greatest distances between forest and fire, food, and price/value, and men also consider these concepts to be comparatively distant from forest. The distance of price/value from forest is

Table 5-1. *DSWR Men's and Women's Perceptions of the Distance between Forest and Other Study Concepts*

Concept	All female	All male
Fish	3.50	3.81
Wood	1.75	2.17
Rattan	2.42	2.58
Honey	2.96	3.24
Garden	4.12	3.06
Price/value	6.09	5.36
Good	4.22	3.94
Future	4.45	4.69
Spirit	4.05	4.54
Animal	2.29	2.44
Food	6.48	5.10
Earth/soil	1.65	1.72
Water	2.78	3.10
I	4.67	4.71
Man	4.96	4.76
Woman	5.85	5.21
Village/home	5.95	5.24
Fire	6.96	5.36

Table 5-2. Men's and Women's Perceptions of the Distance between Forest and Other Study Concepts

Concept	Nanga Kedebu'		Danau Seluang		Kelayang		Wong Garai	
	Female	Male	Female	Male	Female	Male	Female	Male
Fish	3.73	3.04	3.72	3.71	4.08	4.06	3.93	3.64
Wood	2.00	1.32	1.67	1.73	1.73	1.84	2.64	2.32
Rattan	1.77	2.35	2.06	2.27	2.03	3.13	3.68	2.86
Honey	1.88	2.35	3.39	3.12	3.89	3.47	4.39	3.32
Garden	2.23	2.81	4.06	5.36	1.76	1.69	3.18	2.81
Price/value	5.19	5.33	6.39	6.82	5.97	5.59	4.86	5.36
Good	3.96	3.11	3.28	4.48	4.95	3.59	4.37	5.00
Future	5.42	3.89	3.39	4.45	5.35	4.19	4.86	5.14
Spirit	3.85	3.26	4.61	4.29	3.19	3.28	5.15	4.45
Animal	1.62	1.48	2.33	2.15	1.57	1.91	3.30	3.59
Food	5.50	6.33	6.50	7.18	4.32	4.34	3.78	5.05
Earth/soil	1.38	1.07	1.39	1.64	1.30	1.22	2.25	2.36
Water	2.96	2.48	2.83	2.59	3.41	2.91	3.39	3.59
I	4.42	4.04	4.89	5.22	3.86	2.81	4.86	4.18
Man	4.81	4.07	4.28	5.04	4.65	3.66	5.04	5.86
Woman	5.19	5.44	5.22	5.73	4.65	3.97	5.21	6.64
Village/home	5.65	5.48	4.67	5.92	4.84	4.34	5.21	6.59
Fire	4.04	6.15	6.94	6.98	3.81	3.53	5.54	7.00

interesting in light of international timber prices and reflects a genuine difference between the views of locals and the views of outsiders. Both men and women consider **forest** to be closer to **man** than to **woman**, but the difference is not great.[7]

Good, future, and **forest** are concepts that we felt bore special significance for sustainability (see below). The distance between **forest** and these concepts falls in an intermediate range for both men and women. Men consider the **forest** to be closer to **good** than women do, and slightly less closely connected with **future**. These data do not suggest that these people have a particularly close or positive feeling about the forest, despite their extreme dependence on it, considerable knowledge about it, and continual interaction with it.[8] From our general experience, we know that forests also contain many perceived dangers (dangerous or poisonous animals, potential for injury from falls, fast regrowth into places needed for uses such as housing and agriculture, malevolent spirits, and so forth).

The concept **money** provides another lens for interpreting perceptions (Tables 5-3 and 5-4). It is clear that the peoples of the DSWR consider **money** to be more tightly integrated than **forest** into this conceptual domain (Table 5-3).[9] Women consider **money** to be closer to **people (I, man,**

Table 5-3. *DSWR Men's and Women's Perceptions of the Distance between Money and Other Study Concepts*

Concept	All female	All male
Fish	3.17	3.21
Wood	4.74	3.93
Rattan	4.51	3.56
Honey	3.69	3.36
Garden	3.43	4
Price/value	2.85	3.24
Good	3.62	3.18
Future	4.01	3.52
Spirit	9.32	8.2
Animal	5.58	6.01
Food	3.41	4.13
Earth/soil	5.28	4.77
Water	5.07	4.4
I	2.22	3.29
Man	2.75	2.89
Woman	2.25	2.64
Village/home	3.46	2.99
Fire	7.23	6.03

Table 5-4. Men's and Women's Perceptions of the Distance between **Money** and Other Study Concepts

Concept	Nanga Kedebu'		Danau Seluang		Kelayang		Wong Garai	
	Female	Male	Female	Male	Female	Male	Female	Male
Fish	1.92	1.79	2.28	3.82	2.43	1.88	4.96	3.41
Wood	3.48	4.14	2.78	4.73	3.89	2.91	5.07	5.50
Rattan	2.65	3.65	3.06	4.65	3.11	2.88	4.74	5.18
Honey	1.81	2.23	2.39	3.53	3.68	2.64	5.43	5.77
Garden	2.79	2.54	1.94	3.33	4.11	4.21	6.36	4.73
Price/value	2.28	2.67	1.44	2.42	3.59	2.45	5.33	4.09
Good	2.64	2.07	1.67	4.10	2.76	2.12	4.64	4.41
Future	3.19	3.76	1.67	3.50	3.59	4.33	5.07	5.50
Spirit	8.75	10.12	9.11	9.74	7.81	8.30	7.14	7.41
Animal	6.85	4.36	4.00	5.55	5.78	5.03	6.54	7.05
Food	3.77	2.41	2.50	3.55	2.62	1.81	5.50	4.32
Earth/soil	4.85	4.85	2.83	5.27	3.24	3.09	6.00	5.82
Water	3.77	3.62	3.83	5.20	3.84	3.59	5.36	6.50
I	2.65	2.35	2.11	1.54	2.76	1.66	4.64	3.68
Man	2.23	2.77	1.94	2.35	3.30	1.78	4.11	3.64
Woman	2.04	1.85	1.89	2.06	2.49	2.10	3.68	3.18
Village/home	2.85	2.85	3.06	3.67	2.43	1.55	3.07	3.73
Fire	5.77	7.27	6.67	7.27	5.54	4.68	5.86	7.09

woman) than men do. The only concept from which **money** is truly distant is **spirit**, and to a lesser extent, **fire**.

Careful inspection of Tables 5-1 to 5-4 can yield an abundant harvest of insights into the worldviews found in these four villages, but we cannot make such a thorough analysis here.

Ethnic Differences in Conceptual Distances

The two main ethnic groups in the DSWR surroundings, the Melayu and the Iban, have lived in the area for a long time (Harwell 1997, 2000a; Wadley 1999b). Although both groups rely heavily on natural resources for their subsistence, they tend to use very different habitats. The Iban of Wong Garai reflect a habitat use that is closer to the "ideal type" for the Iban than do the peoples of Kelayang (who live closer to fish resources, have more Melayu neighbors, and for whom fishing is more important). Similarly, the Melayu of Danau Seluang have more interaction with Iban than do those of Nanga Kedebu'. Previous experience with cognitive mapping in Sumatra (Colfer and others 1989),[10] as well as our qualitative evaluation of differences in values and lifestyles, suggested that ethnic differences might be quite significant.

The overall ethnic differences are summarized in Table 5-5, with **forest** and **money** as key concepts. Despite the different lifestyles of the two groups, we noted some remarkable similarities along with some not too surprising differences. For instance, the Iban live on dry land and practice swidden agriculture. Their agricultural endeavors, represented in English by **garden**,[11] are intimately connected with the forest (because the forest is regularly cut to make fields and forest often surrounds fields). As expected, the connection for the Iban between **forest** and **food** is markedly closer.

The Melayu, who live by fishing in and around the **forest**, place **water** slightly closer to **forest** than do the Iban (who recognize the importance of water for fish and for their agricultural endeavors). Although the Melayu use far fewer forest products in general than do the Iban, they concentrate their forest product collection on three items: wood, rattan, and honey. Thus, the Melayu indicate closer connections between **wood** and **forest**, **rattan** and **forest**, and **honey** and **forest** vis-à-vis the Iban.

Not surprisingly, the Melayu consider **money** to be closer to the other concepts in general than do the Iban (Table 5-5). Although both groups use money extensively, the Melayu rely on it for daily life (see also Table 5-4). Iban men, on the other hand, go to Malaysia to earn money, with which they buy consumer goods, whereas Iban daily needs are largely supplied by subsistence activities (agroforestry, agriculture, hunting, fishing, and so forth). Kelayang is located closer to Melayu communities than is Wong Garai. It is not surprising that with increasing geographical distance from the

Table 5-5. *DSWR Men's and Women's Perceptions of the Distance between Forest and Other Study Concepts*

Concept	Iban		Melayu	
	Forest	**Money**	**Forest**	**Money**
Fish	3.96	3.06	3.86	2.61
Wood	2.08	4.20	1.69	3.88
Rattan	2.60	3.84	2.06	3.55
Honey	3.79	4.19	2.94	2.65
Garden	2.27	4.79	3.63	2.85
Price/value	5.49	3.75	5.72	2.39
Good	4.50	3.38	4.10	2.77
Future	4.87	4.53	4.27	3.44
Spirit	3.89	7.82	4.47	8.87
Animal	2.43	5.97	2.29	4.99
Food	4.35	3.40	6.39	3.16
Earth/soil	1.69	4.31	1.62	4.64
Water	3.30	4.63	2.96	4.23
I	3.89	3.10	4.93	2.19
Man	4.70	3.18	4.95	2.42
Woman	4.97	2.81	5.65	2.17
Village/home	5.14	2.60	5.46	3.21
Fire	4.72	5.66	6.08	6.82

Melayu and other marketing outlets, the cognitive distance between **money** and the other concepts would increase.

Galileo Concepts Directly Related to Sustainability

Because we hypothesized that the concepts **good, future,** and **forest** were important to consider in assessing social criteria and indicators for sustainable forest management, we prepared two tables that show the distances from those concepts to others in the data sets (Tables 5-6 and 5-7).

Interestingly, **spirit** is comparatively close to **forest,** suggesting a possible mechanism for protecting the forest by encouraging belief in spirits (but see Wadley and others 1997). Iban views of spirits are decidedly ambivalent, including fear; for the Melayu, spirits are considered to be bad. In Wong Garai, the Iban deal with spirits most safely by exhibiting humility in speech and action. Wanton destruction, boasting of fishing catches, and other egotistical behavior are not done where spirits might see or hear (particularly in

Table 5-6. *DSWR Cognitive Distances among Selected Concepts for Sustainability*

Concept	Good	Future	Forest
Fish	3.41	5.20	3.90
Wood	3.73	4.97	1.86
Rattan	3.38	4.96	2.30
Honey	2.96	4.93	3.32
Garden	3.40	4.80	3.03
Price/value	3.25	4.73	5.62
Spirit	7.63	7.83	4.22
Animal	5.60	5.35	2.35
Food	2.97	3.82	5.49
Earth/soil	3.64	4.03	1.65
Water	3.56	3.86	3.11
I	2.95	3.23	4.47
Man	3.04	3.45	4.84
Woman	3.16	3.83	5.35
Village/home	3.42	4.15	5.32
Fire	5.02	5.30	5.48
Money	3.04	3.92	6.26

the forest). This approach seems to have an obvious relevance for resource use and sustainability (see also Wadley and others in preparation; Colfer and others (1997a) found a similar Kenyah aversion to such boasting in East Kalimantan for the same reasons). The whole question of spirits and spirituality is complex. A full assessment would need far more specification, drawing from the discourse of the people studied, than is available using the Galileo method alone (personal communication from Robert Lee in a review of this chapter, November 4, 1999).

The distances from **price/value** and **money** to **forest** are considerably larger than to **future** and still larger than to **good**. These data could suggest the absence of what westerners consider a conservation ethic; however, in our research, we all have noted features of the people's worldviews that suggest a concern about the land and its future capability to produce.[12] We remain uncertain whether this problem may be with the concept of *conservation ethic* itself or whether we have simply not succeeded in finding a way to measure it satisfactorily. Perhaps the inclusion of the concepts **love** and **fear** could provide a useful distance in trying to place different stakeholder groups on some sort of "conservation ethic continuum."

The small distance between various forest products or NTFPs **(wood, rattan, honey, animal, fish, garden)** and **forest** is in marked contrast to the

Table 5-7. Cognitive Distances among Selected Concepts for Sustainability

Concept	Nanga Kedebu'			Danau Seluang			Kelayang			Wong Garai		
	Good	Future	Forest	Good	Future	Forest	Good	Future	Forest	Good	Future	Forest
Fish	2.95	5.15	3.98	3.83	5.27	3.71	2.77	4.93	4.07	4.52	5.58	3.80
Wood	3.34	4.50	1.66	3.60	4.65	1.71	3.36	5.34	1.79	5.10	5.68	2.50
Rattan	2.48	5.43	1.93	3.88	4.51	2.21	2.93	4.23	2.09	4.80	5.86	3.32
Honey	2.44	5.16	2.74	2.88	3.97	3.19	2.41	4.76	3.70	4.70	6.16	3.92
Garden	2.45	4.17	2.49	4.06	4.68	5.01	2.94	4.80	1.74	4.74	5.98	3.02
Price/value	2.55	4.00	4.90	3.25	5.28	6.71	3.04	4.33	5.79	4.68	5.76	5.08
Spirit	7.37	7.85	4.55	8.48	8.87	4.38	7.67	7.69	3.23	6.80	6.54	4.84
Animal	5.14	5.17	2.36	5.42	4.43	2.20	5.66	6.29	1.73	6.52	5.62	3.43
Food	2.82	2.95	5.86	3.07	2.86	7.00	2.35	5.06	4.36	3.98	4.82	4.35
Earth/soil	3.78	3.34	1.66	3.53	3.07	1.57	3.57	5.09	1.26	3.66	5.00	2.30
Water	3.56	3.80	3.22	3.39	2.61	2.65	3.30	4.77	3.17	4.16	4.42	3.48
I	2.52	3.15	4.76	3.10	2.60	5.13	2.21	3.11	3.41	4.46	4.42	4.56
Man	3.08	3.71	5.05	3.25	2.37	4.84	1.71	3.39	4.20	4.54	4.59	5.40
Woman	3.27	4.17	5.70	3.58	2.88	5.59	1.84	3.37	4.34	4.26	5.22	5.84
Village/home	3.71	4.06	5.34	3.81	3.26	5.59	1.97	4.33	4.66	4.42	5.26	5.82
Fire	5.05	5.59	5.31	6.14	5.12	6.97	3.34	4.80	3.69	5.78	5.76	6.18
Money	2.17	3.79	5.59	3.46	3.03	6.79	2.56	4.03	6.04	4.54	5.27	6.90

distance between those products and **future**. This difference may reflect a feeling that the future does not lie with exploitation of natural resources, that it lies elsewhere or with other economic endeavors—despite the comparative proximity of **good** to these products.

In examining the worldview of each village (Table 5-7), we find a greater distance between **forest** and **food** for the Melayu than for the Iban. The Iban obtain almost all their food from the forest, whereas Melayu food either comes from the lakes and streams or is bought. **Woman** in all communities is perceived as farther from the **forest** than is **man**.[13] **Fire**, one of the concepts of particular interest to the Conservation Project at the DSWR, was fairly distant from **good** and **forest** on all sites. This finding is interesting because of the importance of fire in the Iban system of swidden cultivation and in fish processing among the Melayu. People who considered **fire** close to **woman** or to **good**, for instance, typically mentioned its role in cooking food. Considerably more burned forest surrounded Danau Seluang than Nanga Kedebu' (see Chapter 17). These data (and qualitative interviews as well) do not support the Conservation Project's hypothesis that fires were more common in the Danau Seluang area because of different perceptions about fire. Both communities consider **fire** to be distant from **forest** and **good**—even though the people of Nanga Kedebu' see **fire** and **future** as more distant than do the residents of Danau Seluang (which is near the edge of the reserve, in an area that has been more extensively logged and where losses from fire have been greater).

Conclusions and Recommendations

We believe that this study fairly accurately represents the cognitive view that local peoples have of their environment. The differences in the cognitive maps between men and women and between ethnic groups are consistent with our expectations based on long-term ethnographic research: they reflect how local peoples see their environment and their place within it.

Some methodological questions remain. First, we wonder whether the distances are entirely comparable, even with a standard ten-unit measuring rod (the distance between **black** and **white** specified for all respondents). Higher distances in general were attributed in Wong Garai than in Kelayang, even though they are both Iban communities (Table 5-7). We had a similar experience in East Kalimantan, where we found the people of Long Ampung consistently estimated greater distances than those in Long Segar (Colfer 1982). This issue has implications for the process of developing threshold values for making comparable cross-site assessments.[14]

Second, in our cross-cultural comparisons (see Chapter 6), we found what appears to be a core set of concepts that are valid or important in all the humid tropical forests we examined, with empty "slots" (in the questionnaire) for special, locally determined concepts. For example, rattan might be important in one location, palm hearts in another—but NTFPs were important in all areas. One of our concerns remains the different meanings of a concept in different locations and among different peoples. Although we translated **garden** as *umai* (Iban) and *tayak* (Melayu), these concepts are not entirely comparable. Whereas the Iban garden is a rice field that is associated with the people's subsistence base, uses large amounts of their time, and involves major spiritual significance, the Melayu garden is a small, part-time endeavor not practiced by all families; it has no spiritual significance that we were able to ascertain. Another locally important difference relates to the importance among the Iban of varying stages of forest regrowth (of which the *kampong* stage is likely to be much closer to **spirit** than the word we selected, *babas*). A core set of concepts, if achievable, would simplify cross-site comparisons considerably (for more discussion of this issue, see Chapter 6).

Third, after this research, we wanted to investigate relevant concepts in various contexts, along a continuum of sustainable forest management from both human and biophysical standpoints. In the DSWR, the forests are and have been fairly sustainably managed. Comparable data from more diverse stakeholders in Cameroon, Brazil, and East Kalimantan are discussed in Chapter 6 (for related studies in the United States and Australia, see Cary 1995 or Richardson and others 1996). We hoped that such comparisons could provide useful information on the implications of the conservation ethic, emotional proximity to the forest, and the forest–culture link for sustainability.

In this chapter, we have presented illustrative analyses that we believe accurately reflect local people's cognitive views. The data demonstrate intermediate—not particularly close—cognitive relationships between **forest** and such important concepts as **I**, **good**, **future**, **spirit**, **man**, and **woman**. As such, they do not provide strong support for any of the issues we identified as potentially pertaining to a conservation ethic. Yet ethnographic research has demonstrated strong, if ambivalent, links between people and the forests there.

We had hoped that this kind of analysis could help us pin down the idea of a conservation ethic more precisely. That has not been the result. The Galileo method is designed to measure cognition. Perhaps future efforts to measure a conservation ethic should focus on emotional content rather than the cognitive aspects measured by the Galileo method.

Acknowledgements

The research reported in this chapter was undertaken with sponsorship from the U.S. Agency for International Development (AID) and in collaboration with the Indonesian Government (PHPA), Wetlands International–Indonesia Program, and with the informal cooperation of the Sustainable Forest Management Project of the U.K. Department for International Development (DFID) in West Kalimantan, Indonesia.

We thank all these institutions for their substantive contributions and cooperation in all our efforts. Most important, we thank the people of the Danau Sentarum Wildlife Reserve (now National Park) and its surroundings for their patience and kindness in answering our many questions. We created pseudonyms to protect the privacy of individuals in the communities around the DSWR who shared their perspectives with us.

Annex 5-1. The Galileo Method as Applied in the DSWR

Traditionally, a Galileo study requires respondents to report their perceptions of the differences (often called *distances*) among a set of concepts considered central to the definition of a selected topic, for example, **forests**. These estimated dissimilarities are averaged across all respondents in any segment and projected onto orthogonal coordinate axes to produce a perceptual map or space. Within this space, distances are predictive of attitudes, beliefs, and behaviors.

In our study, 277 respondents estimated the pairwise dissimilarities among a set of terms including **forest** and 19 other concepts identified in previous analyses as pertinent to the perception of forests in Kalimantan villages. The resulting square mean dissimilarities matrix was then analyzed in several ways, including perceptual maps (multidimensional scaling), charts, graphs, tables, and advanced artificial neural networks. Perceptual maps were made using Galileo software, which produces very precise representations of the dissimilarities in graphic form and allows transformations (rotations and translations) to common orientations for easy comparisons of data over time and across subsamples. Previous research has shown Galileo to be an appropriate tool when

- holistic models of cognitive structure and processes are required,
- precise results are desirable,
- a standard metric needs to be maintained across time or subsamples (as when time-ordered maps are needed, or when maps are to be compared from sample to sample), and
- the concepts to be mapped are known.

Galileo modeling may be less appropriate (Woelfel and Barnett 1982, 1992; Woelfel and others 1986, 1989; Cary and others 1989) when investigators are uncertain as to which concepts occur in the cognitive model, or when

- light respondent burdens are crucial,
- there is no need to maintain an invariant metric over time and across samples, and
- precise results are not important.

When less is known about the concepts that need to be included—as in preliminary studies—similar results can be obtained from CATPAC, a self-organizing neural network that reads text and uncovers the main underlying concepts. CATPAC makes it possible to work from in-depth interviews rather than quantitative scales and derive similar results (Cary 1995).

Endnotes

1. Muslim Melayu fisherfolk live in the seasonally flooded core of the reserve. Christian and animist Iban live in the surrounding hills; they are swidden cultivators and, to a much lesser degree, forest workers. These two main groups inhabit ecologically very different habitats and have distinct natural resource management systems.

2. Sample sizes are approximate because overall sample size varies slightly from item to item. Complete statistics are available from the authors.

3. Plots were made with TerraVision, an interactive computer graphics program for perceptual mapping.

4. In this case, the probability that the concept will be in its correct position vis-à-vis other concepts on the map is roughly 75%—a high figure, because of the high degree of agreement among respondents at DSWR about the distances among concepts.

5. See also Freeman 1970, Howell 1984, Davison and Sutlive 1991, Roseman 1991, Colfer and others 1997a, and Wadley 1999a. Moreover, these human–environment links extend to assert human responsibilities to each other, including dead ancestors. Many Dayak believe that various forms of injustice or misdeeds must be redressed to avoid adverse effects in both the environment and the human condition, including climatic extremes, harvest failures, illness, and death (Dove and Kammen 1997; Harwell 2000a,b; Katz and others in press).

6. See Colfer 1981, 1982, 1983a, 1985a, 1985b, and 1991 on the Kenyah; Sutlive and Appell 1991 for a thorough coverage of the issue; or Tsing 1993 for an unusual approach.

7. Colfer found similar results among the Kenyah of two locations in East Kalimantan in 1980 (Colfer 1982).

8. In the 1980 study of perceptions of forests in East Kalimantan, **forest** and **good** were considered quite close among both adults and young people in the remote Long Ampung; a little more distant among adults in Long Segar (a resettlement village closer to "civilisation"), and still more distant among Long Segar's young people (Colfer 1982).

9. Table 5-6 allows us to infer the substantial distance between **forest** and **money.** That **forest** and **money** are not directly compared on this table is an artifact of our analysis process. We also conducted an analysis which used these results to predict the outcome of a vote where people had to choose between **money** and **forest. Money** won by a landslide.

10. In the Sumatra study, one ethnic group was indigenous (Minangkabau), the others were transmigrants (Javanese, Sundanese) from very different environments and cultures.

11. "Garden" is not a good translation for **umai,** the Iban word used in the form. However, because the two groups have such different kinds of agriculture, and we

wanted to compare the findings, we have used one English word. **Umai** would be more properly defined as "field."

12. Wadley notes that Wong Garai men make considerable money logging in Sarawak and suggests, based on these data, that the forests at home may be considered more valuable in their existing state than are forests elsewhere.

13. This was also true in the 1980 East Kalimantan study mentioned above (Colfer 1982). From simple interest, we performed an Automatic Message Generator (AMG) analysis on these data, for possible use in extension or awareness programs to encourage people to use the forest more sustainably. Woelfel's communications research experience indicates that concepts lying between **me** and a concept of interest (in our case, **forest**) have the most potential, when used in extension or awareness programs, to bring the concept of interest closer to **me**. This in turn can result in behavioral changes relating to that concept. Using this software, we identified the concepts with the most potential for moving the concept, **forest** closer to **me**, in the community's cognitive map. The two top contenders were three-concept messages, consisting of **fish, honey,** and **woman** and **honey, woman,** and **village/home**. A message that links **forest** to these concepts is likely to result in local people's feeling closer to the forest and behaving more sustainably toward it. Either of these three-concept combinations had the potential to remove all but 5% of the distance remaining between **forest** and **me**.

14. One common practice of Galileo researchers that might alleviate this difficulty is to choose two concepts from within the domain as the criterion pair. Several researchers (for example, Woelfel and Fink 1980) have suggested that this might provide a more accurate model, because the measuring rod does not have to be "transported" such a large psychological distance to be compared with the distances in the domain. A second advantage is that the internal criterion distance can be used to rescale the data after the fact, should differences in scale size appear.

CHAPTER SIX

In Search of
a Conservation Ethic

Agus Salim, Mary Ann Brocklesby, Anne Marie Tiani,
Bertin Tchikangwa, Mustofa Agung Sardjono,
Roberto Porro, Joseph Woelfel, and
Carol J. Pierce Colfer

As indicated by the results reported in the other chapters in this section, the definition of a *conservation ethic* remains somewhat vague. In this chapter, we use a comparative approach to continue to explore its meaning and ways to measure it. Building on our explorations into stakeholder identification with seven dimensions—proximity, preexisting rights, dependency, poverty, indigenous knowledge, culture–forest integration, and power deficit—we focused on some particularly relevant questions:

- Is a conservation ethic intimately connected to a feeling of proximity to the forest?
- Is a conservation ethic identifiably different from a close "forest–culture link"?
- What role does spirituality play?
- How can we assess a conservation ethic quickly and easily?

As in Chapter 5, we do not try to specify how closely linked people's perceptions are to their actions, even though this issue will have obvious ramifications for sustainable forest management (SFM).

The attempt to address the issues reported here involved the use of cognitive mapping. We reasoned that a conservation ethic would have to involve a cognitive element. An accurate measurement of people's perceptions relating to forests might help us better define the idea of a conservation ethic. We chose the Galileo multidimensional scaling method as a measurement instrument (described in the Introduction; see also Chapter 5). One of its strengths (the use of local concepts) is also one of its weaknesses. We were not really free to "toss in" concepts of specific interest to us.[1] Fortunately, sufficient commonality of concepts across sites allowed us to make some reasonable comparisons, although the number of sites differs depending on the cognitive distances compared. We considered the appropriateness of concepts to the context more important than subsequent convenience in analysis.

In the Introduction, we discussed our interests in comparing forest-rich and forest-poor sites and how/whether/which human beliefs and behaviors vary along this continuum. On the basis of concepts relevant to forests in our research sites, we developed several hypotheses related to the issue of a conservation ethic that are easily testable using Galileo-derived data. As part of our attempt to measure the amorphous concept of *spirituality* as it relates to forests in our research sites, we hypothesized first that the concept **forest**[2] would be considered closer to **spirit** concepts in forest-rich areas than in forest-poor areas. If this proved true, we reasoned, it would comprise additional evidence that SFM and spirituality might be causally linked.

We further proposed that forest-rich sites would be characterized by perceptions of **forest** that were closer to **good** and **future** than would forest-poor sites. The possible links (represented by short distances on the cognitive map) between **good, future**, and **forest** seemed to signify a potential proxy for sustainability, with its emphasis on maintaining forests for future use.

Finally, we hypothesized that **me** would be closer to **forest** in forest-rich sites than in forest-poor sites. We were interested in this conceptual distance for two reasons. First, it is a measure of the value placed on the forest by the respondents. Concepts perceived to be close to a respondent are typically more highly valued by that individual (see Chapter 5 for more discussion of this aspect of the method). Second, in our efforts to determine a few significant dimensions in identifying "who counts" in SFM, we identified proximity to the forest as one important dimension. Several researchers have suggested that emotional proximity to forests also should be addressed (for example, Behan 1988; personal communication from Dennis Dykstra, 1995). A close conceptual link between **me** and **forest** seemed to be a potentially good measure of emotional proximity.

We report here our attempts to measure these perceptions at several sites in Indonesia, Cameroon, and Brazil.

Galileo Results

The Galileo multidimensional scaling method (described in the Introduction) has been used for cognitive mapping by extension services and for public opinion polls in fields as diverse as politics, agriculture, energy, family planning, and conservation (for example, Newton 1977; Canan and Hennessy 1981; Cary and Holmes 1982; Newton and others 1984; Cary 1995).

The data can be analyzed in many different ways to address various questions, particularly related to human variation within the populations surveyed. In the following discussion, we compare results from sites where the method was successfully tested and we were able to obtain comparable results (Table 6-1).

The use of locally relevant concepts—often not the case in cross-cultural research—is one strength of the Galileo method. However, because we also were interested in comparing findings from site to site, we tried to make the selected concepts as comparable as possible across sites, without distorting the views of our respondents. We began with a set that had been developed in Indonesia and adapted it to one location in Cameroon. Researchers then modified this set as needed for their own locations. Interestingly, given the wide geographic dispersion of the sites, the concept sets do not vary greatly from site to site, but the variation that does exist has complicated analysis.

The results from this method are averaged across all respondents at a given site. The most fundamental and accurate data are presented in the form of a *means matrix*, which shows the distance between every concept and every other concept. From these data, plots representing cognitive maps can be constructed to give a three-dimensional impression of how the concepts fit together.[3]

Figures 6-1 to 6-5 show examples of these cognitive maps derived from data in five locations. Forest-rich sites are Mbongo and the Dja Reserve (Cameroon) and Bulungan (Indonesia); forest-poor sites are Mbalmayo (Cameroon) and Long Segar (Indonesia).

In addition to our more general questions about a conservation ethic, we were interested in measuring people's spirituality, because Indicator 3.3.6 in CIFOR's *Generic C&I Template* (1999) states, "People maintain spiritual and/ or emotional links to the land." We sought appropriate words to reflect spirituality (*spirit, ancestors, sacred*) but remained dissatisfied that they really reflected the meaning of the phrase "spiritual links to the land."

We hoped that the Galileo plots, by the placement of spiritual concepts in relation to **forest**, might demonstrate a conservation ethic. However, interpreting the integration of spiritual issues (represented variously as **spirit**, **sacred**, or **ancestors**) as reflected in these data sets is not straightforward. The first observation is that the cultural similarities between sites may be a more significant factor than forest quality (see also Chapter 11).[4] Respondents from the forest-rich Dja Reserve (Figure 6-1) considered **ancestors** to be closely integrated (with **village**, **future**, **man**, **woman**, and **land**). In Mbalmayo (Figure 6-2), a forest-poor site inhabited by people of a similar cultural background to many of those in the Dja Reserve, **spirit** was considered close to **me** (which usually indicates the importance of a concept to the respondent), although both were comparatively distant from most other concepts.

Table 6-1. *Description of Study Sites*

Site location	Country	Village(s)	Forest quality	No. of respondents	Main researcher
Bulungan, East Kalimantan	Indonesia	Birun, Paking Baru, Long Iman, Bintuan, Paking Lama	Rich	45	Sardjono
Mount Cameroon Project	Cameroon	Mbongo	Rich	264	Brocklesby
Dja Reserve	Cameroon	Mindouma I, II, III	Rich	46	Tchikangwa
Transamazon	Brazil	Trairão	Rich	39	R. Porro
Porto de Moz	Brazil	São João	Rich	37	R. Porro
Long Segar, East Kalimantan	Indonesia	Long Segar, Kernyanyan, Mara Kenyah	Poor	27	Sardjono
Mbalmayo	Cameroon	Mbalmayo	Poor	202	Tiani
Transamazon	Brazil	Transiriri	Poor	80	R. Porro
Porto de Moz	Brazil	Bom Jesus	Poor	32	R. Porro

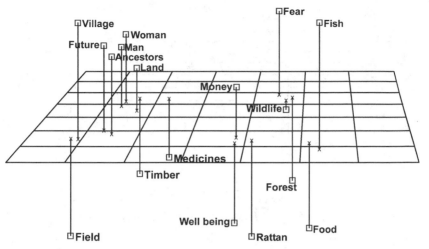

Figure 6-1. *Cognitive Map of Respondents in Dja Reserve, Cameroon*

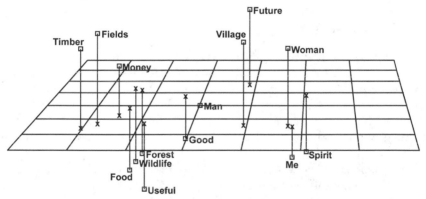

Figure 6-2. *Cognitive Map of Respondents in Mbalmayo, Cameroon*

In Mbongo (Figure 6-3), where the cultural heritage differs markedly from that of southern Cameroon, **sacred** is removed from all other concepts. In both Indonesian sites, **spirit** is distant from all concepts except **fear**; in forest-poor Long Segar (Figure 6-4), **spirit** is farther from **forest** than in forest-rich Bulungan (Figure 6-5). The West Kalimantan data reported in Chapter 5 also show **spirit** as a fairly isolated concept.

Our knowledge of these cultures suggested that the study participants do indeed have important ways in which they link the natural environment and their spiritual beliefs and that we try other approaches in the analysis. In the following discussion, we extract specific distances for the most relevant concepts (sometimes in tabular form) for additional analysis. Such data are (a) straightfor-

159

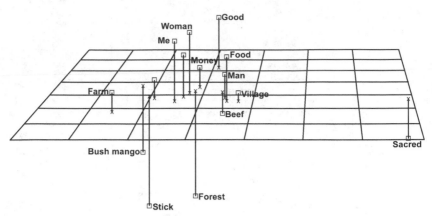

Figure 6-3. *Cognitive Map of Respondents in Mbongo, Cameroon*

ward and apparently accurate representations of the distances people perceive among the relevant concepts and (b) simple enough to compare across sites.

Further Analysis

Remaining with the issue of spirituality as a possible proxy for a conservation ethic, we examined the distances between **forest** and **spirit** in four locations: Bulungan and Long Segar (Indonesia) and Mbongo and Mbalmayo (Cameroon). On average, the distance between **spirit** and **forest** was closer in the forest-rich sites (5.89) than in the forest-poor sites (6.38), confirming our hypothesized relationship. But we found significant national differences.

In Indonesia, **spirit** was considered to be 7.24 units away from **forest** in forest-rich Bulungan, and the distance was 8.22 in forest-poor Long Segar (to get a feeling for the comparative distance involved, see mean distances for concepts reported in Table 6-2). Interestingly, at these two sites, **spirit** was seen as even farther removed from **land** (8.11 and 9.78, respectively) and in the same direction (closer in the forest-rich area and farther in the forest-poor area). In Cameroon, **forest** and **spirit** were considered to be equally distant (4.54) in forest-rich Mbongo and forest-poor Mbalmayo.

The fact that, on average, people see concepts related to spirits as closer to themselves in forest-rich areas than in forest-poor areas is interesting in its technical confirmation of our hypothesis. But even more interesting in a general sense is the greater perceived distance between **spirit** and **forest** compared with **forest** and other concepts related to natural resources. The people who inhabit these forests are precisely those we would expect to have a high degree of spirituality connected with the forest.

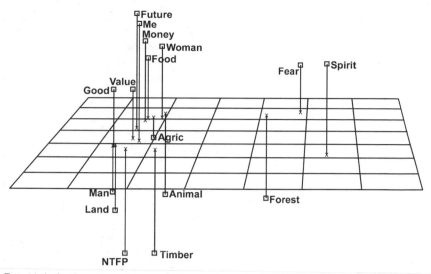

Figure 6-4. *Cognitive Map of Respondents in Long Segar, East Kalimantan, Indonesia*

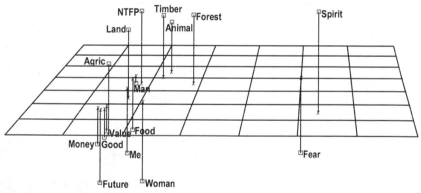

Figure 6-5. *Cognitive Map of Respondents in Bulungan, East Kalimantan, Indonesia*

Can We Use the Galileo Method to Define a Conservation Ethic?

This question emerged in the criteria and indicator testing process. Assessors wanted a simple method to determine who warranted inclusion in the concerns of forest managers. The "Who Counts Matrix" (Colfer and others 1999c)—which provides a simple technique for differentiating among the stakeholders in a particular area, with users scoring each stakeholder on seven

Table 6-2. *Distance from* **Good** *and* **Future** *to* **Forest** *at Nine Sites*

Site name	Forest and		Overall average distance
	Good	**Future**	
Forest-rich sites			
Bulungan	2.47	2.84	3.79
Dja Reserve	2.38	5.23	3.91
Mbongo	3.98	3.23	3.07
Trairão	3.21	3.24	3.62
São João	2.65	2.05	3.36
Mean	2.94	3.32	3.55
Forest-poor sites			
Long Segar	3.41	5.56	4.06
Mbalmayo	1.22	3.19	2.36
Transiriri	4.21	3.71	4.00
Bom Jesus	2.81	3.00	3.35
Mean	2.91	3.87	3.44

dimensions—is the approach we recommended. But we remained intrigued by the idea of using the presence of a conservation ethic (not one of the seven dimensions) to help us identify the most important stakeholders.

In our methods tests, all researchers selected the concepts **forest**, **good**, and **future**. These concepts potentially seemed to reflect the idea of a conservation ethic. The placement of **forest** far from concepts such as **good** and **future** might well suggest the absence of such an ethic.

We emphasize that we approached this research from two perspectives. We wanted to see

- whether there was a link between forest quality and close cognitive distances between these concepts and
- whether, based on our qualitative understanding of local people's views, this method reflected what we understood by the term *conservation ethic* more generally.

The distances between **forest** and the other two concepts, **good** and **future**, were extracted from the cognitive data (Table 6-2). These distances can be compared with the overall average of all distances between all other concepts, in each of the nine sites for which we have comparable data.[5]

Our second hypothesis proposed that concerns about sustainability (a conservation ethic), as manifest in a close conceptual relationship between **good**, **future**, and **forest**, would be more evident in forest–rich areas than in forest–poor areas. This relationship would be reflected in smaller average distances between those concepts in the forest-rich sites. The data show that this is clearly not the case. **Forest** and **good** are slightly more distantly related concepts in the forest-rich sites than in the forest-poor sites. The average distance between **forest** and **future** is consistent with our hypothesis, but this average masks significant differences among sites.

Making within-country comparisons, we find that at the Indonesian sites, **forest** and **good**, and **forest** and **future** are considerably closer in the forest-rich sites—consistent with our hypothesis. In Brazil, we also find the relevant concepts to be closer in the forest-rich sites, Trairão (colonists) and São João (*ribeirinhos*), than in the forest-poor sites of Transiriri (colonists) and Bom Jesus (*ribeirinhos*). However, the reverse is true in Cameroon, where these distances for the forest-rich Dja Reserve and Mbongo are greater than for forest-poor Mbalmayo.

From these conflicting results, it is not possible to make a definitive statement that people living in forest-rich contexts are more likely to have a conservation ethic, as reflected in these measurements.

Can We Measure the Closeness People Feel to the Forest?

Another dimension we considered important in the Who Counts Matrix was people's proximity to the forest. In most cases, we were concerned about physical proximity; however, in many developed countries, what might be called *emotional closeness* to the forest may be important, as well as the significant impact of such feelings on forest management (personal communication from Dennis Dykstra, 1995). It is also possible that a feeling of closeness might signify something akin to the conservation ethic idea in developing countries.

Woelfel and Fink (1980) consider the self-point, or **me**, to be of particular interest. The position vector in the analysis from the self to any other concept, they define as an *attitude* (juxtaposed with vectors among other concepts, seen as *beliefs*). They have found such attitudes to be more closely related to behavior than are beliefs.

Table 6-3 shows the average distances between **me** and **forest**. Unfortunately, the Brazilian and Dja Reserve data could not be used in this comparison, because **me** was not included. The average distance between **me** and **forest** was smaller for forest-rich sites than for forest-poor sites. However, we have the same pattern, where the Indonesian sites support the hypothesis (that people living in forest-rich sites are more likely to feel close to the forest than those living in forest-poor sites), and the Cameroon sites do not. Indonesians from forest-rich Bulungan estimated their conceptual distance to the forest to be considerably less than the distance estimated by forest-poor Long Segar residents.

Additional data are available from other studies (Colfer 1982; Chapter 5), all of which reflect forest-rich conditions. In 1980–1981, Colfer conducted a similar study in Long Ampung (near the border with Sarawak in East Kali-

Table 6-3. *Distance between **Me** and **Forest** at Four Sites*

Site name	**Me** and **Forest**	Overall average distance
Forest-rich sites		
Bulungan	2.51	3.79
Mbongo	4.04	3.07
Mean	3.28	3.55
Forest-poor sites		
Long Segar	4.89	4.06
Mbalmayo	2.62	2.36
Mean	3.51	3.44

mantan) and in Long Segar (where forests remained "rich" at that time), Indonesia. Adults in the remote Long Ampung estimated the distance between **forest** and **me** to be 3.1; those in Long Segar estimated it to be 3.0. Both are considerably closer than the more recent result from a degraded Long Segar: 4.89. The 1980–1981 study differentiated adults and young people. The latter considered the forest to be farther from themselves: 4.6 for Long Ampung young people and 6.1 for Long Segar young people. The Iban of forest-rich Danau Sentarum estimated the distance between themselves and their forest to be 3.89, whereas the local Melayu—a group more dependent on fishing than forests—considered this distance to be 4.93.

Again, the overall results are ambiguous, with respect to the hypothesis posed—confirmed in Indonesia, not Cameroon.

Conclusions

Conclusions relating to our specific hypothesis—that certain concepts (**spirit, good, future, me**) would be closer to **forest** in forest-rich areas than in forest-poor areas—are summarized as follows.

- The relationship between **spirit** and **forest** was closer in forest-rich areas than in forest-poor areas, confirming our hypothesis. However, concepts related to spirits were among the most distant from the other natural resource-related concepts on all sites but one (the Dja Reserve, Cameroon).
- The hypothesized relationship between **forest**, **good**, and **future** (closer in forest-rich areas than in forest-poor) was not confirmed for the overall data set. (Data from Indonesia and Brazil supported the hypothesis.)
- The hypothesized relationship between **forest** and **me** (closer in forest-rich areas than in forest-poor areas) was confirmed, but support was weak.

(The Indonesian data are consistent with the hypothesis; the Cameroonian data are not.)

We feel that the Galileo method has accurately reflected local people's cognitive maps, in the sense that it provided a good summary of how people see the clustering of relevant concepts. This information is of interest to natural resource managers and is helpful in efforts to achieve SFM. The method also allows great flexibility in examining various aspects of people's perceptions, including variations by social category (gender, ethnicity, wealth, and so forth; for a gender-based analysis, see Colfer and others 1997c). Finally, information generated can be analyzed to develop communication strategies tailored to the people surveyed. Clear transfer of information and ideas is very important in conservation efforts (the Automatic Message Generator function is described briefly in Chapter 5).

However, we have been disappointed with the Galileo method as a reflection or measurement of a conservation ethic. Our qualitative understanding of the systems we examined with this quantitative method includes features we consider likely components of a conservation ethic:

- beliefs about spirits that are closely connected to the natural world;
- rituals that link spirits, natural resources, and the social world; and
- taboos that protect (for spiritual reasons) particular areas from human interventions or use.

It is possible that we selected the wrong terms to reflect spirituality. If we had presented similar questionnaires to Cameroonian or Indonesian urban dwellers, or to people from developed countries, we might have received responses that averaged even greater distances between **spirit** and **forest**.

If we were to pursue this approach in efforts to identify who counts in SFM, we would have to do additional field studies to gather comparative data about people less intimately connected with their forest environment. However, we do not think such a strategy is warranted. Instead, we conclude that the Galileo is a useful method for reflecting local people's views of their natural environment—knowledge that is most valuable within particular contexts.

Endnotes

1. Two reviewers of this paper (Bob Lee and Esther Katz) independently suggested that **abundance** would have been a useful concept to include.

2. The concepts used in the Galileo study are in **boldface** (to differentiate them from terms used in a general sense).

3. These three-dimensional plots misrepresent reality to a certain degree because the measurements are actually in a multidimensional space (depending on how many

concepts are being measured). However, a three-dimensional plot typically accounts for around 70% of the variation, so it gives a rough idea.

4. One reviewer suggested that user error could account for this finding. We think this improbable, because the researchers were competent and because the implementation of the method is quite straightforward. The reviewer also wondered what other factors (besides forest quality) differed about the sites. This concern mirrors our own growing discomfort with this kind of reductionist science.

5. The *average distance* is important here, because patterned differences in distances are sometimes apparent, whereby one site will systematically estimate larger distances between sites than another (see also Colfer 1982).

Security of Intergenerational Access to Resources

Research from a wide range of disciplines has produced a wealth of literature focused on access to resources. The number of anthropological studies that document the existence and functioning of local mechanisms that regulate access to resources is legion (particularly weighty contributions include those edited by Kundstadter and others 1978; McCay and Acheson 1987; Poffenberger 1990; Croll and Parkin 1992; Redford and Padoch 1992; Padoch and Peluso 1996). The Land Tenure Center in the United States has been documenting the world's land tenure systems for decades. The Food and Agriculture Organization's series of monographs and manuals on community forestry also addresses these issues (for example, Bruce 1989; FAO 1989), as have many individual scholars (for example, Fortmann and Bruce 1988; Rocheleau 1988; Banana and Gombya-Ssembajjwe 1996; Diaw 1997). Ostrom's (1990) excellent analysis of institutions for collective action represents a growing body of political science literature on local management, including access to resources. A series of essays from Rose (1994) provides useful insights into the meaning of property itself.

The same attention does not appear to have been devoted to determining exactly how security of intergenerational access to resources relates (or does not relate) to sustainable forest management. Nor has there been much scholarly attention to traditional land tenure vis-à-vis logging. Many important actors in forest certification (for example, Rainforest Alliance 1993; ITW 1994; Schotveld and Stortenbeker 1994; Soil Association 1994; LEI 1997; SGS 1997; FSC 1998) list issues pertaining to security of access in their requirements for a sustainably managed forest—though exactly what they mean typically remains unclear.

This section draws on the literature survey published in Colfer and others 1998b.

The international "processes" that have developed criteria and indicators (C&I) for sustainable forest management vary in their treatment of access to resources. The 1997 Montreal Process (reprinted in Prabhu and others 1999) is quite explicit about customary and traditional land rights of indigenous people, employment, and other community needs. The C&I from the African Timber Organization (ATO; still under revision) are, similarly, quite specific on recognition of traditional rights and sharing of benefits (see also Caballe's 1999 version of the principles and criteria produced for the Central African Republic, under ATO sponsorship). The earlier set from the International Tropical Timber Organization (ITTO 1992a) scarcely mentioned this issue, but later versions (ITTO 1997) are more complete. The Helsinki Process (1993) was clearly formulated with a European context and does not address the issue in the direct way necessary in many tropical contexts. It does, however, refer to changes in the rate of employment in forestry, notably in rural areas. The Finnish Ministry of Agriculture and Forestry (1996) made some modifications to the Helsinki Process that strengthen requirements about local access to resources somewhat (for example, "Secures clarity of rights related to forests [for example, ownership ...]," "secures special rights [for example, reindeer husbandry] of the Saami people and local people"). The Tarapoto Process (1998) C&I for the forest management unit level do not discuss rights to land but specify quality of life, impact of the economic use of the forest on the availability of forest resources of importance to local populations, and the amount of direct and indirect employment as important indicators of "local socioeconomic benefits."

Perez (1996) provides a valuable overview of work done on C&I for certification, with a strong emphasis on access-related issues. Copus and Crabtree (1996) suggest a matrix for indicators of socioeconomic sustainability, derived from experience in rural Scotland, that includes several economic issues that may be more crucial in a developed country than in the developing world. Indeed, a wealth of economic indicators and models (for a recent overview, see Becker 1997 or Ruitenbeek 1998) can be linked to security of access issues, but they have not been a focus in our work until quite recently.

The Southern Appalachian Assessment (1996) in the United States dealt with many of the issues that we consider important, phrased as broad questions rather than as C&I. With reference to security of intergenerational access, they identified important issues relating to employment, income, and the relationship of these issues to timber production and other forest-related industries; population; ownership and uses of land; and values pertaining to land and its uses.

In the conservation realm, Borrini-Feyerabend with Buchan (1997) published a two-volume work aimed at improving our ability to achieve "social sustainability in conservation." Many of the access issues we have addressed in the timber management context correspond with issues in these volumes, though from somewhat different perspectives. This work asks key questions, such as, "How do the natural resources of the conservation initiative contribute to the livelihood of local people?" and "Does the conservation initiative affect access to land or resources and the control over them for one or more stakeholders?" They also offer indicators, with "warning flags," pertaining to many of the issues we have also found important.

The International Union for the Conservation of Nature (IUCN) has been involved in several efforts to improve monitoring and evaluation relating to human well-being. Poffenberger (1996) reports case studies from India, Nepal, British Columbia, Panama, and Ghana and concludes that access to forest resources is important in stabilizing resources. Jackson (1997) details some of the tools and methods used by IUCN in southern Africa to evaluate collaborative management of natural resources, although the emphasis is more on participation issues than on access to resources, per se. Prescott-Allen's (1995) idea of a "barometer of sustainability" is particularly appealing, but its only direct, access-related issue is "wealth and livelihood." It includes the idea that "human well-being is dependent on the well-being of the ecosystem" (IUCN 1995a). In this system, indicators are considered context-specific and are selected by the users (IUCN 1995b).

Much of the growing literature on community-based management is pertinent. Stevens (1997) uses C&I to assess the sustainability of a Turkish forest village ecosystem. From the access perspective, he looks at household finances (income sources and amounts, indebtedness). Like us, he recognizes a link between "the state of the natural resource base, and the ... social and financial indicators that depend upon them" (Stevens 1997, 30). The work undertaken by the Asia Forestry Network has included methods for assessing access to resources (for example, Poffenberger and others 1992a, 1992b), and the case studies address these issues as well (for example, Poffenberger 1998; Poffenberger and McGean 1993a, 1993b). Similarly, the Overseas Development Institute Rural Development Forestry Study Guides provide useful methodological insights (for example, Carter 1996; Hobley 1996). A recent review by the Ford Foundation (1998) also stresses the importance of elements of access ("ascertaining actual benefits" and "promoting the equitable distribution of benefits").

Burford de Oliveira's C&I testing in four community-managed forests (two in Indonesia, one each in Cameroon and Brazil) involved attention to access to resources, though phrased differently in the different contexts. One of our current activities is comparing our results with those of Burford de Oliveira, in search of commonalities. (Some of the results of these tests are available in Burford de Oliveira 1997, 1999; Burford de Oliveira and others 2000; Ritchie and others 2000.)

Security of intergenerational access to resources is the subject of Chapters 7–11. Chapter 7 (Günter) serves as a transition because it deals with both the question of stakeholder identification (discussed in Sections 1 and 2) and access issues. Günter's contribution is unusual among this set of papers

169

because the orientation is toward workers and small-scale entrepreneurs in the logging industry rather than communities per se. He also provides a useful comparison with the other studies in that his case was on a small island state in the Americas (Trinidad). As a forester, he has a somewhat different perspective on problems relating to forest use than do many social scientists.

In Chapter 8, Colfer and others discuss the implications of security of access in Kalimantan, Indonesia (see also Chapters 1, 4, 5, 12, and 16). This analysis led to the development of a qualitatively based scoring system for use in assessing human well-being (Salim and others 1999).

In Chapter 9, Gami and Nasi describe results from the use of participatory mapping in assessing access in Gabon. This Gabon context is one of the most forest-rich environments in our repertoire. The similarity in problems between this site and the forest-poor contexts was one red flag, leading us to conclude that our reductionist approach was not working.

In Chapter 10, Russell and Tchamou examine issues directly relevant for security of access in the south of Cameroon. They focus on the intersection of soil fertility, forest cover, and intergenerational equity. They explore some of the same phenomena reported in Chapter 3 but from a more long-term research perspective. They argue that the commodification of natural resources has had deleterious effects on the resources themselves and on the cultural institutions that both give meaning to people's lives and have served management functions in the past.

In Chapter 11, Porro and others compare results from forest-rich and forest-poor contexts. They report the comparative results of two pebble-sorting methods, with the inconclusive results we would now predict.

CHAPTER SEVEN

Intergenerational Equity and Sharing of Benefits in a Developing Island State

Research in Trinidad

Mario Günter

When addressing sustainable development in island developing states in general—and sustainable forestry in the Caribbean in particular—Trinidad provides a valuable focal point as one of the region's few countries with substantial yet endangered forest resources remaining. Compared with other countries of similar size, Trinidad is blessed with a diverse range of flora and fauna and several spectacular natural habitats.[1] The island originally was covered almost completely by a heavy tropical rainforest. Only the dry westernmost end of the Northern Range has a sparse, drought-tolerant vegetative cover. The highest peak of the Northern Range, El Cerro del Aripo (940 meters), is covered with forest to its summit. The forest types include evergreen seasonal, semievergreen seasonal, deciduous seasonal, dry evergreen, seasonal montane, montane, and swamp forests.

This chapter contains extracts from my Ph.D. thesis, Social Criteria and Indicators (C&I) for Sustainable Forest Management in Small Island Developing States (SIDS), where little research on social sustainability has been carried out to date. SIDS of the Caribbean are not the focus of international forestry research, although deforestation rates and socio-economic issues are major problems. These circumstances constitute a major concern for sustainable development. New approaches to investigating environmental and human interrelations by means of C&I are urgently needed.

Available data concerning the existing forest area over the past few decades seem to be inadequate. Even most recent publications refer to the extent of the indigenous forest as it was first documented by Beard in 1946, when more than 50% of the total land area was covered with forest. The most up-to-date survey of existing natural forests was carried out by the Forest Resource Inventory and Management Section of the Forestry Division in 1980. Unfortunately the inventory was based on aerial photographs dating from 1969, when approximately 44.8% of Trinidad's land area was under forest cover, of which roughly 7% consisted of teak and pine plantations (Chalmers 1992, 26). Illegal shifting cultivation, fire, and overexploitation during recent decades have increased concern about the quality and species composition of much of today's forest resources.

The actual extent of privately owned forests—which for some years has been quoted as 56,000 hectares (based on the Agricultural Census of 1963; Forestry Division 1995a, 1)—also appears to be in question. The number of private landowners who have converted their forest areas into agriculturally productive land has been steadily increasing. Whereas uncontrolled slash-and-burn practices are generally undertaken during the dry season by landless people, private landowners tend to make use of the rainy season for planting sugarcane or other profitable crops, such as fruits. Using a fell-and-sell method, they usually generate an additional income by negotiating the timber's value with a sawmiller, who may then process the raw material. As a result of this development and illegal squatting, the forest area under private ownership fell sharply by 84.6% to 8,470 hectares between the Agricultural Censuses of 1963 and 1983 (Smart 1988, cited in Ministry of Agriculture, Land, and Marine Resources 1992, 3-1). These lands have been alienated and converted without legal authorization to residential, commercial, or industrial uses. Very few landowners pursue any form of forest management or reforestation. The long rotation period for most timber species is considered to be a deterrent to their sustainable management, whereas primary cash crops (cocoa, coffee, and citrus), once producing, bring in a regular annual revenue.[2]

According to Food and Agriculture Organization (FAO) statistics, the forest cover in 1990 was between 30% and 33%, and an additional 3.4% of land was forestry plantations (FAO 1993; 1995b, 4; 1995c, 37). This figure roughly coincides with information provided in interviews with forestry officials, who estimated that the forest cover would be 35% by 2000; however, only 15–20% of this total was identified as primary forests. Furthermore, Trinidad loses about 10–15% of its forest resources per decade. The peak annual deforestation between 1980 and 1990 reached almost 4% (FAO 1995d, 330). More pessimistic outlooks suggest deforestation rates of up to 20% per decade (European Communities Commission 1996, 4). However, especially in remote areas with low population figures, we still find sites that can be deemed "forest rich"—primarily the Northern Range and southeastern forest reserves.[3]

Deforestation rates correlate with a steadily growing population and a demand for fuelwood and charcoal that increased by 38% throughout the 1980s (World Resources Institute 1992). The production of sawnwood over the seven years from 1985 to 1992 almost doubled (Statistisches Bundesamt 1996, 83). With a population of more than 1.3 million, Trinidad's population density (255 people per square kilometer) is one of the highest in Latin America and the Caribbean. Another decisive pressure that affects forest resources is the percentage of people living below the absolute poverty level, which reaches almost 40% in the rural forest-covered areas; in the Caribbean region, this figure is exceeded only in the Dominican Republic, Haiti, and Jamaica (UNICEF 1994).

Additional threats to the forest's ecological soundness are huge oil and other lucrative mineral resource deposits. They are found in the forest-rich Northern Range areas, in the southeastern parts of the island, and offshore. So far, the Northern Range has escaped intensive exploitation because the less mountainous districts have been more accessible. However, the extraction of inorganic minerals (the blue limestone of the Northern Range and the sands and gravel of the Northern Basin) poses a severe threat to some areas.

Trinidad and Tobago are still largely dependent on the oil and petrochemical industry. Serious attempts to diversify the economy have not yet achieved expected results. In 1997, the petroleum sector accounted for 26% of the country's gross domestic product (GDP) and 72% of its foreign exchange (Economist Intelligence Unit 1997, 23). Oil spills and seepage are evident throughout the areas of the oil installations, destroying important flora and fauna and thereby reducing biodiversity. This kind of pollution also poses a serious threat to the recreational potential of affected areas. It has been estimated that in 1995, 250 hectares of forested land of the southeastern region alone were adversely affected by oil pollution (Forestry Division 1995b, 38). The problem is created by petroleum operations through routine and accidental oil spills, the intrusion of oil exploration, and production operations in protected areas. Wide corridors are often bulldozed through virgin forests for the construction of roads that serve isolated wellheads. The Forestry Division has no control over the activities of the oil companies, which drill where they wish and remediate little of the environmental damage they cause.

A 69,000-hectare system of wildlife sanctuaries and other protected areas (approximately 14% of the total land area) was established in 1980. Of this area, only 43% is presently afforded some protection status, of which 31% is forest reserves and 12% is game sanctuaries. The remaining areas are privately owned or other state lands (Chalmers 1992). The major problem has been the lack of financial resources required to meet the costs of establishing the system.

Squatting[4]

The major threats to the forests are firmly linked to societal problems. The dominant issue of the forestry sector in Trinidad has become the almost uncontrolled, widespread destruction of public and private forest areas by squatters. Especially in the Northern Range, large areas of hillside have been deforested, causing the growth of savanna-type vegetation that is very flammable during the dry season. Reforestation of such areas has turned out to be labor-intensive, and fire protection in the early years involves high expenditures. Although squatting in Trinidad occurred in the past century, circumstances during the oil boom and the subsequent economic recession led to increased squatting. As a result, serious damage has been done to the environment. Whereas residential squatting also has a negative impact on the environment, fundamentally, the agricultural squatters cause the most damage. With the downturn in the economy, former workers from rural areas (and even neighboring countries affected by rising unemployment) returned to remote areas, invading private and public lands because they were unable to obtain land through legal channels. Most of these areas—even those proposed for protection—contain fragile ecosystems or important biodiversity values.

The main socioeconomic factors responsible for squatting are the high cost and unavailability of land, a shortage of housing, high levels of unemployment, the inability of most unemployed people to acquire or rent accommodation, a preference for earning a living from agriculture, and the low levels of policing and control of squatters' activities. With a growing population and increasing demand for food, even more pressures will be placed on forested areas. A report on the squatting situation by the Forestry Division concluded that throughout all districts, 1,940 squatters occupied 4,429 hectares of land in 1982. To date, approximately 8,357 hectares of forested land are occupied by farmers who possess no legal documents (Forestry Division 1995b, 37). This figure probably has increased drastically as a result of ongoing squatting. The following figures speak for themselves: Among house parcel holders, the Landholder Survey of 1991 showed that 12.4% on privately owned land and 48.5% on state-owned land possessed no documents. Among agricultural parcel holders, 22.6% on privately owned land and 33.3 % of parcel holders on state-owned land possessed no legal documents (Stanfield and Singer 1993).

The Forestry Division itself does not have the capacity or power to evict squatters. Although the division has the authority to police the unauthorized use of forest resources, personnel and transportation limitations severely restrict its effectiveness. Squatting and its consequences are very closely linked with tenure insecurity. The combination of poor management in the Forestry Division and the high incidence of squatting on private lands has resulted in a situation where any rights that may have been documented (in land leases) at one time are now all but worthless; those rights cannot be defended within a system where the supposed policing organization is ineffective. Additionally, confusion at the administrative level exacerbates the already complex problems of meeting existing demands for land by forest users.

Outright squatting is found primarily on state land where no lease has ever been issued. However, institutional weaknesses have resulted in apparent widespread illegal possession of land that is under lease to or owned by a third party. The virtual institutionalization of squatters' rights has brought a "subjective" land tenure security and has affected the ways that people deal with land they claim as their own. People behave more sustainably toward their "property" when they are in a social and economic position that allows them to do so; landless dwellers lack such an incentive. Additionally, the rural poor do not consider forested areas as being particularly attractive as reserves or savings, and so many occupied tracts are converted for agricultural use to satisfy people's basic needs.

Deforestation as a contributing factor in uncontrolled squatting might be weakened by channeling the demand for access to land in the direction of state-owned lands other than forest reserves. However, people whose livelihoods are already closely connected to forest resources must use the forest sustainably to reduce degradation of the forest and ecosystem (Chalmers 1992). Management methods are needed that will use a wide range of services and goods from the forests and turn the potential destroyers into a true rural population that gains economic and social benefits from this improvement.

Achieving this goal would pose legal and political problems, because only when national policies are announced and implemented can forestry-specific problems be addressed at the state level. The forestry sector alone cannot introduce effective ways of dealing with squatters. The complexity of the issue requires a multisectoral approach. The lack of regulated land tenure, government support, employment, and economic alternatives has resulted in not only environmental destruction but also apparent loss of self-esteem and emotional instability.

Various strategies for the legalization of squatters' claims have had little success in rehabilitating communities and individuals. They were considered to be piecemeal in approach, legally inapplicable, and ignorant of the social and technical dimensions of the problem (Mooledhar, cited in Stanfield and Singer 1993, Part II, 493). Integrated development plans need to be established that acknowledge the squatters' participation and meet their needs for infrastructure, health services, education and training, economic development possibilities, and so forth. Furthermore, the squatters have to be involved in nature and forest conservation efforts, to learn to value their own environment and to relate to it in such a way that it will guarantee not only their living standards but also their quality of life.

Forest–Society Links

Logging and various other timber-related industries have played an important role in the socioeconomic well-being of many inhabitants in Trinidad. Exact numbers of those who are either directly or indirectly employed in these industries do not exist. Forestry and forest-related activities played an important role in community life in rural parts of the country long before agroforestry and social forestry became popular. Earning a living from forest resources is still the principal objective for various groups ranging from Forestry Division laborers to game wardens, charcoal burners, woodcutters, hunters, agroforestry (*tangya*) gardeners, and sawmillers. The private sawmill industry alone consists of 63 enterprises, fairly well distributed across the country. The current employment level of perhaps 350 in sawmills and 1,500–2,000 in woodworking industries could even be increased (Nagle 1991).[5]

Logging operations carried out on public land are controlled through the granting of a conservator's license, which takes into account social, silvicultural, and industrial considerations. Timber is sold to registered licensees who supply the industry with local raw material. However, from a peak of nearly 1,600 licensees in 1980, numbers have steadily declined to around 400 because of the conditions under which they have to operate. On a girth license, the maximum number of trees that can be cut is ten, whereas sale on a cubical license is limited to 500 cubic feet. The license period may not exceed three months without authorization. Two visits by foresters from the Forestry Division are required. On the first visit, they select and mark the trees, estimate the approximate royalty, and fell the trees. On the second visit, logs are measured accurately and marked as released, when the royalty has been collected. Removal permits are issued, and then the logs can be transported from the forest. To ensure the ecological stability of the forest, these operations are controlled under a periodic block system.[6] The results of my stakeholder analysis indicated that a closer assessment of selected criteria and indicators (C&I) would be highly desirable for these private licensees (PLs) operating in a block system. Among all stakeholders, the PLs were identified to be most important in terms of issues of social sustainability (see Table 7-1). With regard to the study's focus on human well-being, only social C&I were applied.

In the future, the value of all these small forest industries may even be overshadowed by the enormous potential of forest-based ecotourism. On the basis of existing trends and the remaining healthy natural environment, a significant ecotourism market clearly could contribute to rural development.[7] On the other hand, this development can meet all expectations only when the proposed national parks and wildlife sanctuaries are fully operational in a reliable way. Currently, these forest-based industries are dwarfed by the petrochemical sector. However, they represent a self-sufficient, profitable industry with strong development linkages between rural and urban areas. Fur-

Table 7-1. *Identifying Stakeholder Relevance*

Stakeholder	Dimension						
	Prox-imity	Pre-existing rights	Depen-dency	Indigenous knowledge	Culture–forest integration	Power deficit	Value (mean)
Private licensees	1.33	1.89	1.22	1.44	1.56	1.67	1.52
Forest workers	1.44	2.11	1.33	1.67	1.89	1.67	1.69
Sawmillers	1.67	1.67	1.11	1.78	1.89	2.00	1.69
Squatters	1.22	1.78	1.56	2.44	2.11	1.33	1.74
Hunters	1.78	1.56	1.89	1.89	2.00	1.67	1.80
Forestry officials	1.89	1.89	1.44	1.56	1.78	2.78	1.89
Tanteak workers	1.67	1.78	1.33	2.33	2.44	2.56	2.02
Conservationists	2.11	2.11	2.33	2.00	2.22	2.22	2.17
Ecotourism	1.89	2.89	2.11	2.22	2.44	1.89	2.24
Consumers	2.56	2.44	2.33	2.67	2.56	2.33	2.48
National citizens	2.44	2.44	2.56	2.78	2.89	2.44	2.59

Notes: 1 = high; 2 = medium; 3 = low.

thermore, they can generate additional revenues—especially for rural people—and could enhance stewardship of the forest.

In Trinidad, timber is rarely the sole source of income, and in large families, several members contribute to the household income from various sources. Activity in the forestry sector is closely related to the performance of agriculture (cash crops) and other major natural resources (gas, petroleum, and oil). Each region's economy is diversified just enough to avoid dominance by the forest sector—even in "remote" regions. People are not in a position that they need to rely on the forest sector alone for income, and distances between regions are very small, so no forest community is truly isolated. The quantity as well as quality of timber in Trinidad also limits complete dependence on the resource itself. Extra income is needed in the rainy season, when work in the forest is extremely difficult.

Especially in small island states, sustainability problems related to the demands of a growing society arise because of limited natural resources. There is no logical inconsistency in the theoretical concept of sustainability as it applies in these countries, but development based on economic, ecological, and social coherence is constrained by the complicating factors that face the inhabitants every day. Only when all involved disciplines accept these constraints can incremental steps be taken toward preserving the natural assets and biological diversity of ecosystems for the benefit of societies.

Sustainability efforts should stress not only ecological goals but also how individuals, groups, and communities can contribute to maintaining resources and, more important, how and to what extent those efforts can affect and increase the well-being of the people involved. Implementing steps to promote sustainability has to counteract the socioeconomic causes of deforestation without posing a threat to the needs of the people—needs that only too often are the very reason for the exploitation of resources. These aims, contributing to social stability and welfare, are additionally supported by various recreational functions. The most desirable result would be to regard people not as a major problem but as a potential resource. In this socioenvironmental equation, people are not yet widely considered a resource for their own development, generating mutual benefits. Current local practices that do not contribute to sustainability should not be seen as insurmountable obstructions but as symptoms of a deteriorating situation in need of amelioration.

Because of the various problems associated with forest resources, forest management in a small island state can be as complex and diverse as that in a country with vast forested areas, huge populations, and a national economy that relies on forest resources to a much greater extent. Small island states therefore are microcosms of larger societies, which makes them particularly interesting in terms of a case study; they allow the observer to view the situation in its entirety. Clearly, achieving sustainable forest management in Trinidad will be as complex as the situation itself. We need to identify the most important stakeholder groups and examine their dependence on forest resources.

Stakeholder Identification

To apply C&I as tools for measuring any trend toward or away from sustainable management, we first had to identify the most relevant interest groups. Using C&I to measure the forest–society relationship of a particular group improves the clarity of the sustainability concept itself. The interests of the stakeholders and the conflicts within and among their groups can be seen as reference points on which C&I should focus. When concentrating on social aspects as one of the subsystems for achieving sustainability in forest management, we must emphasize the ways that stakeholder groups rely on forest resources to understand the extent of forest dependence. This aspect is more important when addressing access to forest resources, because a group's vulnerability is linked to such dependence. We used the information obtained from stakeholder analysis to clarify the various groups and determine their vulnerability. Our intent was to determine whether access to forest resources incorporates the groups' demands for participation in forest use and how the groups perceive this access to be changing over time, by generation.

Despite its complexity in terms of interest groups, the limited forest area of a small island state allows the observer to regard it as one particular forest or almost as one complex management unit with categories of relevant populations. Because the main task of the fieldwork was testing socially relevant

C&I, our initial identification of stakeholders had to be quick and easy, yet reliable. We used the "Who Counts Matrix" (see Colfer 1995),[8] with six dimensions by which stakeholder groups are judged (proximity, preexisting rights, dependency, indigenous knowledge, culture–forest integration, and power deficit), to identify relevant stakeholders.

All six dimensions apply for Trinidad. Unlike in large countries, physical proximity plays a minor role in a small island state where substantial forest still covers most of the island. Physical access (proximity) is relatively easy for all stakeholders. Therefore, *proximity* as used in this chapter includes the notion of emotional closeness. It does not mean that physical and emotional proximity are automatically correlated; emotional proximity does not have to be high only because people live close to the forest. It also can be high among conservationists who live far away from the forest. Emotional proximity depends on the quality of a group's or an individual's interest. Even a direct dependence on forest resources does not necessarily have to result in an emotional closeness.[9] Real indigenous knowledge rarely exists anymore. Local experience, on the other hand, varies widely among stakeholders, replacing the indigenous knowledge of indigenous peoples.[10] This study complies with the original intent of the method, focusing on the experiential knowledge of long-term residents.

Dimensions such as *preexisting rights* and *power deficit* turned out to be of higher importance in Trinidad, with its uncontrolled squatting problem, linking both dimensions closely. Today, even squatters claim preexisting rights as a consequence of their widespread and continuous activities; in terms of power deficit, their situation is the worst of all groups. The most difficult problem with squatters is their lack of homogeneity as a group. Squatting can be found almost throughout all stakeholder categories because it has nearly become a normal feature of Trinidad society. The squatters can be referred to as subgroups within the identified stakeholders. Even among the questioned interviewees, squatters were identified. Residential squatting was confirmed to be practiced by 26% of PLs and 24% of workers at the state-owned Trinidad and Tobago Forest Products Co. Ltd. (Tanteak). Nevertheless, all squatters were treated as one group to establish the way they are perceived within the human–forest relationship. *Preexisting rights, power deficit, and dependency* are at least three dimensions that are directly linked to Trinidad's main problem of squatting. All dimensions of the Who Counts Matrix apply for Trinidad, and they are interrelated. We therefore applied equal weighting for all six dimensions, as in the original method.

We consulted "neutral" local experts on forestry and sustainability matters in Trinidad and asked them to fill out a Who Counts Matrix with all the stakeholders and dimensions.[11] On the basis of numerous field trips, my own estimates also were discussed and led to a further understanding of present stakeholder traits. These two assessments were considered when identifying stakeholder relevance (Table 7-1). Involving more people in this method increased reliability of estimates, made use of insider knowledge of local expertise, and tested the applicability of the matrix. The mean score for each column in Table 7-1 is given. The score of the stakeholders increases moving down the columns; the higher the score, the less direct the interest in the use of forest resources and the less vulnerable a specific group can be characterized. Field visits and meetings with local experts led us to conclude that, in terms of C&I for social sustainability in forest management, the groups most worth focusing on were self-employed woodworkers who work on a license (that is, PLs) and local sawmillers.

The identification of PLs as suitable candidates for research was confirmed by the matrix technique. On the basis of social sustainability, this group is the most relevant agent for participation in managing forest units and thus will be most affected by these issues. PLs are closely followed by forest workers and sawmillers, whose reliance on the resources and consequent vulnerability are also shown in the matrix results (see also Günter 1998). In previous applications of this method, extreme figures (1.00) were found only among indigenous groups. The apparent closeness in the results of the groups in our sample may be the result of two factors: indigenous groups no longer exist in Trinidad, and the calculation of an arithmetic mean of the scores given by several respondents. A scale with more points (1 to 6) would improve the comprehensiveness and meaningfulness of the matrix results without affecting the rankings of the stakeholders.

The conceptual problems related to weighting the dimensions and the scoring method were partially compensated by improving the estimates by means of local experts' assessments. The experts' positive feedback on the achieved results confirmed this assumption. One disadvantage of these consultations was the additional amount of time required; it took almost a month to complete this process. However, the results proved to be worth the extra time. The method accurately reflects the views of stakeholders about who is directly affected by a management unit's state of sustainability—and in the case of a small island state, it shows how to focus on a set of stakeholders in a broader context.

Intergenerational Access

After identifying "who counts" in sustainable forest management, we assessed the security of intergenerational access to resources. The objective of focusing only on the most important group and the limited amount of time available did not allow research on all groups. Accordingly, we focused on PLs. This group is important when addressing intergenerational equity because the licensing system in Trinidad dates back to colonial times. These individuals have been helping their fathers or grandfathers in the forests—a circumstance that is of special interest to their assessments.[12]

The state-owned Trinidad and Tobago Forest Products Co. Ltd. (Tanteak) allowed us to interview some of its workers, an ideal contribution because forest workers were perceived to be of high importance as well.[13] The following investigation identifies whether access to resources, as estimated by PLs and Tanteak workers (TWs) themselves, can maintain or even enhance the flow of benefits from the forests. Our goal was to find out how access to resources of the affected stakeholder groups is perceived to be changing from generation to generation. *Intergenerational access to resources* implies the notion of benefit sharing among generations. It is one of the main obligations of sustainability and therefore is also of considerable importance for sustainable forest management. Only by creating conditions under which lasting access to resources is assured can equity be achieved among generations.

We initially intended to investigate this aspect by using CIFOR's pebble-sorting method (see Colfer and others 1999b). However, testing this method proved problematic. The interviewees had no problems understanding the method or its purpose, but they had difficulty expressing their views in terms of pebbles.

Access to resources by generation was one part of a questionnaire that contained various other C&I. We found the responses of stakeholders to the sorting method to be very time-consuming. One possible reason is level of education; the higher a respondent's level of education, the less the method was accepted. Age also appeared to have negative effects. Especially among PLs, who have a much higher average age than TWs, the acceptance decreased with increasing age.[14] Possibly the idea of a complete stranger asking them to distribute pebbles made them feel humiliated or offended. However, the use of rating scales—also used with other questions on the questionnaire—caused no problems. Therefore, we presented a scale for each generation (grandparents, self, grandchildren) on which the respondents had to estimate the access to forest resources per generation. We first asked the respondents about the current situation, then about the situation of their grandparents, and about the future. In addition to age groups, Trinidad's various ethnic and religious groups were taken into consideration. Only one woman was present among all PLs, and she refused to be interviewed. All TWs were men. The heavy labor aspect appears to exclude women from both groups.

Unlike the pebble-sorting method, in which 100 pebbles had to be distributed among all three generations, the scale was applied for each generation separately to estimate perceptions of relevant access. The responses per group therefore were not documented by means of an arithmetic mean but by a median, serving to illustrate the distribution's midpoint on the scale (Figure 7-1). This measure increased the meaningfulness and the comparability of the figures not only among the generations but also among the stakeholders. The results are easy to interpret: each PL or TW perceives that, compared with his own situation and the situations of his grandparents, access to forest resources is hardly an option for future generations. Sustain-

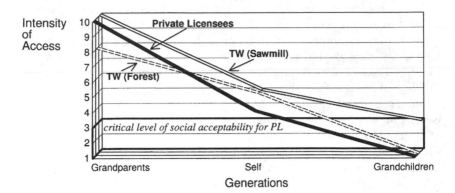

Figure 7-1. *Access to Forest Resources as Perceived by Private Licensees (PL) and Tanteak Workers (TW)*

ability (defined as meeting the needs of the present generation without threatening the needs of future generations) is clearly not achieved. Intergenerational aspects of sustainability, equity, and benefit sharing among generations as achieved through access to resources is perceived to be unrealizable, especially by those who are most dependent on forest resources. The sustainable flow of benefits, as it is projected, is even impossible.

PLs generate additional income by making use of nontimber forest products as hunters and food crop farmers or by collecting medicinal plants. Access to forest resources for this group does not imply the idea of access to timber alone. Their perspective must be seen in a broader context of resources, whereas the TWs are primarily focused on timber. TWs and PLs project the future situation to be equally bad. The reason for the slightly more optimistic future projection of TWs might be correlated with their actual distance from the resource and from conflicts in the forest.[15] Presumably, other stakeholders who are less dependent on forest resources and less aware of the day-to-day struggles for forest resources might have a less pessimistic view of the future.

The periodic block system in the naturally grown Mora forests, where PLs operate, is very well managed by the Forestry Division and leads to sound forest management. A situation has been created, dating back to colonial times, that allows local individuals to participate in and benefit from forest use. However, the present-day conditions under which those PLs have to operate are neither socially nor economically acceptable or sustainable. Although their numbers have steadily declined, the number of licenses each person receives per year has remained the same or even decreased. Furthermore, the requirement to use low-quality timber, which the licensing policy includes only too often, negatively affects their income. At present, access to resources for PLs is about to reach a level where their use of the forest is no longer socially sustainable. The feedback from PLs and the experience of their living and working conditions suggest that they are approaching the minimum level of social acceptability (Figure 7-1). The respondents considered the critical level to be close to 3 for the applied scale. PLs report that present forest management restrictions that affect their access to resources already negatively influence their earnings. Therefore, the critical level of acceptability for PLs may even be adjusted to the perceived current access of 4. For other, more distant and less dependent stakeholders, this level of access to forest resources might still be sufficient.

One further aspect that supported the idea that a minimum level of access—to generate a satisfactory livelihood—has nearly been reached was the PLs' current living conditions, which were reported to have declined in recent years, especially among the older licensees. This finding led us to conclude that access to resources and state-of-living conditions are closely linked to each other. Without a doubt, this is the case especially among groups who depend on direct use of resources. We therefore particularly wanted to know whether the groups perceive that a change of living conditions had taken place since they started to work in the forests and whether this trend is negative or positive. Based on the estimates in Figure 7-1, we expected a steady or slightly negative development for PLs as well as TWs.

The scale from "much worse" to "much better" (Figure 7-2) was presented in exactly these terms to the PLs and TWs interviewed. The distribution clearly shows that the situation is reversed for the two groups. Whereas on a scale from 1 ("much worse") to 6 ("much better") a median (3.0) for PLs would indicate "slightly worse," TWs' median (5.0) would represent "better." Access to resources for PLs really has reached a critical level. For the majority of interviewees (68%), working as a PL negatively affected their living conditions. On the other hand, 94% of TWs had improved their living conditions since their employment with Tanteak. For TWs, the reduction in perceived access to resources shown in Figure 7-1 does not have any immediate effect on their living conditions. TWs receive a fortnightly regular income, paid vacation and sick or casual leave days, traveling allowances, protective gear, retirement benefits, medical checks, and housing loans—to mention a few of the benefits granted by Tanteak. PLs, in contrast, are completely dependent on the revenues they achieve per license. In terms of social sustainability, the welfare system granted to TWs is highly desirable.

One can ask about the basis on which TWs estimate reduced access to resources as in Figure 7-1: What makes them project such a negative trend although they profit dramatically from their work? Trinidad's forest plantation estates of teak and pine, for which Tanteak has had a monopoly concession agreement until recent times, are characterized by inappropriate plantation management and illicit removals. Inefficient management practices, which result in harvest losses and material waste, have been criticized by various FAO consultants (Shand 1994; von Bothmer 1994; Young 1994).[16] The

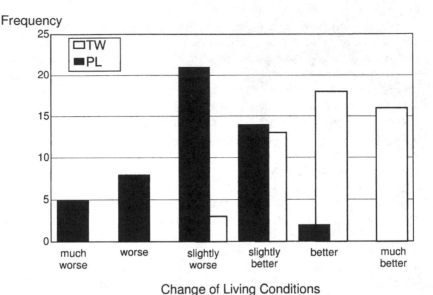

Figure 7-2. *Change in Living Conditions since Working as Private Licensees (PL) and Tanteak Workers (TW)*

plantations are not properly managed according to the rotation plan and cannot be deemed environmentally sustainable. These problems are accompanied by fire damage, medium quality of the standing volume, squatting by gardeners, clearfelling without replanting, and encroachment on the plantation estates. These factors must be seen in relation to the TWs' pessimistic estimates on access to resources as shown in Figure 7-1. When asked the reasons for their skeptical conclusions, 52% of TWs said they based their estimates on reduced forest and plantation areas, 28% explained their assumption with official management restrictions by the Forestry Division, and 20% of the interviewees gave both reasons. The explanations given by PLs mirror these main fears and represent an almost opposite perspective: for 58%, official restrictions by the Forestry Division concerning their participation in forest resources use was the main reason, and only 14% named reduced forest areas—both reasons were given by 28%.[17] These results show that the present conditions under which PLs have to operate constitute the primary reason for their perceptions about future access to resources.

Exchange of Information

The constraints imposed by forestry officials mostly affect access to resources by PLs. This group is vulnerable to and dependent on decisions determined externally by forestry officials without proper consultation or mutual agreement on possible effects of resource use restrictions.

Communication between PLs and the Forestry Division and between TWs and Tanteak management is perceived to be insufficient by PLs and TWs, respectively. Although the opportunity exists for both groups to express their interests—mostly by means of formal meetings and representatives—the results are highly unsatisfactory. The mere demand for better communication among relevant stakeholders is of no use until qualitative effects are achieved. It was therefore of interest to ask the groups to what extent their concerns and interests are taken into consideration. A scale from 1 ("low consideration") to 6 ("high consideration") produced an equal mean of 2.4 for both groups. This result clearly reflects a desire for better communication by PLs and TWs.[18] Stakeholders such as Forestry Division and Tanteak officials, who are in positions to make decisions that directly affect the well-being of other groups, are neglecting this issue. The people's role in sustainable forest management currently receives little attention. The voices of all stakeholders are not of equal interest to officials whose decisions are significant. Unfortunately, PLs, the group that counts most, suffer most from this unbalanced situation.

Trinidad's squatting problems have created a situation that indisputably puts the forest's ecological integrity at the forefront of sustainable forest management. However, these problems should provide the very reason to enhance communication among interdependent stakeholders. Without effective mechanisms of communication, even "legitimate" stakeholders may be forced to act illegally. On the other hand, interaction and negotiation among stakeholders may not exclude squatters or other unlawful cash crop farmers. Only a forest management practice that integrates all aspects of sustainability will improve socioeconomic conditions for the relevant target groups.

The limits of sustainability in this context are approaching. Measures have to be implemented to reach mutual consent, which requires hearing the voices of all stakeholders. Furthermore, only by enforcing extensive actions aimed at solving the present social dilemmas can the ecological condition of the remaining primary and secondary forest areas be maintained or enhanced.

Conclusions

The problems addressed in this research were familiar to all groups involved. With regard to deforestation rates, overall management under the guidance of the Forestry Division serves to maintain the forested areas first. Reforestation and afforestation operations and the protection of the virgin forests of the Northern Range are of primary importance. At present, social principles of sustainability are overshadowed by the need to control negative human impacts on the forests. However, sustainable forest management—which for

PLs is designed to give local individuals the chance to participate and benefit from resource use—has to grant social principles an equal status to ensure the well-being of present and future generations. The same is true for TWs. Although their working conditions are satisfactory, in terms of intergenerational equity and sharing of benefits, they also are affected by inadequate management of plantations and the company itself.

From silvicultural and socioeconomic perspectives, the license system leaves a lot to be desired. The Forestry Division argues that it has no control over illegal felling and that increased exploitation would lead to further irreversible degradation. On the other hand, the restricted flow of raw materials that the division imposes results in the ongoing illegal cutting and felling of trees. However, since the introduction of the periodic block system, PLs have not overexploited the resource. It is important to reaffirm the effectiveness of the block system on a 30-year felling cycle with regard to the maintenance of the forest's floristic composition and its wildlife.

The uncomplicated methodical application and interpretation of C&I has proved to be an efficient tool to study current developments. Furthermore, when used and discussed adequately, experience showed that C&I also hold potential to increase awareness among officials responsible for forestry's social consequences. The priorities for C&I research on socially acceptable forest management are largely short term to solve immediate problems. By highlighting the main restrictions and problems with present resource use as well as the consequences for stakeholders, C&I can simultaneously direct long-term management and research aims. Equity and benefit sharing among generations are important guidelines for any long-term forest management if conflicts are to be reduced and human well-being to be attained. The capacity to contribute to local peoples' well-being on a small island developing state with limited resources and a steadily growing population is clearly limited. Nevertheless, it may not lead to management patterns that undervalue sustainability's demand for persistent security of access to resources by dependent stakeholders.

The applied C&I identified negative impacts for the realization of social sustainability and thus also have shown the most important obstacles, as well as starting points, that should be addressed for the sake of the well-being of local individuals involved in forest management. The success of such actions primarily depends on the cooperation of the dominant stakeholder groups and their readiness to acknowledge that ecological and economic sustainability and social sustainability cannot be separated but that they are interdependent factors that commonly contribute to sustainable forest management in its true sense. Ecological and social sustainability can be threatened by the same factors on a national level. Actions to address the identified problem areas that hinder social acceptability also will contribute to ecological stability.

On a national level, the interdependence of factors that influence sustainability is more easily understood when studying small areas. Therefore, on small island states such as Trinidad economic and social factors can have a direct impact on the country's ecological stability, which can be more immediately perceived and measured—for example, using C&I—than on larger geographical areas. Ecological degradation is influencing both economic and social conditions. As the study has shown, forest-dependent people, who have earned their living by sustainable use of the forest's resources for generations, are directly affected by this development, having to fear for their own and their children's well-being. To counter this development, it will become necessary to link various political activities and decisions more closely. Leaving responsibilities to single departments without proper coordination with other political decisionmakers and affected interest groups will result in further deterioration of present conditions. Drastic measures will increasingly be needed to address the serious ecological and social problems discussed in this chapter.

On a national level, these efforts can be supported by the creation of C&I information networks, summarizing the work and results of various interdisciplinary studies on the ecological and social aspects of sustainability. For forest management, the mere forest development figures can be enriched by social indicators such as population, poverty, or squatting rates—even the specific stakeholders themselves, and their demands on forest use. Additionally, the applied methods and tools of investigating sustainability should be incorporated in such information networks.

For small island developing states of a particular region (for example, the Caribbean), such a C&I information pool can serve as a reference source for decisionmakers and interested scientists from various fields. Numerous national society- and forest-relevant components can be considered in relation to others so that positive or negative trends can be examined, and by successively applying C&I, new decisions can be taken. Socioecological experiences of various islands can be shared, leading to advantages for national and subnational forest policy matters. They also serve to deepen our understanding of forest–society relationships and their consequences and, moreover, help to continually expand the content of C&I and their underlying methodology as tools of sustainable forest management. The quality of the decisions made will directly depend on the relevance of the C&I and the methods of application. Therefore, it is necessary to make C&I and the methodology available for additional adaptations and improvements.

Acknowledgements

In November 1997, a second Caribbean Ministerial Meeting on the Implementation of the Program of Action for the Sustainable Development of Small Island Developing States (SIDS) took place in Barbados, convened by the U.N. Economic Commission for Latin America and the Caribbean

(ECLAC). Because of my focus on society–forest relations, I was invited by ECLAC's Environment and Development Officer, Eric Blommestein, to join the ECLAC Headquarters for the Caribbean as an associate researcher and to carry out my fieldwork in Trinidad.

The help of my supervisor, Prof. Mikus (University of Heidelberg [Germany], Geography Department); the support of Trinidad's Forestry Division; and the methodological assistance I received as a collaborator in the Center for International Forestry Research's Assessing Sustainable Forest Management: Testing Criteria and Indicators project were essential for the completion of my fieldwork.

Endnotes

1. The nation's geographic location as a continental island separated only in historical times from the South American coast, has fostered a biodiversity that is richer than that of any other Caribbean nation. Trinidad alone (4,828 square kilometers), without its sister island Tobago (300 square kilometers), possesses 2,160 flowering plants, of which 110 are endemic. The fauna of the island is no less impressive, with 108 mammals species, 426 bird species, 70 reptile species, 25 amphibian species, and 617 butterfly species; again, a vast number of these species are unique to the island. The shoreline is similarly rich with marine life (UNDP 1985; CFCA 1994). A recent study by the Organization of American States deemed 61 sites worthy of protection. However, the existing legal and institutional framework for the conservation and management of the nation's ecosystems is inadequate (World Bank 1995).

2. In Costa Rica, a new system has been introduced that not only aims to stop deforestation by private landowners but also supports reforestation on private land. Under the program, which is based on growth and quality of the newly planted trees, the state periodically pays consistent returns to address people's short-term interests. The problem of long rotation periods is minimized by granting a steady income.

3. In Trinidad, a *forest reserve* is a logging area that has been designated to be under permanent forest cover. A *nature reserve,* on the other hand, has the status of a preserved area.

4. The negative connotations of the term *squatting* are slightly different in Trinidad. Officially, it is mainly used as a political term that expresses the illegal activities of the people involved. The actions are the same whether they are called *squatters, landless, rural poor,* or some similar term. However, the circumstances in Trinidad have generated a special conflict potential. Squatting is not only related to effects on the environment but also is closely linked with the authorities' inefficiency and lack of action to handle this social dilemma. The uncontrolled practices mirror the official helplessness toward the problem. Squatting, therefore, is also a reflection of the societal disparity and a characterization of the quality of present solutions (Chalmers 1992, 243).

5. Furniture is the most important secondary forest industry. During the oil-boom years, the furniture market reached a peak value of US$60 million in 1982.

6. Since 1960, forests have been divided into blocks of approximately 400 acres (160 hectares), and felling is controlled by girth limit. A block is opened for exploita-

tion on a 30-year cycle. The sale is confined to 12 months after the opening of each block. Both foresters and the people who work in the forests consider that this method has allowed controlled systematic exploitation of the forest, adequate stocking of young and immature trees, and the natural regeneration of commercially valuable species.

7. The Asa Wright Nature Center, for example, currently employs about 30 people who have more than 100 dependents.

8. This method has been updated (Colfer and others 1999c) taking into account input from additional field trials; "poverty" was added as a seventh dimension.

9. Asked if they would prefer higher production and therefore higher personal income, even if the regeneration of the forest was endangered, 64% of Tanteak's sawmill workers, only 34% of Tanteak's forest workers, and 30% of the private licensees answered "Yes." Even though these groups all live in very close proximity, these responses show that their attitudes toward the forest differ markedly.

10. The last descendants of a Caribbean tribe in the Arima area have been fully absorbed by modern Trinidad society and have almost completely lost their cultural identity.

11. The experts consulted represented the following organizations and institutions: U.N. Economic Commission for Latin America and the Caribbean; U.N. Development Programme, Sustainable Economic Development Unit; University of the West Indies; Caribbean Forest Conservation Association; Eastern Caribbean Institute for Agriculture and Forestry; Interamerican Institute for Cooperation in Agriculture; and the Forestry Division.

12. 86.5% of Private Licensees with previous working experience in the forest had gained their experience by helping family-member licensees.

13. Tanteak's forest workers are estimated to be much more important than the company itself (Table 7-1). Altogether, 50 private licensees, 25 of Tanteak's forest workers, and 25 workers of the Tanteak sawmills were interviewed. The two groups of Tanteak workers held slightly different estimates on future access to resources (Figure 7-1). This relatively large number of interviewees was chosen to increase the overall significance of the questionnaire in which the C&I were included.

14. The average age of the private licensee interviewees was 49; 17 were older than 60. Tanteak workers' average age was 33.

15. This observation argues for the relevance of the dimension *proximity* in the "Who Counts Matrix." In this case the influence of physical closeness cannot be ignored.

16. These studies also refer to the economic and social climate within the company, which has had negative effects on the worker's job satisfaction and work output as a result of poor management practices.

17. "Restrictions by the Forestry Division" are responsible for declining numbers of licensees, the utilization of low-quality timber and unmarketable trees and unfair treatment among licensees, as well as unfair treatment between licensees and other groups (for example, sawmillers) or exclusion of private licensees from the use of teak plantations and so forth.

18. This is also highlighted by the fact that 78% of the interviewed Tanteak workers asserted that they are not informed of the decisions made by the company's management.

CHAPTER EIGHT

Assessing Intergenerational Access to Resources

Using Criteria and Indicators in West Kalimantan, Indonesia

Carol J. Pierce Colfer, Reed L. Wadley, Emily Harwell, and Ravi Prabhu

In this chapter, we present data from research conducted in 1996 in and around the Danau Sentarum Wildlife Reserve (DSWR) in West Kalimantan, Indonesia. Our initial purpose was to contribute to the development of principles, criteria, and indicators for sustainable forest management. The particular topic investigated here is intergenerational access to resources in sustainable forest management.

After describing our methods,[1] we provide the case materials that form our results, within the organizational framework provided by Center for International Forestry Research (CIFOR) criteria and indicators (C&I) relating to the security of intergenerational access to forest resources.[2] Our purpose was to determine why, and by what means, intergenerational access to resources is important for sustainable forest management. In other words, what are the causal links—as evident in one forest-rich location—between these issues and sustainable forest management?

We conclude with a discussion of our scoring of the qualitative cases pertaining to each indicator[3] (for a fuller treatment of this scoring system designed for users, see Salim and others 1999). Besides contributing to our understanding of how security of access relates to sustainable forest management, this method allows quantification of our C&I assessments and helps us

---

This chapter is a revised and shortened version of a Center for International Forestry Research working paper (Colfer and others 1997b).

decide on the sustainability of the management of a particular forest, including its peoples' access to resources.

Theoretical Context

The fundamental impetus to the research reported here was a widely recognized need to be able to assess intergenerational access to resources simply, inexpensively, and reliably. As we worked on developing methods, we realized that there was a more fundamental question: why, and by what means, is security of intergenerational access to resources important for sustainable forest management?

The importance of local people's security of intergenerational access to resources was consistently identified in CIFOR's various field tests of C&I for sustainable forest management (Prabhu and others 1996, 1998; see also Colfer and others 1995). The issue also has been debated in numerous other scholarly studies (for example, Fortmann and Bruce 1988; Ostrom 1990; Lynch and Alcorn 1994; Rose 1994; Besley 1995; Grigsby 1995; Lueck 1995; see also the introduction to Section 3). Yet real dissatisfaction with our ability to effectively assess the CIFOR C&I remained, along with uncertainty about the causal links between such access and sustainable forest management. Access to resources seemed too difficult to determine reliably in the short amount of time typically available. Here we suggest some ways to address these issues.

Definitions

What do we mean by *security of intergenerational access to resources*? The most common examples cited by research teams included security of land tenure, use rights to forest products, and fair distribution of forest benefits.[4] The meaning of *intergenerational* is quite clear in the Indonesian context;[5] the resources in question are for the benefit of both the present and subsequent generations (see Becker 1997 for a brief philosophical discussion of this issue). *Security* refers to a reasonable certainty that the future will not involve a significant reduction in people's access.

Access implies three qualities: that the resource remains (sufficient quantity and quality);[6] that the people can use it, as needed or to the same extent as in the past;[7] and that "fairness," or equity, exists in regulations governing its use and distribution. By *resources,* we mean natural resources—forests and their products, streams, lakes, agricultural lands, fisheries—anything in nature that has or could have a productive potential and/or provide ecological or cultural services in forested landscapes.

Research Site

A preliminary step in the pursuit of understanding causal links involved a test of three access-related methods in and around the DSWR in West Kalimantan, Indonesia (Figure 8-1; the DSWR is also discussed in Chapters 5, 12, and 16).

This research site was chosen for several reasons. Three of the authors have spent a total of six years conducting ethnographic research in the area (1992–1997), and we had access to more than 130 reports from a Conservation Project in the reserve, along with results of our own studies. The area represented various purported managers, including local people, conservation managers, and timber concessionaires. Finally, we anticipated considerable variation in local people's security of access on the basis of their different resource use, different lengths of residence in their communities, and different potential conflicts with other stakeholders.

The primary forest actors in this area include Muslim Melayu fisherfolk, who live in the seasonally flooded core of the reserve; Christian and animist Iban swidden cultivators, who live in the surrounding hills; and to a much lesser degree, forest workers. The two main groups inhabit ecologically very different habitats and have distinct natural resource management systems. Other important stakeholders include residents of the larger Melayu "mother villages" along the Kapuas River, traders, timber concession holders, timber workers, the Conservation Project, and local government. We were also cognizant of the potentially different concerns of men versus women, old versus young, rich versus poor, and newcomers versus old-timers (Colfer and others 1999b; or Nurse and others 1995).

We decided to focus on four communities: the Melayu communities of Nanga Kedebu' and Danau Seluang, and the Iban communities of Wong Garai and Bemban. Because of logistical problems in Bemban, we added the Iban community of Kelayang as a partial replacement (see description in Chapter 5).

Methods

We selected two methods and developed a third to test: a history form, participatory mapping, and the iterative continuum method (ICM). The first two are described in the following paragraphs; the third, in the Introduction. The order in which the methods are presented reflects the increasing expertise needed by the assessor.

History Form

Tainter (1995) and Vayda (1996), among others, argue for the important role that history must play in any attempt to address sustainability issues. Sustainability, by definition, has a temporal component. Similarly, intergenerational access has a built-in time frame. By using the history form, we wanted to gain some sense of the sweep of history within the area; we hoped that look-

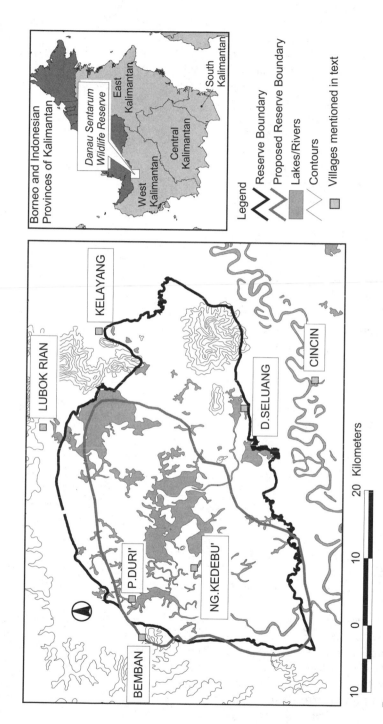

Figure 8-1. Map of the DSWR Showing Traditional Territories

ing backward might be helpful to us in the much more difficult task of look-ing forward. We also hoped that looking at the current situation with some understanding of past events would help us to understand some of the "causes" behind the present situation (Vayda 1996). Finally, we hoped that the history form would serve as a catalyst to discussions, which in turn would clarify the dynamics of factors affecting people's access to resources.

The method involved filling out a form with dates, starting in 1920 (see also time lines described in the *Participatory Rural Appraisal Handbook* 1990). We then asked individuals and groups in the study villages to tell us impor-tant events in the community's history. As we learned more, we were able to prompt people with known dates and to help them estimate unknown dates, because the use of dates is not common in this area. The results, recorded on the form, provided a historical perspective on natural disasters, warfare, the arrival of significant outsiders (such as the timber concessionaires and the Conservation Project), adoption or development of important new technol-ogy, and so forth.

The form served as a catalyst to discussion, and it provided useful dates, such as that of the timber companies' arrival. However, people's inexperi-ence in dealing with dates was a significant constraint to its utility. Trend lines, as described in the *Participatory Rural Appraisal Handbook* (1990), were subsequently tested and found more useful for this purpose.[8]

Participatory Mapping

We selected participatory mapping as a research method because we sensed that residents were more likely to be forthcoming about boundaries, regula-tions, sanctions, and conflicts—often sensitive topics—when confronted with a visual image than they might be when asked direct questions. We hoped to be able to elicit this kind of information over informal discussions about the maps and in walks through the area with local residents.

An important activity within the U.K. Department for International Development (DfID) Conservation Project was the participatory mapping of traditional use zones in and around the DSWR (Dennis and others 1997b; see also Chapter 16), but these maps proved too complex for our purposes. We simplified and combined them with sketch maps, hand-drawn in collab-oration with local people (see also Sirait and others 1994 or Momberg and others 1996).

We first asked people to identify locations where they gathered forest products. In the course of the discussions, information about indigenous management practices, access and use rights, historical trends, and conflicts emerged as well. We used the maps in various contexts, with different users and stakeholders. Local people were interested in our simplified base maps (partly in finding errors!) and in many cases enjoyed pointing out areas with different uses, different histories, conflicting claims, and so forth. Both the maps and the accompanying excursions into community territories to see the resources about which access was to be assessed contributed to the cases we present below.

In Danau Seluang, for instance, we got a fairly clear view (Figure 8-2) of logged and burned areas, areas where rattan grew abundantly, and areas of comparatively "good" forest (including locally protected areas). Excursions into the forest to check the maps prompted discussions of conflicts among adjacent villages, different perceptions of boundaries, and the bases of historical claims to land and other resources. Indeed, the inclusion by Danau Seluang residents of the bamboo and protected areas to the east in their territory reflects differing perceptions (without apparent conflict) by this community and the adjacent Iban community.

In Nanga Kedebu', we focused on nontimber forest products (NTFPs). In Danau Seluang, we had found only rattan mentioned consistently as an important NTFP. Three outings in search of other important NTFPs convinced us that the Melayu were not using many other NTFPs. This conclusion is consistent with 1992–1993 household record-keeping data from there (Colfer and others 2000a).

In Bemban, we elicited historical data on the settlement of the community, locations of timber camps and logging activities, and local land use (Figure 8-3) as well as discussion of conflicts between the community and various outsiders (plantation owners, other villages, and timber concessionaires).

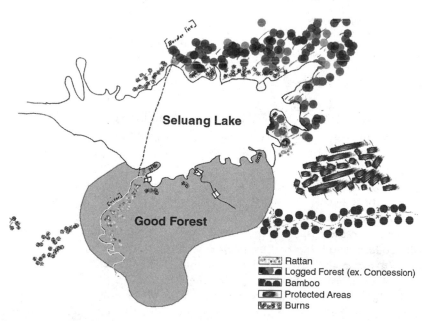

Seluang Lake

Good Forest

Rattan
Logged Forest (ex. Concession)
Bamboo
Protected Areas
Burns

Figure 8-2. *Sketch Map of Danau Seluang*

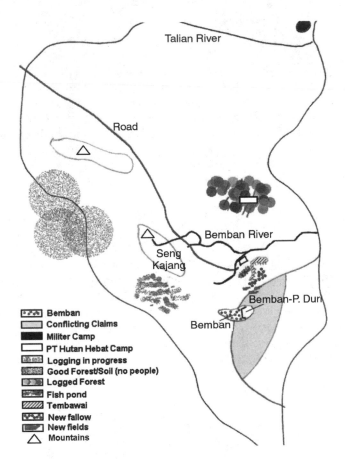

Figure 8-3. *Sketch Map of Bemban*

The people of Kelayang, sparked by Conservation Project interest, had made their own map, on which we were able to build. Again, the mapping exercise elicited areas of resource use and conflict as well as different land uses and histories.

In Wong Garai, because of Wadley's long experience there, we were able to elicit more detailed information on the extent of the traditional use area and the present area of the longhouse's effective control (Figure 8-4). In reviewing the list of specially preserved forest and old longhouse sites originally collected (see the box on page 198), he unearthed 103 named plants from 46 forest reserves and 26 old longhouse sites. These plants included food, construction materials, and medicine.[9]

For the other communities we studied, we had access to satellite imagery and the cooperation of a remote sensing specialist (see Chapter 16). Using satellite maps (such as those used by Dennis in Chapter 16), we were able to

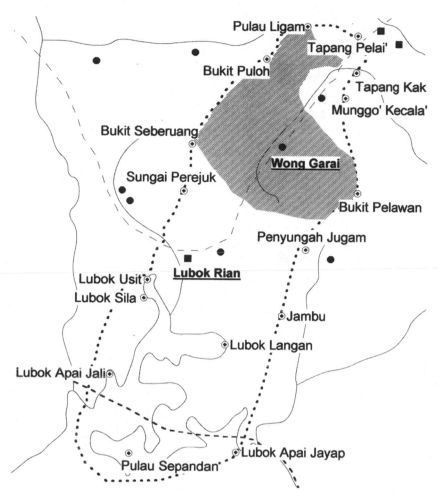

Figure 8-4. *Areas of Traditional Use and Effective Control in Wong Garai*

surmise that little or no dramatic forest loss had occurred between 1973 and 1994 in the three study villages.

For Wong Garai, we used an indirect method for assessing sustainability of land use, based on farming information collected during Wadley's initial research. Four years of farming data were selected (1979, 1983, 1988, and 1993) to show forest types being farmed, field sizes, and the length of fallow used—all practices that have implications for forest cover (see also Table 8-1). The longhouse increased in size from 7 households in 1979 to 14 households in 1993. The data show a consequent increase in the number of fields and an uneven increase in the total area farmed. No decline in fallow length (indi-

Iban Management

Wadley's research clearly shows the contrast between plants reported in two forested indigenous land use types: *tembawai* (old longhouse sites) and *pulau* (preserved forest areas). Although both categories of sites show similar counts containing fruit trees (domesticated, semidomesticated, and wild), 11 *pulau* have trees used for lumber, but only 3 *tembawai* do. Whereas 65% of the *tembawai* contain useful fruit trees, only 46.8% of *pulau* do. This difference illustrates the different nature of management for these two categories: even *tembawai* more than one hundred years old contain many fruit trees because people plant fruit nearby while they occupy a longhouse; after the people move on to another nearby long-house, they promote the succession of saplings from the original trees, thus producing a forest patch dominated by fruit trees. *Pulau*, in contrast, are patches of forest that have been preserved from felling for various reasons—as sacred sites, as places to collect rattan or wild latex, and as places to cut lumber for longhouse construction. The succession of useful tree species is also promoted in these sites (see, for example, Sather 1990; Padoch and Peters 1993; Wadley and others 1996).

cating land shortage) or increase in fallow length (indicating the opening of old growth forest) is apparent. What these data reveal is the annual cycling of fallowed forest—in some years, young fallow; in other years, older fallow—with an average fallow length of 22.7 years. Forest cover, as in the other communities where satellite maps were available, has remained fairly stable over the years. Subsequent and more detailed analyses of the Wong Garai situation support these findings (Wadley 1999b).

The Iterative Continuum Method

The ICM is described in the Introduction to this book and in the box on page 200. The ICM provided a rich source of case material for analysis.

Principles, Criteria, and Indicators for Assessing Security of Intergenerational Access to Resources

The goal of developing clear and relevant principles, criteria, and indicators for assessing sustainable forest management has been hotly pursued in recent years (see Upton and Bass 1995 for an overview). No element in this process has been more controversial or more difficult to attain than the development of good social C&I. But the potential gains are significant if successful: the

Table 8-1. *Forest Type, Field Size, and Fallow Length of Hill Swiddens in Wong Garai, West Kalimantan, Indonesia*

Forest type		Year			
		1979	1983	1988	1993
Young fallow	No. of fields	6	4	5	0
	Area (hectares)	19.28	14.77	15.39	0
	Avg fallow length (years)	7	5.25	8.6	0
Young secondary	No. of fields	1	2	6	10
	Area (hectares)	2.87	9.45	16	32.79
	Avg fallow length (years)	30	13.5	19	20.8
Old secondary	No. of fields	0	4	2	4
	Area (hectares)	0	20.53	11.09	14.41
	Avg fallow length (years)	0	45	39.5	38.75
Total	No. of fields	7	10	13	14
	Area (hectares)	22.15	44.75	42.48	47.2
	Avg fallow length (years)	10.3	22.8	18.2	25.9

existence of simple assessment tools, the potential of influencing forest managers to attend meaningfully to resident communities (in all their variety), a greater share of the forest's products and a greater "voice" for those currently disadvantaged, improved understanding of the causal links between human behavior and beliefs and sustainable forest management, and improved management of resources through better information, to name a few.

We described our involvement in the testing of C&I in the Introduction. The C&I pertaining to security of intergenerational access to resources[10] are shown in Annex 1 (see the Introduction to this book). They represent modest progress toward answering the questions posed at the beginning of this chapter: why, and by what means, is secure access to resources important for sustainable forest management?

In the subsequent pages, we try to make these links clearer, using case studies from the DSWR area. Our hope is that these case studies can spur other researchers to add to our growing fund of case materials pertaining to this topic. We also suggest a simple scoring technique (elaborated in Salim and others 1999). We hope that, if our attempt to develop more *quantifiable* methods fails, we will have made some progress in outlining relevant *qualitative* features.

Commentary

One prerequisite for forest people achieving intergenerational access to forest resources is the maintenance of the forest resources. That is, if the forests and

Using the Iterative Continuum Method: An Example

When Colfer began the 1996 research, she was uncertain about people's feelings of security about their tenure and use rights in the area. On her first visit, in 1992, a group of villagers had explained that they had no rights to the land on Bukit Kedebu', that they were "really" residents of Selimbau, a larger town on the Kapuas River. Based on this assertion, Colfer and her husband proceeded with their plan to build the Danau Sentarum Wildlife Reserve (DSWR) Field Center there. Much to their surprise, the local governmental triad (police, military, and district government) called a meeting at which a man from yet another village asserted most aggressively that he and 14 other people "owned" Bukit Kedebu'. Although this disagreement was resolved eventually to everyone's apparent satisfaction, Colfer concluded over the next 15 months that there was a very confusing mélange of ownership and use rights in the area.

How could she better understand the importance of such use rights to sustainable forest management? One important issue identified in the literature seemed to be the presence and operation of regulations. She began looking for further evidence of regulations. She knew they existed for fisheries, but what about for forests? She found evidence that the Melayu considered rattan harvesting to be subject to regulation by local communities and that permission had to be asked before one could harvest it. She found regulations among the Iban about collecting forest foods. Did these rules seem to be regularly applied? Were there sanctions? One question led to another, always keeping in mind the link to the state of the forests in the area and the likelihood and direction of change.

Mechanisms for conflict resolution represented another feature that previous research had identified as important for sustainable forest management. Colfer began listening for stories of conflicts and trying to understand how people resolved them. In conflicts between timber companies and local communities, some were resolved by negotiation, others by violence. Neighboring communities clashed over appropriate fishing gear, boundaries, and regulations; the various methods of resolution were duly noted.

The iterative continuum method (ICM) process requires the researcher to stay alert and to be open to many kinds of evidence—because of the huge variety in human uses of forests. One observation leads to others, following the connections among human values and behavior on one hand, and sustainable forest management on the other. Experience in participant observation techniques is helpful in this process.

their resources have been depleted or destroyed, it is impossible for the current or next generation to have access to them. This concern results in the first criterion (Criterion 3.1: "Local management is effective in controlling maintenance of and access to the resource") and its indicators, the development of which has been influenced by Ostrom's work (1990, 1994; Wollenberg and Colfer 1996). Clearly, we ignore a host of complementary ecological C&I here.

The C&I discussed here overlap with other social C&I. For instance, Indicator 3.1.1 ("ownership and use rights to resources … are clear and respect preexisting claims") has implications for the question of voice in forest management or co-management. Without a firm economic base, forest actors may remain comparatively silent and powerless (see Gatuslao 1988 and Canuday 1996 for some counter examples).

The distinction between Criterion 3.1 and Criterion 3.2 ("Forest actors have a reasonable share in the economic benefits derived from forest use") has been confusing to some biological scientists. It may help to think in terms of input and output variables, with Criterion 3.1 as an input (the basis on which access to resources rests) and Criterion 3.2 as an output (the products that come from that resource base).

Finally, our work has been influenced by a perhaps unwarranted assumption that C&I can be organized into hierarchies. We have made considerable progress toward improving our definitions of principles, criteria, indicators, and verifiers (Prabhu 1995; Lammerts van Bueren and Blom 1997), but we have become more skeptical that these hierarchical connections are as immutable as they appear on paper. An indicator in one context can, in our view, function as a criterion in another—and vice versa. Income levels, for instance, may be considered (and phrased as) an indicator for a criterion on the state of people's health, or conversely, adequate incomes could be conceived as a criterion for human well-being, with human health as an indicator.

The hierarchical approach has its appeal, but we wonder also if there may be more hierarchical levels applicable for social phenomena, as Young (1992, 144) implies. Young quotes Thoreau (1957, 197), who said "the imagination, give it the least license, dives deeper and soars higher than Nature does." Is it possible that human systems, more directly affected by human imagination, may require more levels than we are allowing in Lammerts van Bueren and Blom's interesting formulation? Or, must we ultimately recognize a certain arbitrariness in our hierarchies? We cannot answer these questions with certainty here. However, we do view the use of the hierarchical formulation as a means rather than an end. We initially thought it had utility insofar as it could further our understanding of these problems; we now think it has utility insofar as it can further our ability to communicate significant human and environmental issues to others. We are moving away from the static approach implied by the C&I framework, even though we recognize its utility as a communication device and an assessment tool.

Illustrative Evidence Relating to Security of Access

In the following section, we supply illustrative cases, or evidence, that we found useful in assessing conditions at the DSWR. Although we do not discuss all the cases, we did use them all in an attempt to quantify the qualitative; each author made a comparatively independent assessment of all the cases in Table 8-2. For each criterion or indicator,[11] a qualitatively determined score between 1 and 10 was assigned, where 10 represents the most sustainable value. All three field researchers shared common "anchoring" points on the continuum (used in the ICM method) when they filled in their forms, but scoring was a personal judgement based on the evidence presented under each indicator. The final column in Table 8-2 provides the average scores for each criterion or indicator.

The numbering that we use is the same as that used in CIFOR's *Generic C&I Template*. Each indicator is numbered consecutively within each criterion. Numbered cases are presented under each indicator. Colfer, Wadley, and Harwell separately scored each case with regard to its significance for security of access over time. These scores ("Col," "Wad," and "Har," respectively) were then averaged, as presented in Table 8-2.

Criterion 3.1: Local management is effective in controlling maintenance of, and access to, the resource.

Indicator 3.1.1: Ownership and use rights to resources (inter- and intragenerational) are clear and respect preexisting claims.[12]

1. Residents of Nanga Kedebu', Bukit Rancong, and Danau Seluang have permission to reside in the lakes area from their respective "mother villages" on the Kapuas River. No village has been permanently inhabited for more than a few decades, and many residents of these three communities are seasonal. On the other hand, each community has a clearly, albeit extralegally, defined territory (as shown in Figure 8-1).
   C = 7; W = 7 (Melayu)

2. In 1989, Wong Garai saved a significant tract of old growth forest from being logged. Wong Garai territory falls within the P.T. Militer concession, but the people appealed to district and regional governments and received important help from one of their own who was a member of the regency legislature at the time. The forest was declared a protected area by the regency head (see Chapter 12). (The box on page 203 is a case study with a less positive outcome.)
   C = 7; W = 7 (Iban)

Indicator 3.1.2: Rules and norms of resource use are monitored and successfully enforced.[13]

1. Nanga Kedebu' residents expressed "righteous anger" at other nearby

Communities and Loggers in Competition

In April 1996, the Melayu of Bakakak burned down a logging base camp in territory they considered theirs, where P.T. Hutan Hebat had begun logging. The people claimed this area as a "protected area" from which they expected to harvest wood for their current and future building needs. The Regional Forestry Office in Pontianak had given Hutan Hebat special permission to cut in this area (which was, in fact, outside the company's current annual work area). In discussion between the company and the community, the community had requested company contributions that the company felt were excessive. Estimates of the requests ranged from 10 million to 30 million rupiah (Rp; US$4,300–12,800). The community had not yet agreed to Hutan Hebat's cutting when the company began its logging operation. The burning appeared to be a spontaneous, villagewide reaction that reflected people's feelings that Hutan Hebat was infringing on their legitimate rights.

An investigation involved the police, the military, and the regional government as well as the company and the community; Hutan Hebat stopped cutting temporarily. However, our last understanding was that the Regional Forestry Office had stuck by its original permission and that the other governmental agencies were supporting Hutan Hebat. Hutan Hebat agreed to improve the boardwalks and to build a religious school in the community. A local Forestry Department official said this action was not a requirement but a "token of good will." The people, it appeared, had lost their rights to the area they had been managing for their own future use.

[This case was assigned a 2 by Colfer and Wadley, for the Melayu management.]

---

communities whose members came and collected rattan or caught fish in their territory, contrary to Nanga Kedebu' regulations.

C = 7; W = 7(Melayu)

2. A group of Bemban children and young women went out to a previous longhouse site to collect ferns for supper. They explained that only people from the community could collect ferns in this area. A young girl took the jackfruit Colfer was carrying, saying that Colfer might be fined for taking the fruit, whereas she was allowed to do so (see Sandin 1980).

C = 8; W = 8 (Iban)

Indicator 3.1.3: Means of conflict resolution function without violence.

1. In the late 1980s, Wong Garai had a land dispute with a neighboring longhouse. In years past, Wong Garai had allowed members of the other longhouse to farm land within its territory, which the other longhouse

Table 8-2. *Authors' Scoring of "Security of Intergenerational Access to Resources" Cases in DSWR and Environs, West Kalimantan, Indonesia (June 1996 Conditions)*

Case/evidence	Iban				Melayu				DSWR
	Col	Wad	Har	Avg	Col	Wad	Har	Avg	Avg
Criterion 3.1: Local management is effective in controlling maintenance of, and access to, the resource.									
Indicator 3.1.1: Ownership and use rights to resources (inter- and intragenerational) are clear and respect preexisting claims.[a]	5.6	5.3	8.0	6.3	4.0	4.0	4.0	4.0	5.2
Indicator 3.1.2: Rules and norms of resource use are monitored and successfully enforced.[a]	6.5	6.5	7.5	6.8	5.5	5.5	4.5	5.2	6.0
Indicator 3.1.3: Means of conflict resolution function without violence.[b]	6.0	6.0	8.0	6.7	5.0	5.0	4.0	4.7	5.7
Indicator 3.1.4: Access to forest resources is perceived locally to be fair.[c]	4.0	4.0	5.0	4.3	2.3	2.3	3.0	2.5	3.4
Indicator 3.1.5: Local people feel secure about access to resources.[c]	5.3	5.3	4.0	4.9	6.0	6.0	4.0	5.3	5.1
Criterion 3.2: Forest actors have a reasonable share in the economic benefits derived from forest use.									
Indicator 3.2.1: Mechanisms for sharing benefits are seen as fair by local communities.[d]	4.0	4.0	2.0	3.3	2.7	2.7	2.0	2.5	2.9
Indicator 3.2.2: Opportunities exist for local and forest-dependent people to receive employment and training from forest companies.[d]	3.0	3.0	2.0	2.7	2.0	2.0	2.0	2.0	2.3
Indicator 3.2.3: Wages and other benefits conform to national and/or International Labor Organization (ILO) standards.[e]	2.0	2.0	3.0	2.3			3.0	3.0	2.5
Indicator 3.2.4: Damages are compensated in a fair manner.[f]	3.5	3.5	3.0	3.3			3.0	3.0	3.3

Criterion 3.3: People link their and their children's future with management of forest resources.

Indicator 3.3.1: People invest in their surroundings (that is, time, effort, and money).[d]	8.0	8.0	8.0	8.0	8.0	8.7	4.0	6.9	7.5
Indicator 3.3.3: People recognize the need to balance number of people with natural resource use.[b]	6.3	6.3	7.0	6.5	5.0	5.0	2.5	4.2	5.4
Indicator 3.3.4: Children are educated (formally and informally) about natural resource management.[d]	5.7	6.0	4.0	5.2		4.0		4.0	4.9
Indicator 3.3.5: Destruction of natural resources by local communities is rare.[d]	4.0	4.0	5.0	4.3	4.0	4.0	5.0	4.3	4.3
Indicator 3.3.6: People maintain spiritual or emotional links to the land.[c]	8.7	8.7	9.0	8.8	5.0	5.0	3.0	4.3	6.6
Grand mean				5.3				4.0	4.7

Notes: Col = Colfer; Wad = Wadley; Har = Harwell. Indicator 3.3.2 was not evaluated.

[a] Based on average scores (Colfer and Wadley) for six cases.
[b] Based on average scores (Colfer and Wadley) for five cases.
[c] Based on average scores (Colfer and Wadley) for four cases.
[d] Based on average scores (Colfer and Wadley) for three cases.
[e] Based on average scores (Colfer and Wadley) for one case.
[f] Based on average scores (Colfer and Wadley) for two cases.

then claimed as their own. They brought the case before the *temenggong* (traditional law leader) for a hearing, and the *temenggong* decided that the two disputants should divide the land. Wong Garai refused to accept the decision (locally agreed to be their right), arguing that the other long-house has no *tembawai* (old longhouse sites) on Wong Garai territory that would mark their claim to the land.

C = 8; W = 8 (Iban)

2. Nanga Kedebu' residents frequently disagreed with P.T. Hutan Hebat, a timber company that regularly towed log rafts through Nanga Kedebu' territory. One community member served as a tugboat pilot for the company and as an informal mediator in resolving these disputes. Despite grumbling with regard to levels of compensation for damage to local fishing gear, there was acceptance of this system both by the community and the company.

C = 8; W = 8 (Melayu)

Indicator 3.1.4: Access to forest resources is perceived locally to be fair.

1. In Nanga Kedebu', logs had recently been quietly removed from passing P.T. Hutan Hebat log rafts. The logs were to be sawn into lumber and used to build a mosque. This action was agreed to by community members and justified with reference to the profits being gained by timber companies, vis-à-vis local benefits from local resources.

C = 2; W = 2 (Melayu)

2. Forest fires occurred extensively in 1992 (the last really dry year prior to the 1996 fieldwork reported here) in Danau Seluang's territory. They significantly reduced the availability of rattan and timber and destroyed about 500 wooden *tikung* (artificial bees' nests). Burning was variously described as purposeful and related to outsiders' envy or anger because they were denied permission to harvest, or entirely due to carelessness.

C = 3; W = 3 (Melayu)

Indicator 3.1.5: Local people feel secure about access to resources.

1. Wong Garai shares access to some forest and riverine land with another longhouse. There is some concern that this increase in use is leading to overexploitation, particularly of fish. People are also concerned about their future ability to collect fish in the lakes area during the dry season—something they have been doing for at least 150 years and to which they make traditional use claims—given the increasing presence of Melayu in traditional Iban use areas and the possibility that the government will begin to enforce its own very different boundaries in the future.

C = 5; W = 5 (Iban)

2. Throughout the DSWR area, concern is expressed that others (for example, timber companies, other ethnic groups, and transmigrants) are encroaching on their areas of traditional use, threatening their ability to

use those resources in the future. For the Iban, the concern focuses on forest resources; for the Melayu, fisheries.

C = 4; W = 4 (Iban; Melayu)

Criterion 3.2: Forest actors have a reasonable share in the economic benefits derived from forest use.

Indicator 3.2.1: Mechanisms for sharing benefits are seen as fair by local communities.

1. In Nanga Kedebu' and Bukit Rancong, people felt that funds made available to the Conservation Project from ecotourists and payment of salaries and other in-kind help from the project were unfairly distributed.
   C = 2; W = 2 (Melayu)

2. Payment of royalties to local communities, in recognition of their prior rights, has been suggested as a mechanism for sharing benefits more fairly. No royalties are paid to DSWR communities or to those in the surrounding area. Various taxes are paid by companies to the Kapuas-based forestry agent, but they go to Pontianak (and to Jakarta; see Ascher 1993).
   C = 2; W = 2 (Melayu)

Indicator 3.2.2: Opportunities exist for local and forest-dependent people to receive employment and training from forest companies.

1. Very few residents within and around the DSWR work for the timber concessions. Kelayang is in the P.T. Panggau Libau concession, partially owned by Iban from the Lubok Rian area, some of whom are related to Kelayang residents. Although Kelayang economic involvement with this company is greater than that found between other companies (such as P.T. Militer or P.T. Hutan Hebat) and local communities, conflicts still arise (see Chapter 12). Conflicts have resulted from employment opportunities that were perceived to be inadequate, promised but unpaid rent on land, requests for rattan that was then not bought, and unfair recompense when a community member was killed by a company speedboat.
   C = 3; W = 3 (Iban)

2. In Nanga Kedebu', only one person is considered to have had a long-term relationship with the timber company. Young men occasionally work for a while with timber companies, but some people perceive that when fishing is good, the young men will leave the company. This perception may mean that incomes from fishing (and related economic endeavors) are better than incomes from the company.
   C = 2; W = 2 (Melayu)

Indicator 3.2.3: Wages and other benefits conform to national and/or International Labor Organization (ILO) standards.

1. The workers Colfer spoke with—a mix of locals and newcomers—considered themselves to be adequately paid, with reasonable benefits, working

conditions, and safety standards.[14] On the other hand, Wadley found that Iban who have worked for Indonesian logging companies generally complain about the low wages locally (compared with what they could earn for comparable work in Malaysia), dangerous conditions, and poor equipment. Quite a few Wong Garai residents who had worked for P.T. Panggau Libau said they had never been paid and would never work there again.

C = 2; W = 2 (Iban)

Indicator 3.2.4: Damages are compensated in a fair manner.

1. In 1992, a subcontractor with P.T. Militer/P.T. Hutan Hebat paid the community of Bemban one portable, 500-W generator for the right to harvest an unknown number of hectares in Bemban's traditional area. (The situation outlined in the box on page 209 describes another example.) Local people are becoming more savvy, though they often lack power and voice in demanding justice.

C = 2; W = 2 (Iban)

Criterion 3.3: People link their and their children's futures with management of forest resources.

Indicator 3.3.1: People invest in their surroundings (that is, time, effort, money).

1. For the Melayu, enforcement of local regulations to protect resources involves protecting special areas as fish nurseries, prohibiting small mesh sizes and the harvesting of fish under a certain size, restricting access to rattan and valuable wood, and outlawing burning. For the Iban, it involves maintaining special forest preserves (*pulau*) and old longhouse sites (*tembawai*) and prohibiting farming the peaks of mountains to allow for the forest regeneration of swiddens (see Wadley and others 1996).

C = 10; W = 10 (Iban); C = 8 (Melayu)

2. Increasing educational levels (with significant sacrifice and investment by both parents and children) have recognized and profound negative consequences, such as loss of traditional ecological and ritual knowledge, devaluation of traditional work and knowledge, and increased consumerism.

C = 6; W = 6 (Iban; Melayu)

Indicator 3.3.2: Out-migration levels are low.

1. People—primarily relatives who live along the Kapuas River—migrate into the reserve seasonally. Close economic and kinship ties between DSWR communities and their "mother villages" along the Kapuas would make control of this seasonal influx difficult. Many people who started as seasonal fishers in the reserve have settled and built permanent homes there. Many also express a commitment to staying and making the community better for their children.

C = 5; W = 5 (Melayu)

Compensation Paid to Communities

In early 1996, an irrigation project was started on Wong Garai land that would feed downriver into the fields of other communities and a planned transmigration project.[a] Wong Garai had successfully lobbied to get the main irrigation dam built within its territory, but when the site was surveyed and work started, project workers did not notify the longhouse. Banana trees and cassava plants in one garden at the site were cut down, and some graves in an old forest cemetery were disturbed with digging and tree felling.

Several Wong Garai women were first to see the work. They directly challenged the workers, forcing them to stop. After holding a traditional dispute hearing at the longhouse, the construction company was fined more than Rp 500,000 (US$212),[b] which was divided with two other longhouses that had ancestors buried in the disturbed cemetery. The company also was required to pay for local rock and sand used in the dam and canal construction.

[This case was assigned a 5 by Colfer and Wadley, for Iban management.]

[a] *Transmigration* is an Indonesian government–sponsored program to move people from densely populated areas (such as Java) to less densely populated areas (such as Kalimantan).

[b] Some fines and costs to the company included Rp 300,000 for disturbing the orchard, Rp 200,000 for not reporting their activities, court costs (for example, Rp 3,000 per kilogram for 63 kilograms of pork, Rp 151,900 for bought food), Rp 200 per cubic meter for stones and sand, Rp 25,000 per tree for each full grown rubber tree destroyed (with prices varying depending on the size), Rp 90,000 for each fully grown tengkawang tree, and so forth. (*Note:* The 1996 exchange rate was US$1 = Rp 2,350.)

2. Iban men are regular circular migrants to Malaysia, where they work for higher wages. They normally return home, bringing welcome booty with them, at harvest time.

   C = 5; W = 5 (Iban)

Indicator 3.3.3: People recognize the need to balance numbers of people with natural resource use.

1. Birth control has been widely accepted, often linked to resource use issues. Iban women, however—recognizing that families are better able to provide for fewer children, that they are freed from the real risk of death in pregnancy and childbirth, and that they can be more economically productive—worry that low or stable fertility levels among indigenous people

like themselves may provide an excuse to move transmigrants into the area who may overwhelm them numerically.

C = 7; W = 7 (Iban)

2. Migration into the reserve appears to be considerable, with no effort or means to control it. Indeed, the ethic of hospitality makes such control difficult without outside support.

C = 3; W = 3 (Melayu)

Indicator 3.3.4: Children are educated (formally and informally) about natural resource management.

1. We met several young people who had been selected by their parents to pursue various disciplines (within one family) and then return home to share their knowledge with the family and community.

C = 5; W = 5 (Iban)

2. The Iban have a still functioning system of land tenure and tree ownership rules and practice (see Wadley 1996) and maintain many rituals connected to farming. But they fear these "old ways"—the ritual chants, the rich ceremonial language, and farming and forest knowledge—are being lost to the youth. Competition from national education and television is constant. In June 1996, for example, a set of important longhouse rituals (making of offerings, chanting of invocations to ancestor gods) was being performed at 1 A.M. At the same time, the young people had set up a stereo system to play Indonesian pop music at high volume, to which they danced at the other end of the longhouse. The resources of cultural and ecological knowledge (integral to sustainable management), which their immediate and distant ancestors had acquired, were being lost (see also Chapter 5).

C = 4; W = 5 (Iban)

Indicator 3.3.5: Destruction of natural resources by local communities is rare.[15]

1. Recurrent poisoning of fish with commercial pesticides—largely by a minority of Iban merchants, but also by some Melayu—was reported.

C = 4; W = 4 (Iban; Melayu)

2. Current supplies of timber species available to local people are significantly reduced (*tembesu'*, *kawi*, *kelansau*, *medang*, *menyawai*). There is a widespread perception of overharvesting of swamp forest (*rawa*) by local people. (Those used by the Melayu are mostly swamp species.)

C = 5; W = 5 (Melayu)

Indicator 3.3.6: People maintain spiritual or emotional links to the land.

1. During Colfer's four-day stay in Kelayang, three resource-related religious ceremonies were observed, all of which included the active involvement of the young (one to "feed" a crocodile spirit in the river whose hunger had been revealed in a dream to constitute a threat to a community mem-

ber, one to "feed" the soil before beginning to clear a rice field, and one to "feed" the soil in preparation for planting).
C = 10; W = 10 (Iban)

2. Iban refer to the forest as *seput menoa,* "the breath of the land," and recognize the hydrological consequences of too much forest cutting—for example, drying up of water sources[16] (see Wadley and others 1996).
C = 8; W = 8 (Iban)

The fact that the forests in and around the DSWR are in relatively healthy shape suggests that these scores may be high on a global scale. The low average score (2.5) for Criterion 3.2 suggests a possible flash point; indeed, feelings of unfairness about local people's shares in forest benefits that they felt should be their own were both a recurring complaint and a rationale for examples of violent confrontation.[17] Our comparatively high assessments of the strength of people's feelings of security about access to resources (7) and their clear conceptual link between their own and their children's well-being and the forests (5.4) seem likely to contribute to sustainability by confirming their "stake" in the forest and by providing motivation for protecting it against potentially destructive new endeavors in the area.

Conclusions

In this chapter, we have described illustrative results obtained from using three methods designed to assess security of intergenerational access to resources quickly, inexpensively, and reliably. We used the principles, criteria, and indicators identified in CIFOR's C&I process as a framework, and we presented and scored cases that provide evidence of causal links between such access and sustainable forest management. We see this case material as contributing to an illustrative "library" of cases from different contexts, building ultimately to a fuller understanding of the causal links between people's access and sustainable forest management.

Our attempt to understand the causal links between access to resources and sustainable forest management is a long-term goal, of which this study formed a small part. Evidence from one forest-rich site cannot prove that maintenance of fair intergenerational access to resources and economic benefits is always important for sustainable forest management. However, the evidence (in the form of cases) we accumulated for the DSWR has given us a better understanding of the kinds of links between the social C&I (or the conditions they reflect) and sustainable forest management more generally. The kinds of links we have identified support the conclusion that *best practices* in forest management—whether by local people or by timber concessionaires—will require that

- resources be maintained if people now and in the future are to continue to have access to them (Criterion 3.1),
- local people must share in the economic benefits from forest use (Criterion 3.2), and
- people (in this case, also managers)[18] link their own and their children's futures with good management of the resource (Criterion 3.3).

Endnotes

1. See Colfer and others 1997b, 1999a, and 1999b for more complete descriptions of these methods.

2. In this chapter, we shorten *security of intergenerational access to resources* to *access to resources* or even *access*.

3. We are cognizant of the pertinence of Becker's (1997, 32) criticism of scoring systems. She says, "The problem with scoring systems is that they pretend objectivity and uniformity, whereas the choice of components and their assigned weights is highly subjective, and the aggregation of different spatial, temporal and sectoral dimensions is often not meaningful."

4. The issue is complicated by the many interpretations of *fairness* that reflect real differences in people's perceptions and understandings (for fuller discussions, see Farmer and Tiefenthaler 1995; Prakash and Thompson 1994).

5. The meaning of *intergenerational* has proved more complicated in Cameroon, where competition is rather dramatic, even antagonistic, between generations—something rare or understated in the Indonesian context.

6. One could argue that this issue can be left to the ecologists. However, our own perspective is that local people are likely to have important responsibilities in maintaining that resource. Where people have developed mechanisms for maintaining a resource, its condition is likely to be better.

7. As with many criteria and indicators, there are potential conflicts here. If the population has grown drastically, for instance, the same resource base may no longer support previous levels of use. This change in turn will affect the first quality of *access,* resource availability. It is also a red flag relating to sustainability. This element of access also ignores the important issue of changing aspirations among local populations, who may no longer want a particular resource.

8. This type of method is described in Bruce 1989, Carter 1996, Momberg and others 1996, and Panday and others 1997, among others. The Asia Forest Network also has put out a series of cases, many of which have excellent examples of the uses of this kind of map (for example, Chatterji and others 1996; Poffenberger and McGean 1993a, 1993b; Poffenberger and others 1995, 1996). See Lightfoot and others 1991 for other approaches with similar goals.

9. Hanne Christiansen (not dated) has documented an Iban lexicon of some 2,000 plant species and reports that in one longhouse, at least 127 families of plants are known and regularly used (see also Bernstein and others not dated).

10. We felt comfortable following Lammerts van Bueren and Blom's (1997) requirements for *principles* and *criteria,* but the indicators we have developed join their *indicators* and *norms* and are, in the case of this principle, almost exclusively qualitative.

11. Colfer and Wadley were in communication by e-mail, but Harwell was in the field when these estimates were made. Our communication problems resulted in her assessing only the criteria.

12. Our emphasis here is on local ownership and use rights, but there is considerable difference of opinion about actual rights to resources. Local people feel that the resources belong to them, and the government considers the resources to belong to the nation.

13. Again, our emphasis is on local rules and regulations. But a host of rules and regulations from different parts of the Ministry of Forestry are not normally monitored or enforced (for example, the government forester who knew neither the regulations on timber harvesting nor who was supposed to enforce them, and Conservation Project personnel who manage the wildlife reserve but regularly ignore purple herons and storm's storks tied to Melayu rafts and macaques and small birds kept as pets by the Iban).

14. As with the perception of security of tenure, the perceptions of local workers may be different from those of outside assessors. Local working conditions would not, for instance, comply with those proposed by previous Center for International Forestry Research teams or with International Labor Organization standards.

15. Other criteria and indicators deal with destruction by other stakeholders (for example, the harvesting of timber by concessionaires without regard to regulations, the transmigration of large numbers of families into already occupied forest areas, and the conversion of natural forest areas to industrial timber estates or oil palm or rubber plantations).

16. This hydrological knowledge has a spiritual component; the Iban contend that if they do not take care of the land both ecologically and ritually, it and they will be threatened with supernatural "heat" (*angat*) that manifests itself in people's poor health and in social disruption.

17. In early 1997, in another area of West Kalimantan, a confrontation occurred that was so violent that many people were killed, and the military intervened. Although its causes are widely debated (often attributed to ethnic or religious conflict), we feel with some confidence that inequitable access to resources and benefits played a significant role in this sad occurrence. Similar problems recurred in 1999 in the same and other provinces of Indonesia (partially also related to other political issues). See Harwell 2000a for an indepth analysis.

18. The role local people play in management in and around the Danau Sentarum Wildlife Reserve is examined in Chapter 12.

CHAPTER NINE

Sustainability and Security of Intergenerational Access to Resources

Participatory Mapping Studies in Gabon

Norbert Gami and Robert Nasi

In this short chapter, we provide some background information—both ecological and human—about Gabon. The site discussed in this chapter is the most forest-rich[1] site of those reported in this book. We then report the results from a partial social science methods test related to the more systematic and comprehensive studies conducted in Cameroon, Indonesia, and Brazil and reported elsewhere in this book.

This research was part of an interdisciplinary assessment of criteria and indicators (C&I) undertaken in 1998 (Nasi and others 1998). Fieldwork took place during a very short period (two weeks) under difficult conditions, and it reflects the conditions under which social assessments are more typically conducted vis-à-vis the longer-term methods tests reported elsewhere in this book.

The results presented in this chapter are particularly interesting given our initial reductionist hypothesis (that is, that intergenerational access to resources would be more secure and clear in forest-rich contexts). Our findings lead us to propose explicitly collaborative management of these forests.

Gabon and the Congo Basin

A visual inspection of the Tropical Ecosystem Environment Observation by Satellites (TREES)[2] vegetation map of Central Africa reveals a very different context from the degraded West African landscape (Mayaux and others 1998). Except for some fragmentation along the main access routes (roads, rivers) and around the main population centers, the dense forests remain relatively well preserved. The extensive forest exploitation in the subregion has not significantly affected the forest cover (logging is very selective and primarily interests foreign markets; destruction and degradation resulting from logging represents generally less than 10% of forest cover).

Data confirm this visual diagnosis (Table 9-1). However, some observable differences between the Food and Agriculture Organization (FAO) and TREES data are mainly due to the different definitions adopted to characterize the forests. Note that the FAO data for the Central African Republic include the mixed woody-herbaceous open formations (dry forests, savanna), even though they are not included for the other countries (Cameroon, Democratic Republic of the Congo) (Table 9-2).

In summary, the Central African region is made up of low-population countries; with a few local exceptions, it is still largely covered by natural forests, and the countries are all wood exporters. In this context, the priority actions to preserve forests and local people's livelihoods should be

- the extensive management of large production forest areas through forest management plans based on the natural regeneration capacity of ecosystems and
- the consolidation of the existing network of protected areas.

The development and implementation of these forest management plans should take into consideration existing and traditional knowledge.

Table 9-1. *Forest Area in Central Africa, 1997*

Countries	Population density[a]	Forest area (× 1,000 hectares)		
		FAO[a]	TREES[b]	FAO[c]
Cameroon	28.4	46,540	17,378	19,582
Central African Republic	5.3	62,298	6,037	29,924
Equatorial Guinea	14.3	2,805	1,811	1,778
Gabon	5.1	25,767	20,677	17,838
Congo	7.6	34,150	23,916	19,500
Democratic Republic of the Congo	19.4	226,705	114,147	109,203
Central Africa		398,265	183,966	197,825

Note: Population density is measured in inhabitants per square kilometer.

[a] According to FAO 1997.

[b] Dense evergreen or semideciduous forests, including some fragmented forests (corrected data).

[c] Natural forests (minimal tree cover of 10%).

215

Table 9-2. *Vegetation Cover Distribution in Central Africa*

	Forest area (× 1,000 hectares)					
Class	Cameroon	Central African Republic	Equatorial Guinea	Gabon	Congo	Democratic Republic of the Congo
Dense humid forest	18,074	4,424	1,928	21,328	23,900	110,608
Secondary forest/ fallow	4,911	321	531	1,722	1,574	14,346
Mosaic forest/ savanna	2,177	2,519	2	18	272	25,228
Dense dry forest/ Miombo	18	615	0	0	0	47,992
Woody savanna	17,010	50,773	0	0	0	9,927
Shrub savanna	3,425	3,708	0	0	0	0
Herbaceous formations	261	410	32	2,651	8,275	22,690
Herbaceous swamp	86	0	0	152	322	892
Hydrographic elements	68	0	0	94	330	3,132
Mangroves	242	0	32	155	0	0

Note: All values for dense dry forest, woody savanna, and shrub savanna are mixed ligno-herbaceous vegetal formations.

Source: Data from Mayaux and others 1998.

Traditional Knowledge and Forest Resource Management

In the preamble to their report on the situation of indigenous populations living in dense forests, Bahuchet and de Maret (1993) state, "There is no virgin forest. The luxuriance of the equatorial vegetation, which amazes western travelers so much, must not make us forget that they result from millennia of human history." Yet reality is that the forest populations are either not consulted at all or inadequately consulted when their lands are allocated to a logging company or designated as protected areas (Joiris 1996; Ondo 1997; Gami 1998).

But what evidence can we present from the region's ethnographic literature on traditional natural resource management? There seems to be a burgeoning interest on the part of the various conservation projects to take into consideration local populations' knowledge to promote sustainable overall forest management. Such interests have spawned a whole new vocabulary: participatory management, co-management, patrimonial management, and so forth. Colonial policy, which involved moving entire villages and regrouping land ownership accordingly, contributed to weakening local customs. Mbot (1997) sees that policy as the first stage of breaking down customary forest management, hunting, and fishing rules.

However, evidence shows that some such natural resource management remains, especially relating to the magico-religious aspects of people's lives. Pagezy (1996) explains how the *bilima,* supernatural beings, intervene in the regulation of natural resources extraction—particularly of fish and game—among the Mongo people in the south (Ntomba, Ekonda) of the Democratic Republic of the Congo. Similarly, the Lumbu and the Vili, who live in the south of Congo-Brazzaville, believe that fishing in lakes and lagoons is regulated by spirits called *muissi.* Fishing in lakes and lagoons must respect several rules so as not to make the *muissi* angry. Access to sea resources falls under the authority of the chief of the clan or of the lineage, the *mfumu kanda* (Gami 1998). Such traditional management modes, or patterns for regulating access to resources, are found among all the populations that live in the forest of the Congo basin (for example, Bahuchet and de Maret 1993; Joiris 1996).

Various authors are increasingly recognizing the management capabilities of local people. Joiris (1996) suggests, "Given that ecosystem biodiversity in protected areas did not undergo significant degradation due to local economies, such populations should be regarded as possessors of a certain know-how about the rational use of forest ecosystems." Princet (1994) reinforces this view: "For quite a long time and generally speaking, the knowledge of local populations had been unknown to or ignored by almost all scientists who instead taught and disseminated their own truths as being universal."

In *Rites et Croyance des Peuples du Gabon,* Walker and Sillans (1962) emphasize the importance of certain rites in the sustainable management of natural resources. Unlike the Bwiti, the Mwiri and Mangongo (widely occurring men's secret societies in Ngounié and Bas-Ogooué, Gabon) do not seem to be intended to honor ancestors. According to the authors, it is a sort of league for protecting the environment and preserving public places, helped by a "secret police" that seeks and punishes guilty people, no matter who they are. Excessive cutting, hunting, fishing, and harvesting of forest products causes the scarcity of game, fish, fruits, and so forth. Thanks to these secret police, the Mwiri can intervene quickly to stop the abuse by creating "local natural reserves" where it will be forbidden to hunt, fish, cut trees, or harvest fruits for many years. Blaney and others (1997) indicate the persistence of this practice in southeastern Gabon, where blocks of sacred forests still belong to some clans or lineages. People also avoid certain sites because of magico-religious beliefs.

The Mboko people of northern Congo-Brazzaville traditionally have strict rules regulating access to sea resources. In fact, fishing ponds that belong to various clans or lineages are only used during the dry season, with the blessing of the chief of the lineage (Gami 1995). The role of the chief of the lineage in controlling access to resources appeals to the notion of *village territory* or of *village borders* as defined by Karsenty and others (1997). Traditional areas are demarcated, reserved for various village activities. These areas are under the authority of the chief of the clan, who controls users.

However, in Central African countries, forest land is state land by law, and typically forest regulations neither take into consideration local populations' knowledge about management nor their customary laws. Formal regulations also have not, in practice, provided means for involving stakeholders (particularly local people) in the sustainable management of resources. Forest concessions are given out in areas without taking into consideration the views of local people. At best they are "granted" some sort of rights to use certain forest products. However, according to local customary law, the land belongs to the community and cannot be alienated. It is the patrimony of the clan or of the lineage. Therefore, problems and conflicts arise among the population, the state, and loggers. They arise from the opposition of "legality" (state and logging companies) and "legitimacy" (local people considering that the forest is theirs).

Some signs indicate that things are changing. Drouineau and others (1999), for example, report a hopeful clause in a new law on logging in Gabon, which requires logging companies to obtain a logging license from the state by submitting a forest management plan that recognizes villagers' customary forest rights. Some of the newly designed forestry laws allow for community forests or require that local people be involved in setting the boundaries for production forests. But this process is slow and painful because there is reluctance to change (from the administration and logging companies' standpoints); in addition, serious power and economic differentials as well as a lack of access to information inhibit the local peoples' ability to act on these new laws.

Methodology

Evidence of conflicting interests and perceptions on resource ownership, access, and availability in the forested areas of Gabon is ample. One important first step in resolving the conflicting perceptions involves the use of some methodological tool to enable populations to define the various territories used by their villages for hunting, fishing, gathering, cultivating, and cultural or spiritual purposes.

Participatory mapping is such a technique, and we had the opportunity to try it during a test to develop C&I for sustainable forest management in Gabon. This test was carried out by a multidisciplinary team and addressed several aspects of C&I (policy, forest management, ecology, and social issues; see also a description of the testing process in the Introduction to this book). In this chapter, we consider only the results obtained from participatory mapping and their usefulness.

Study Site

Vegetation Characteristics

The vegetation of the study site can grossly be divided into three main types.

- A forest with *okoumé* (*Aucoumea klaineana*), *soro,* and *beli* is characterized by the appearance of *beli* (*Paraberlinia bifoliolata*), new *andoung* species (*Monopetalanthus* spp.), and the abundance of *soro* (*Scyphocephalium ochocoa*). It is the vegetal formation that covers the largest part of the study site area, spreading from the western boundary to the southeastern Olounga-akieni savanna contact and from Okondja toward the northeast. It presents some variants; the most spectacular ones are the forests of the Alele zone (in which the understory is very rich in oil palms) and those from the zone west of Okondja, locally very rich in *azobe* (*Lophira alata*).
- A forest without *okoumé* differs greatly from the previous forest by the absence of *okoumé*, a reduction in the amount of *soro* and *alep* (*Desbordesia glaucescens*), and the possible appearance of species usually associated with the transition forests of southern Cameroon or northern Congo: *wengue* (*Milletia laurentii*), *Pteleopsis hylodendron,* and *Celtis mildbraedii*. This forest is mainly found in the region east of Okondja.
- In the Akieni-Olounga region, an important zone of approximately 100,000 hectares of forest/savanna mosaic is very rich in young *okoumé,* the result of serious invasion of the forests into the savanna when the area is protected from fire.

Population

The population of 16,170 inhabitants—distributed among 43 villages and 4 towns—lives within or in the immediate surroundings of the study site. The population density in the region is approximately 1.7 inhabitants per square kilometer, and the mean population of the villages is more than 100 inhabitants.

Three ethnolinguistic groups are present: Kota, Saké, and Obamba. These groups form segmented societies of the clan type, patrilineal for the Kota and Saké and matrilineal for the Obamba. The Kota and the Saké, who live within our study area, seem to have originated from the central part of the Sangha River region in Cameroon, the Central African Republic, and Congo-Brazzaville. They immigrated to the area in two stages: before the seventeenth century and during the eighteenth and nineteenth centuries (Kwenzi Mikala 1997).

Economic activities are essentially subsistence-based (traditional agriculture by clearing, hunting, and gathering). The previously very developed craft practices (wood carving, raffia and wickerwork, pottery, and metallurgy) have totally disappeared.

Traditional land tenure is quite elaborate and presents easily identified levels.

- At the village level,
  —the village land is a variable surface area, with known boundaries, generally delimited by natural references (rivers, hills, and forests);
  —the origin of the rights of the villagers on these lands comes generally from a peaceful occupation by one of them, considered as the founder; and
  —the name of the village land is designated by the name of the clan founder.

- At the lineage level,
  —the lineage land is a fraction of the village land, with known boundaries and with uses restricted to village members, and
  —lineage right of use finds its origin in the agreement previously reached between the village founder and a chief of any new lineage joining the group.

- At the individual level, the right to use land has two origins:
  —family connections and
  —the physical act of land development (first clearing gives the exclusive right of use of land to the initial clearer and then to his descendants).

The Logging Company: Compagnie Equatoriale des Bois

The Compagnie Equatoriale des Bois (CEB; a branch of the Thanry Group, which owns logging and processing businesses in several countries of tropical Africa) has been logging since 1987 in the Lastourville area after having worked for several years in the Nyanga region. It currently owns eight licenses (PTE and PI)[3] for an area of 505,000 hectares and logs on behalf of another operator, paying fees on the number of cubic meters cut, representing approximately 68,000 more hectares.

The central unit, Bambidie, 30 kilometers east of Lastourville, oversees all the operations (logging, planning, workshop, rolling, and road maintenance). It is composed of offices for the administration and several technical services (for instance, planning), a well-equipped garage, a health center, and lodgings for workers and salaried staff. Approximately 250 people are employed on the site (including in the sawmill), which represents a community of more than 1,000 people including families.

A sawmill with a capacity of 1,200 cubic meters per month was installed at the central unit in 1996. It works mostly with wood recovered from export logs, which would have been abandoned in the forest without the processing unit. It has been completed with two driers (with four 150-cubic-meter rooms) operated by computer, and working with the sawmill waste, which allows it to produce approximately 1,000 cubic meters per month of dried sawn timber. Forty-five people are employed in this unit.

Three logging areas are in operation (Olounga, Okondja, and Ndambi). Each location represents a *base camp,* a miniature replica of the central unit of Bambidie. The staff of each area, approximately 40–50 people, is directed by two salaried staff members (logging chief and garage chief). The total staff for all three areas plus the central unit at Bambidie is approximately 300–350 people.

Several points are notable:

- The quality of the executive staff, Gabonese (6) and expatriate (11), is excellent. The staff functions under the direction of a highly experienced logging director who is really the cornerstone of all the work.
- The company boasts a real internal competence in planning activities, with an expatriate forestry engineer and two Gabonese engineers on staff. Currently, the CEB has carried out the planning inventory on all its licenses, and all the data are integrated within a geographic information system (GIS) database; the planning map is currently being produced.
- The company has a clear vision—unusual in such enterprises—and is oriented toward a better long-term management of the forest heritage through integrated planning of licenses and a coherent industrialization policy. For the Bambidie sawmill, two units are envisaged for the near future: an operational peeling unit for export-quality logs in Libreville and a peeling unit for logs intended for domestic consumption in Bambidie. This management plan would allow the company to challenge the Société Nationale des Bois du Gabon (SNBG) monopoly.

Production by CEB since 1990 is summarized in Table 9-3 and Figure 9-1.

Sampling Sites

We surveyed inhabitants in five villages, all located in the vicinity of or within the CEB logging concession area: Ndambi, Baposso, Ndékabalandji, Mékouka, and Bakoussou 2. Ndambi village, discussed in this paper, is located inside the CEB logging area, but its customary territory (used for hunting, harvesting, and fishing) extends within the concessions of two other logging companies: Société Forestière du Gabon (SOFORGA) and Société des Bois de Lastourville (SBL). Figure 9-2 is a map of the village.

Participatory Mapping

The mapping exercise in Ndambi village was conducted with five adult men and three adult women. In rural forested areas such as those in Gabon and Congo, it is very difficult for farmers to read a printed map. The farmers felt more at ease drawing the map of their village on the ground with a piece of chalk or charcoal rather than using an external map as a starting point. We found it useful to have them draw the territory of the village and mention

221

Table 9-3. *Compagnie Equatoriale des Bois Production and Area Logged, 1990–1997*

	1990	1991	1992	1993	1994	1995	1996	1997
Logged area (hectares)								
Licenses	28,650	17,500	22,000	28,550	35,070	31,050		
Rent licenses				4,800	5,600	16,385		
Total	28,650	17,500	22,000	33,350	40,670	47,435		
Volume produced (cubic meters)								
Okoumé	118,504	70,961	91,971	137,279	121,168	95,417	179,076	185,663
Other woods	21,046	29,289	34,034	18,284	28,238	29,716	21,839	23,881
Total	139,550	100,250	126,005	155,563	149,406	125,133	200,915	209,544

Source: Compagnie Equatoriale des Bois data.

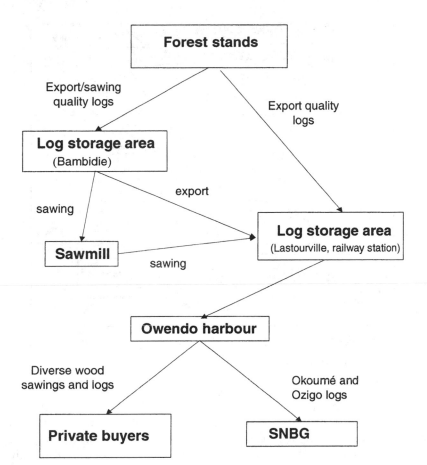

Figure 9-1. *CEB Timber Production Cycle: From Forest to Consumer*

the names of rivers or other characteristic places. We later used an external map to establish the borders of the village based on the farmers' knowledge. The groups chose one person among them to draw the map of their village. That person was always assisted by the others, who watched carefully.

The groups indicated on the map present and past areas meant for hunting and harvesting. They identified the reasons why they were abandoned, conflicts with foreign hunters and loggers, the decline in the availability of game in the area surrounding the village, and the methods advocated for solving conflicts. The other method consisted of assessing, from a drawing, the distance between the village and the place where some animal species were hunted.

The method was effective for eliciting people's views on these subjects. Whereas one of the people being interviewed drew the village domain, the

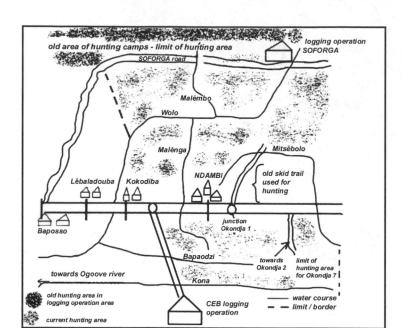

Figure 9-2. *Map of Ndambi Village and Forest Territories in Gabon, Drawn by Villagers on April 11, 1998*

others spontaneously talked about the various problems related to its management, the conflicts with loggers and foreign people, the distance to hunting sites—in short, about access to resources and management techniques of the various actors involved. As in Cameroon, it was difficult to work with a prepared map. The field drawings often began with a main road that cut across the village or spread out along it.

This method was also previously tested in the forests of northern Congo-Brazzaville with PROECO, a project in North Congo (Gami and Mavah 1997) to determine village borders as part of the implementation of a zoning plan in a protected area. In that plan, there was an interest in taking the needs of local populations into consideration, as they related to the use of natural resources. The results there also were satisfactory.

Results

We have divided the results into three categories: hunting, the impact of roads and logging operations, and conflict management. In our mapping exercise, the farmers of Ndambi set the borders of their village and their forest along rivers or other natural boundaries. Figure 9-2 clearly shows the borders of the village of Ndambi and the restrictions imposed by loggers concerning access by villagers to some of their customary lands.

Hunting, Roads, and Logging Operations

The village hunting area straddles two logging concession areas: CEB and SOFORGA. It seems that hunting areas located within logging concessions are now minimally exploited. In fact, before the arrival of logging companies, hunters used to go to those places that were reserved for hunting camps. Villagers complain that SOFORGA and CEB settled in the area without consulting them beforehand, and they consider the workers of the forest company "foreigners" who have no land or land rights in the region. This is partly true for CEB, because the company has brought all its previous workers, who are from the Punu or Nzébi ethnic groups, from Doussala (in the extreme south of Gabon). However, CEB also has recruited several local people to work in the forest (fellers' aides, prospectors, and so forth).

When the map of the village was being drawn in our discussions, farmers expressed hostility to the workers who reportedly decimated the wild animals of their area. They described the workers as beneficiaries of a double salary (from both hunting and logging) and considered them as the primary cause for the reduction in game near the village. Although hunting during working hours or transporting meat with company vehicles is technically not permitted, workers (especially prospectors) hunt with cable snares or guns and somehow manage to transport the meat with company vehicles. In fact, relatives of the workers or retrenched workers living in the logging camp do most of the hunting. This problem is likely for the villages near the logging camp (such as Bambidie) that are experiencing a sharp decrease in game availability. The inhabitants of villages such as Ndambi are angry because workers take advantage of their easy access to game within logging areas. As a result, local hunters do not have the opportunity to consume or sell game from lands that they consider theirs.

The local communities accused loggers of opening roads that facilitate the access of outsiders to hunting areas. People also complained that logging roads were responsible for the increasing distances they had to travel to hunt for wild animals and the reduced availability of game near the village. Farmers declare that "even if all the animals are there, we have to walk a long distance to find them"; "animals have gone away because of the noise of machines." They also mentioned the slump in sales of game in the villages and in the neighboring town of Lastourville (see Figure 9-3).

The participatory mapping process brought about a debate over the destruction of the people's cultural patrimony (traditional knowledge) by loggers. The people accused the loggers of cutting important trees, especially in the areas surrounding the village. For the villagers, these trees represent very important cultural and medicinal interests. Some include the wild *atanga*

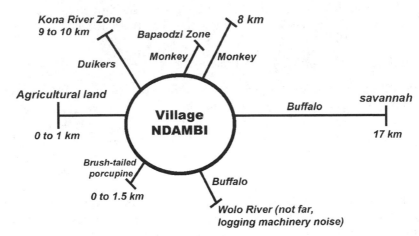

Figure 9-3. *Impact of Logging Roads on Reduced Access to Wild Animals*

(*Dacryodes buttneri*), *moabi* (*Baillonella toxisperma*), the *boudzomon* tree (in the Saké language) or *ndouma* (in the Kota language, which is very often used in traditional medicine), and the *bundéla* tree (Saké language).

All these aspects are causes of conflict between loggers and the local populations. Obviously, this discussion represents the perceptions of local people about the problems and must be weighed against the perceptions of other stakeholders. However, perception alone is powerful and can affect management directly.

Management of Conflict

The participatory mapping method revealed conflicts that we discussed during the focus groups organized in the various villages: Baposso, Ndékabalondji, Mékouka, and Bakoussou 2. Villagers accuse the other partners of not involving them in the management of natural resources. Although the loggers have set up an administrative unit in charge of relations with villagers, and the villagers have set up a committee headed by local administrative and traditional authorities (the divisional officer, the village chief, the canton chief), the villagers feel that the committee does not represent them. Therefore, conflicts between loggers and villagers are solved individually.

In response to these findings, we proposed, for the C&I test, several indicators concerning the criterion "rights and duties of all the stakeholders involved are clearly defined and acknowledged." They are intended to reflect the importance of involving all the forest actors, including local populations, in the sustainable management of natural resources.

Some of the indicators selected by the interdisciplinary Gabon test team follow.

- All the actors involved have their rights of use, tenure, or property clearly defined and guaranteed.
- Details of access to natural resources are clearly defined and respected by all.
- The awareness of all the stakeholders involved of their rights and duties is guaranteed.
- All the partners participate in the management of forest resources.
- Damages are compensated according to current regulations or after negotiation.

Conclusions

In this chapter, we have reported the results of a participatory mapping exercise. This method is a very concrete way to clarify how local populations feel about the problems they encounter in the management of natural resources. It also helps to assess, in an inexpensive way, the surveyed population's knowledge about the forest.

This method helped us identify serious differences in opinion and recurring conflicts between the timber companies and local people in Gabon. Besides providing descriptive information about the forests and people of Gabon, we identified serious conflicts about access to resources, even in the most forest-rich site where we tested social science methods (contrary to our initial hypothesis; see also the inconclusive results in Chapter 11). The kinds of findings reported here have led us to propose more contextualized, process-oriented studies for the future.

The forests of the Congo Basin are among the richest in the world, as far as biodiversity is concerned, and their good management may be important to humankind as a whole. The dense rainforests of Central Africa (Cameroon, Gabon, Equatorial Guinea, and part of Congo-Brazzaville and the Central African Republic) are among the richest on the African continent. Brenan (1978) found that there were 26 genera and 600 species of endemic plants in Cameroon. There are likely 28 genera and more than 1,000 species of endemic plants in Gabon and Equatorial Guinea, among which are many trees. Bahuchet and de Maret (1993) report about 10,000 plants (3,000 endemic), in the Democratic Republic of Congo, plus 409 mammals and 1,086 species of birds. Such unique and abundant biological representation is in danger of being disrupted by various forest actors (local people, loggers, outsiders coming from town for poaching, urban demands for firewood, and so forth). Therefore, we believe it is urgent that all the actors come together, in a cooperative spirit, to think about a better way to sustainably manage forest resources. Sustainable management of the biodiversity of the tropical forests found in the Congo Basin is a chimera unless all relevant parties, including local populations, are involved in the process.

We hope that results such as those presented here can contribute to a greater receptivity to the notion of co-management or participatory management and that a dialogue among the actors who intervene directly or indirectly in the forest will be favored to ensure the access of present and future generations to forest resources.

Endnotes

1. As discussed elsewhere in this book, *forest-rich* areas are defined as "islands of people in a sea of forest"; in contrast to *forest-poor* areas, which are defined as "islands of forest in a sea of people."

2. The TREES project was set up by the European Commission Joint Research Center and the European Space Agency in the early 1990s as the starting point for production of the first global "reference map" of the world's tropical forests from satellite images.

3. PTE stands for Permis Temporaire d'Exploitation (temporary exploitation license), and PI is Permis Industriel (industrial license).

CHAPTER TEN

Soil Fertility and the Generation Gap

The Bënë of Southern Cameroon

Diane Russell and Nicodème Tchamou

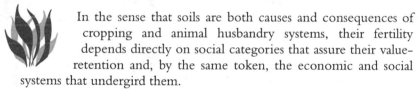

In the sense that soils are both causes and consequences of cropping and animal husbandry systems, their fertility depends directly on social categories that assure their value-retention and, by the same token, the economic and social systems that undergird them.

—C. Reboul (1977, 97; translated by Diane Russell)

In this chapter, we recount a drama of generational conflict within a wider discourse concerning soil fertility and changing values among the Bënë people of southern Cameroon. Studying this discourse allows us to examine the relationships among natural resource management practices, beliefs, values, and social institutions. Retaining or improving soil fertility is usually seen as a technical challenge. The story we present helps researchers to see that this challenge is social, institutional, and perhaps moral as well.

This chapter is based on ethnographic and ethnobotanical research carried out from 1991 to 1993 in the Mbalmayo area of southern Cameroon (see Figure A2 in the Introduction). This research was part of an effort by the International Institute of Tropical Agriculture's Humid Forest Station (IITA/ HFS) to diagnose agronomic problems and characterize resource management systems in Africa's humid forest zone. The goal of this and associated research was to aid in the design of "adoptable" and "adaptable" agroecological systems to retain soil fertility and improve crop productivity (*IITA/HFS*

Medium Term Plan 1990–95). As such, our work was planned as a garden-variety descriptive study. We did not aim to do more than provide a baseline on resource use practices for agronomists and agroforesters who would be designing systems.

But the study went beyond this initial goal because of the unique opportunity that we had to witness the effects of the "cocoa crisis" on households in Mbalmayo, where the local economy was largely shaped by cocoa. The crisis involved the abrupt—almost overnight—collapse of the parastatal organization that bought cocoa and provided inputs and other services—such as delivering chemicals, marketing crops, and even maintaining roads—to cocoa farmers in Cameroon. The collapse was in part caused by the drastic decline in cocoa prices in the early 1990s. More broadly, it was the result of corruption and poor management, which resulted in bankruptcy of the parastatal organization. This crisis was heightened by the tension surrounding the campaign for national elections within the context of "multipartism."

Our research gave us insights that we think can be generally useful in social research for agriculture and forestry. Social science research has long been integrated into agriculture and agroforestry technology and system design at international institutes (for example, Rhoades 1984; Dvorâk 1993). This kind of social research typically focuses on characterization of farming or forestry practices, farm economics, and other social topics directly related to natural resource management and technology adoption. Topics such as the political economy of resource degradation, generational conflict, commodification, and belief systems are usually left to the academics.

The study of social dynamics and institutional change is, in our view, essential to the design of sustainable systems, which are grounded in the notion of intergenerational equity. *Intergenerational equity* refers to the ability of different organizational units (countries, communities, clans, lineages, villages) to maintain a resource base for the use of future generations (Goodland 1991; Norgaard 1992). This resource base should include belief systems, institutions, inheritance, and investment strategies that sustain the resource base as well as technologies and the natural resources themselves.

Many African societies had—and to some extent still have—very long-term, integrated strategies for the transfer and maintenance of resources across generations. These strategies include close links between the ancestors and the inheritors of land and assets, and attainment of statuses and roles that govern the sharing of and control over resources. In the situation we describe, soil fertility was part of a discourse on deteriorating community relations in the wake of the cocoa crisis as well as a complex and changing conception of trees, crops, soils, weeds, and agroforestry practices.

One particular drama that unfolded during our fieldwork illustrates the clash of reciprocal elder-dominated ideology and "modern" principles for natural resource management. The situation has implications for the design of agroforestry systems because the modern view—which emphasizes privatization, free markets, and commodity production for food supply and foreign exchange—underlies the development of many new technologies. As local systems, based in part on reciprocity and community-based sanctions, break apart under numerous pressures, it is unclear how communities in marginalized regions of the world will be able to maintain their resource base.

The difficulties of retaining community-controlled systems are immense. Yet it has been shown that community-based systems are often better suited for natural resource management, conservation, and household food security (Ostrom 1999). The story we tell here illustrates the shifts and risks involved for the community when the market collapses and the resource base deteriorates.

The Bënë

Our study was carried out among a people living in the Mbalmayo area who call themselves the Bëně. The Bëně are a subgroup of the Beti whose clans (*mvog*) descend from the ancestral marriage of Née Bodo and Amombo Kunu or Ndzie Manga (Laburthe-Tolra 1981; see also Chapters 2 and 3). The Beti are an agglomeration of related peoples who are said to have migrated from the northeast, crossing the Sanaga River on the back of a giant serpent. They consider themselves to be nobles who conquered other forest peoples of south-central Cameroon.[1] The Beti are part of the group designated by Francophone ethnographers as "Pahouin" and include the Bulu (southernmost Cameroon) and the Fang (Gabon).

The ethnic identity of the Bëně as part of a Beti–Bulu coalition played an important role in their participation in national politics, which in turn was a factor in the study. The nature of power within Beti society is also a complex and changing factor in the discourse. There is a tension between the concept of the superiority of and delight in successful people as well as the egalitarianism stemming from the notion that each free man is a noble (*nti*) (Alexandre and Binet 1963; Laburthe-Tolra 1981, 374).

Gender relations also shifted in the context of the cocoa crisis. Women's access to agricultural and forest land is based on their relationships with men—fathers, husbands, and sons—and women's labor is largely in the service of male heads of household. However, we found that by the early 1990s, the actual situation of many women differed from the stated norm of male control. Many households were female-headed, at least temporarily, because of divorce or nonlegitimacy of the liaison (for example, men living with widows are not recognized as family members and do not have rights to land). Women were becoming more politically active and in control of the household economy, because less income came from cocoa and more from the sale of food crops (a woman's domain).

231

The Study

We chose four villages as sites of intensive research: two on the main road from Ebolowa to Yaoundé, and two in the interior. All four villages could be said to be part of the same resource management system based on physical and social characteristics (Table 10-1), but they differed in their access to markets and in their social organization. Two villages were managed by (male) chiefs and two by committees that were dominated by women. One village was the site of a sawmill that was abandoned in the early 1980s, and another is a small town with some services such as a post office and a secondary school.

The study method is detailed elsewhere (Russell 1993). The length of time available for the study (two years) and our excellent relations with people in the villages provided an opportunity to expand the depth of the study

Table 10-1. *Characteristics of the Bënë Agricultural System and Environment*

System (physical)	
Rainfall	1,513 mm/year Bimodal (two rainy seasons)
Insolation (sunlight)	1,645 h/year
Soils	Typic Kandiudult (ferralitic, strongly acid, high aluminum saturation)
Topography	600–700 m above sea level with rolling hills and broad valleys
Vegetation	Humid semideciduous adult and young secondary forest characterized by Sterculiaciae and Ulmacae
Leading crops	Cocoa, groundnut, cassava, plantain
Sustainability indicators	
Population density	Overall: 27.4 people/km^2 Rural: 16.6 people/km^2
Average fallow time	11 years (3–50 years depending on field type, status of household, and location)
Cropped land/total land/ arable land (%)	11 (>12% is degraded)
Households sometimes buying staple food (%)	34
Households deriving revenue from bush food sales (%)	53
Households sometimes buying groundnut seed (%)	44

continued on next page

and to verify our findings. Key informants included two village chiefs, the officers of three women's groups, an entrepreneurial farmer and his family, several government officials, and agricultural extension agents.

Because the IITA/HFS was devoted to agricultural research in the forest zone, our aim was to integrate the study of agricultural and forest resource management practices. We thus wanted to look at not only farming but also the collection, use, and marketing of forest products; the use of fallows; knowledge of trees; and decisions about cutting forest. Out of this study came the proposal for work on forest products that has been carried out since 1993 (see also Ndoye and Russell 1993; Ndoye and others 1997/98; Ndoye 1998; Ndoye and Kaimowitz 1998; and Chapter 2).

It was in-between the actual research—during conversations and socializing in the villages—that we learned about the social significance of soil fertility and the generational conflict that both stemmed from and exacerbated

Table 10-1. *Continued*

Region (sociopolitical)	
Infrastructure	Two paved roads
	Feeder roads and bridges in poor repair
	Markets undeveloped in interior
Income-generating opportunities	Artisan 7% (occasional)
	Fishing 15% (9% regular)
	Hunting 18% (10% regular)
	Trade 30% (15% regular)
	Salary 5%
	Distilling whiskey 50%
Migration patterns	Out-migration from villages except those closest to town/city
	Population growth only in towns
	Very little migration of strangers/outsiders to region
Taxation and regulation	Head tax higher in rural than in urban areas
	Based on cocoa income
	License for cutting trees
Land tenure system	Land registration minimal
	Tenure based on inheritance and family claims in village
	Few registered farms
State–rural relations	Privileged (president's home is nearby) but unproductive
Social structure	Ethnically homogeneous
	Weak chiefdoms
	Landholding becoming more important than lineage

Sources: NRI 1992; FAO 1984; field data from authors' survey.

declines in soil fertility. Forces within and external to Bënë society were shaping resource management practice and economic opportunity, which in turn affected beliefs as well as the range and quality of knowledge about soil fertility.

During our fieldwork period, a flamboyant female "exorcist" was circulating in the villages, promising to locate and remove evil spirits that were causing various misfortunes such as the decline in cocoa prices, the collapse of the cocoa parastatal organization, and the resultant deterioration of infrastructure (see Geschiere and Nyamnjoh 1998 for another example of a female exorcist in Cameroon). The campaign preceding national elections was also under way within the context of (supposed) multipartism. Many proposals were being bandied about concerning road building to feed into the new international airport at Yaoundé, new or revived services for villages and towns, and regulations concerning the cutting of trees and status of different forests. It was a very volatile time, when people's economic and political notions were shifting, exacerbating generational tensions.

Before discussing the case in depth, let us look briefly at soil fertility and the Bënë natural resource management system.

Soil Fertility and the Bënë Natural Resource Management System

The precolonial Beti system associated forest clearing with the attainment of adult married status. Soil fertility was maintained through a long fallow cycle, enforced by rituals that involved sowing certain plants at key steps in the cycle and linking soil fertility with the fertility of the woman who maintained the field (Laburthe-Tolra 1981). Food crops were not produced for sale, but processed food was sold on trade routes. The goal of food production was to maintain a food supply for the household and for guests. A vast array of forest products was used, processed, and traded, including bark, fruit, wine from palm, raffia, mushrooms, insects, seeds, and nuts.

Diversification and local food self-sufficiency were sound strategies, because Bënë agroecosystems and markets are both fragile. Soil in the area is "acidic with low to very low cation-exchange capacity, most of which is dominated by aluminum. Most of the useful capacity is associated with organic matter which is concentrated in the shallow topsoil" (NRI 1992, 1). Soil fertility is maintained by building up biomass in the fallow period. Agricultural production cannot be sustained indefinitely with fallow times (that is, the period in which the soil accumulates organic matter between cultivation) of less than seven years.

Fertilizer was unavailable to the vast majority of farmers—and that which was available apparently was misused (see later). Intensive measures such as the use of manure or leguminous trees were not economically viable because of natural constraints (for example, animal diseases), labor insufficiency, low prices for crops, or the higher opportunity costs of other activities (cutting wood, hunting, and fishing).

When fallow length times drop below seven years, the soil can start an irreversible cycle of degradation. By the 1980s, the Mbalmayo area was on the verge of joining this unsustainable category due to lack of land available for agriculture under the fallow system, although each village and household situation was different (Leplaideur 1985). A comparison of 1951–1952 and 1983–1985 photo sets, along with field observations, indicated that the forest had receded significantly as a result of selective logging without regeneration and the expansion of agricultural land. Thenkabail and Nolte (1995) compared forest cover along a "benchmark" strip stretching from the southernmost area of Cameroon to north of Yaoundé:

> These data illustrate a striking decrease in primary forest area from south (Ambam) to north (Yaoundé). Secondary forests dominate in the Ebolowa [south of Mbalmayo] and Mbalmayo sectors and were considerably less in the Yaoundé and Ambam sectors. Area coverage by farmlands was significantly higher in the Yaoundé and Mbalmayo sectors as compared with the Ebolowa and Ambam sectors. Cultivation is primarily concentrated in 2- to 4-km large corridors along the road network. Along with the decrease in primary forest coverage goes a remarkable increase of young fallow areas dominated by *Chromolaena odorata*, whereas fallows with *Imperata cylindrica* as principal vegetation cover only occur in significant portion (10.6% of the geographic area) in the Yaoundé sector. Old fallows with forest regrowth cover 6.3% of the entire study area with slightly higher figures in the Ebolowa and Mbalmayo sectors.

Elements of the food production, plantation, fallow, and forest system in Mbalmayo are shown in Table 10-2 and Figure 10-1. The Bëné name many fields depending on the crops grown and the kind of forest cut for planting. Forest crops such as banana and plantain, yam, and melon (*Cucumerops eduliis*) perform best when planted in the fertile soil of fields cut from forest areas typically left fallow for more than 15 years (*esep*). Thus, these crops "open" the planting cycle. Crops such as cassava and cocoyam can be grown in shorter fallows and often are introduced in the groundnut field (*afub owondo*), which is either cut from short fallow brush or planted within the *esep* after a few years of production. Maize is planted either in the *afub owondo* as an intercrop (after the first harvest of groundnut) or in marginal, swampy land (*asan*).

Farmers in the area know that certain tree species affect soil fertility; indicator species vary with the type of field and season. Tchamou (1994) ranked these species in terms of frequency of use as soil fertility indicators. The Bëné have a sophisticated tree classification system that considers economic

235

Table 10-2. *Mbalmayo Resource Management System Components*

Location or area	Average area or size	Major crops or products	Minor crops
Afub owondo (mixed-crop food field [women])	10.16 ares	Groundnut (16 plants/m^2), cassava, maize, cocoyam	Yam, sugarcane, solanum, amaranthus
Esep ngon ("low maintenance" forest field [men])	16.04 ares (range 3.2–28.9)	*Cucumeropsis manii* or *eduliis* (melon), plantain, maize, cassava	Cucurbita (melon), tobacco, amaranthus, yam, sugarcane
Esep bananeraie (banana/plantain field planted in forest)	23.2 ares	Plantain, cocoyam, banana, cassava, melon, maize	Yam, pepper, amaranthus
Asan (marsh field [interseason])	10.1 ares	Maize, groundnut, cassava, cocoyam	Amaranthus, tomato, tobacco, yam, sweet potato
Specialty fields (certain crops for sale)	12.7 ares	Maize, cassava, pineapples, cocoyam, onions	
Garden	Plants and trees around the house[a]	Plantain, banana, greens, cassava, spices	
Plantation	2.2 hectares	Cocoa and other trees	
Livestock[b]	Chickens (5), goat (1), pig (1) Sheep, cats, dogs, and others (<1)		
Water source		Water, fishing, raffia collection, borders of land	
Fallows[c]	12 per household	Firewood, insects, mushrooms	
Forest[d]	Use restricted (household members and those with permission) in 70% of households	Fruit, firewood, bark, insects, mushrooms, game	

Notes: Labor force is 1 man, 2 women permanent; 4.4 temporary, mostly during school vacations; 14% paid for clearing food fields, 8% paid for clearing plantation; 53% work in groups.

[a] 23% of gardens are enclosed or partially enclosed.   [b] 14% of livestock are sometimes enclosed.

[c] Main vegetation is Chromolaena, *Aframomum* sp., and *Triumfetta* sp. Trees include *Alstonia hoonii*, *Terminalia superba*, and *musanga*.

[d] 70% are used primarily by men; 56% conserved some trees in forest. Most forest was logged extensively 20 years ago. Dominant trees are *musanga*, *triplochiton*, and *Terminalia superba*.

236

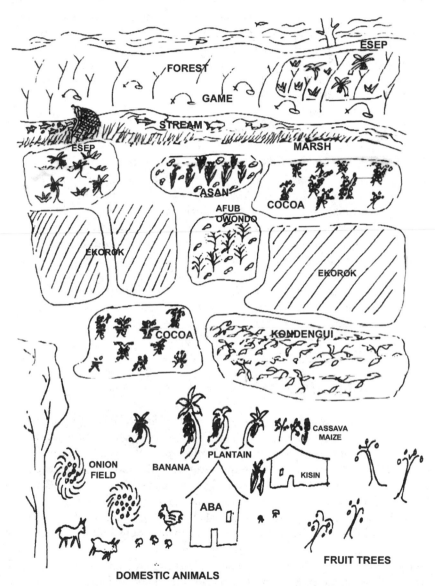

Figure 10-1. *Spatial Representation of the Bënë Natural Resource Management System*

value; food value of the fruit; medicinal value of the fruit, leaves, and bark; leaf biomass; seasonality; attraction of insects; moisture retention; hardness; and ability to harbor sorcerers. The relationship between trees and soil fertility also has a gender dynamic. Women are said to know more about trees found in or near the *afub owondo*, whereas men know about trees in the *esep* fields and plantations. Trees that are considered good for soil fertility are not always saved when forest is cut for food production, whereas trees that have high economic value—especially fruit trees—are routinely protected (see also Munyanziza and Wiersum 1999).

Patterns of tree cutting may have developed relatively recently in relation to changing economic conditions. Forest concessions in the area were exploited for valuable hardwoods for approximately 50 years, and few valuable trees remained at the time of our research. Efforts to replant hardwoods met with limited success because of lack of extension capability in the area (personal communication from Andrew Roby, then of the Overseas Development Institute forestry project in Mbalmayo, Cameroon, 1992). After about 20 years of operation, the sawmills and lumber companies had left little in the way of lasting investment (NRI 1992). However, forest regulations remained that prohibited or impeded local people from cutting trees, while poorly run or bankrupt concessions were still legally permitted to cut them. Artisanal production and exchange systems that would promote the economic value of a greater diversity of trees were disappearing.[2] In addition, knowledge systems that enabled people to use trees for a wide variety of medicinal, food, and artisanal uses were being lost.

The Cocoa Economy

Cocoa cultivation was introduced from Trinidad in 1887, spread during the German era, and expanded after World War II. It created opportunities for increased cash income, although the development of a viable peasant economy was retarded by the food requisition system.[3] Cocoa became deeply integrated into social and economic life. A man's goal was to become a "planter"; to increase his plantation, his cocoa revenue, and his status through marriage and house building (Weber 1977). Outside labor as well as household labor was widely employed in maintaining and expanding the plantation, and in harvesting the cocoa.

By the late 1970s, "improved" cocoa varieties and planting methods were widely disseminated through the Société de Développement de Cacao (SODECAO). These varieties further tied Cameroonian farmers into the world market because they require fungicides and other chemical inputs to perform well. SODECAO provided materials and tools at subsidized prices and also took care of marketing and even road maintenance; consequently, farmers became highly dependent on the organization.

Despite these ties to the world market, cocoa acted as a buffer against both physical and social degradation in the Mbalmayo area. It was integrated with other types of plantations and wild trees, thus maintaining forest cover.[4] Cocoa revenue was spread around the community to fund children's education, marriage payments, house building, and ceremonial occasions such as baptism and first communion. Land could not be bought and sold, but cocoa plantations could be controlled exclusively by the person who planted them, and that person's heirs could inherit the plantations. The spread of smallholder plantations together with the collapse of the food requisition system made Bënë society more egalitarian. Women benefited as well with the decline of polygyny and child marriage and increased access to education. They were usually compensated for their labor, even within families, often by being given low-quality (*hors standard*) cocoa to sell.

In the early 1990s, cocoa prices began to plunge on the world market. SODECAO went bankrupt due to corruption, and farmers suddenly lacked access to the chemicals on which they had come to depend for their improved varieties. Road maintenance (previously funded by SODECAO) ground to a halt, and bridges started to collapse.[5] For example, during our fieldwork, the main bridge to the two isolated villages collapsed; the detour added dozens of miles to the journey. Farmers used old, long-expired fungicides or none at all, and most maintained their plantations only minimally. Table 10-3 describes a range of changes in the Beti resource management system over the past 20 years, based on comparisons between the authors' data and other studies.

Impact on Bënë Villages

To the Bënë elders, soil fertility was a symbol of community coordination and harmony. The soil is "ruined" by disputes and by sole use for economic gain that is not shared throughout the community. Sorcerers can "steal" others' soil fertility through the use of their animal counterparts, such as hedgehogs and wild pigs, that do their work at night. When the community starts noticing declines in productivity, it is important to locate the cause and act. The causes of declining soil fertility are seen to be primarily social: disharmonious social relations. Depending on one's point of view, greedy people want to profit at the expense of others, or sorcerers wish to destroy others' chances of getting ahead.

It is not that people do not see the technical causes of soil fertility declines—lack of biomass, short fallows, and erosion. They do. But these technical problems are seen as *symptoms* rather than root causes. These symptoms are caused by human action (cutting trees, clearing forest, planting larger fields without inputs) and, even more profoundly, by human value systems. When profit is the motive for clearing land, the web of reciprocity is

239

Table 10-3. *Changes in Mbalmayo Resource Management System*

Hired and paid laborers to work their cocoa plantations	55% of households (1985) to 8% of households (1992)
Afub owondo	Second season increasingly important (88% of households plant) Average field size same, number of fields increased, thus total area increased (see below)
Esep	Decline of *esep ngon* (from 10% to 4% of households) Rise of *esep bananeraie* (<10% in 1970s to 33% of households, 13% of fields in 1992)
Asan	30% of households clearing 1991–1992; 21% only to sell crops No reports on *asan* use in past except as rice fields
Specialty fields	Appearance of guava and selected fruit trees, pineapple, and sugarcane in specialty fields for villages on the road Concentration on maize and *macabo* for community fields in isolated villages; plantain for individuals
Plantations	63% of plantations expanding or redensifying and 80% seen to be correctly maintained in 1985 vs. 12% of plantations expanding or redensifying and 70% minimally maintained in 1992 100% of plantations using fungicide in 1985 decreased to 76% (old stock) in 1992 Price of cocoa fell from 420 CFA/kg in 1988 to 135 CFA/kg in 1992
Forest	0% of households had no forest in 1970s; increased to 3% in 1992 Valuable trees removed by loggers until the late 1970s
Avg field size	Households had 50.8 ares under cultivation for two seasons in 1992; 38 ares were average in 1985 for a wider region (Center and South)[a]
Avg area for women in two seasons	From 22 ares (Lékié Division in 1984)[b] to 37.5 ares (Mbalmayo in 1992)
Fallow time	More forest land (long fallow) brought under cultivation, fallow time diminished on *afub owondo* to avg 7.3 years in 1992 (from 10–15 years in 1970s)
Other changes	Disbanding of cocoa cooperatives, liberalization of market Deterioration of roads in interior Multipartism introduced Women's groups gain political strength based on community fields and individual sales of food crops Conflict increases between older and younger generations

Note: Household size is stable except for a village close to Yaoundé, which reports a net increase.
[a] Per Leplaideur 1985.   [b] As cited in Guyer 1984.
Sources: Mainly Leplaideur 1985; Guyer 1984; and authors' research.

broken. Most Bënë believe that the only way to get people to work for others is to enslave them through sorcery.[6] Those who seek to adopt new technologies are suspected of being sorcerers.

The fear that some people or groups might profit at the expense of others could be interpreted as superstition or a no-growth philosophy (see also Foster 1965), but the reality is much more complex. As Taussig (1980, 117) vividly describes a situation in South America, the loss of reciprocity is deeply felt:

> Ideals ... concerning the sharing of wealth and labor increasingly diverge from the practices of everyday life. The Golden Age of plentiful land and food, mutual aid, labor exchanges, and fiesta work parties is invoked all the more heartrendingly as the ideals of equality and reciprocity are subverted. But it is these ideals that give force to the moral outrage and censure of the community.

Key informants reported that underlying the death of the reciprocal system was pressure on the land. The ability of members of Bënë clans to use and lend land with little or no cash changing hands—a gift to the chief, food and drink for the key landowners—was an ideology supporting other tenets of the reciprocal system. Land registration was still relatively rare in villages (less than 3% of the holdings we surveyed), but trends toward more exclusive use of and access to land were evident. The Bënë were concerned first and foremost that no outsiders get title to land. This sentiment included not only other ethnic groups, such as the Bamileke (a Cameroonian ethnic group from another area, renowned for their commercial skills), but also city-based Bënë who wanted to invest in their own villages (see also Geschiere and Nyamnjoh 1998).

Although people were opposed to land sales and to outsiders coming in, many recognized that without external investment, the local economy could not revive. One informant told us in confidence that he hoped the Bamileke would come and invest in his village, but of course he could not express this sentiment publicly. In secret, many people lent land for profit or hoped to do so. The inability of the Bënë to go backward to the reciprocal system or forward to the modern system made the tension over land severe. As one village chief remarked, "I wake up every morning thinking I may get a machete in the back over a land dispute."

The Generation Gap

In Ekwan, the large isolated village where we lived for several weeks as we conducted our survey, we found ourselves in the midst of a generational drama concerning the causes of soil fertility decline and general economic collapse. The young people of the village confided to us that older people, some of whom were sorcerers, did not want them to get ahead. Specifically, the elders kept the cost of bridewealth too high—beyond the reach of a young man who could no longer tap into the cocoa largesse. Without making the gifts of cash and goods to the bride's family required in bridewealth, the young people could not marry legitimately and take part in decision-making. This generational stress, a recurrent theme in African society (for example, see Meillassoux 1981), was exacerbated by the collapse of the cocoa economy. This collapse actually may have given the younger people more power than they would have had ordinarily because the elders could not control the flow of money as before. In addition, young people traditionally had a moral role in the maintenance of soil fertility through fertility rites that were still remembered and occasionally practiced.

The story is further complicated by the political history of the Beti. In previous generations, during the colonial era of food requisition, the Beti were more hierarchical. Elders could control not only labor and investment but also the number of women available to young men, because some accumulated dozens if not hundreds of wives. Few wanted to return to the days of paramount chiefs, but the tension over control of bridewealth harked back to this appropriation of women by the elders.

The political situation also played a role in keeping the young people's frustrations focused on internal conflict. Support for Paul Biya (the president of Cameroon, and a Bulu) was almost mandatory despite "multipartism"; dissent was punished as traitorous to Beti unity. Therefore, the young people had no external political outlet for disappointment with the dismal economic performance of the country. A young chief often remarked to us how dissatisfied he was with the government policies but also said clearly that any move he might make toward dissent would lead to his removal, if not his demise.

The generations could be characterized as the elders (not necessarily chronologically) who believed in the principles of reciprocity (see Discussion). The young people might also believe, or want to believe, but were skeptical and ambitious. In addition, some people, as indicated by their actions, were either "agnostics" or able to manipulate different principles interchangeably. Some—indeed, perhaps many—of the older folks manipulated the reciprocal principles and respect for elders for their own gain, and this manipulation was in part what concerned the young.

To deal with these tensions, an exorcist—for want of a better word to describe Madame A, who was flamboyant, outrageous, highly theatrical, and totally confident in her powers to root out sorcerers and their effects—was brought in by the young people to find the evil objects hidden around the village by sorcerers and destroy them. Three times the police from Mbalmayo tried to arrest Madame A, and each time she escaped. At one point, the young people threatened the chief of Ekwan if he collaborated with the police. We observed a ceremony where Madame A sent her minions, who were ostensibly in a trance, out to find the evil objects in trees and in people's houses and yards. These objects were then placed in front of the village, and the sorcerers were judged. Some confessed, and others did not. The evil objects were heavily associated with the forest and the wild. They were found notably in the *dum* tree (*Ceiba pentadra*), whereas other large hardwood trees were seen to harbor sorcerers and be resistant to the chainsaw (they "turn the chainsaw back"). Wild onions (*meyang*) both ward off sorcery and are a sign that one is a sorcerer.

The major issue was land and the income and livelihood derived from it.[7] By 1991, cocoa plantations already took up 10% of total land in the Nyong and So'o Division, which encompasses the Mbalmayo area (NRI 1992). As cocoa brought in more cash, a higher percentage of bridewealth came from cash. Bridewealth payments became an important source of cash for a family. If cocoa could no longer bring in the revenue needed for bridewealth— as well as head tax, school fees, matches, kerosene, and other necessities— people had to find alternatives that came from exploiting the land. With the economy in decline, few alternatives were available (see Table 10-1, Region).[8]

Without paying bridewealth, a man cannot fully establish himself in the village by claiming land for a house and fields. Thus, young men started cultivating bananas, plantains, and other forest crops; tapping and distilling palm wine; hunting (to a very limited extent, because there was no surplus of game); and clandestine revenue-generating activities. Young people were more eager to try "new" agricultural technology, perhaps because they hadn't yet discovered its risks.[9] Chainsaws were highly prized and growing in use, but their number was impossible to determine (because use was highly restricted or illegal in most cases).

Knowledge about trees, plants, and soil fertility changes with generation, occupation, and gender. Men and women have their areas of expertise in terms of knowing how certain trees and plants affect soil fertility because they specialize in different kinds of fields and crops as well as different kinds of traditional medicines. Although we found the Bënë knowledge base to be extensive, two of our most respected informants—an older man and woman—died during the study, leaving gaps in knowledge about resource management—forestry and soil fertility maintenance, respectively—keenly felt by their kin and neighbors. Informants noted that with universal schooling, it was not possible to pass on the depth of knowledge about trees and plants to the younger generation; children no longer spent enough time with their parents in the forest and field (see also Chapter 8).

Discussion

Like soil fertility, the reciprocal society conjured up by the elders was in decline. Young couples were living together without getting married, the old trees that harbored the sorcerers were being cut down with chainsaws, and knowledge about the use of trees and plants was disappearing. The Bënë were strangely ambivalent about the forest. It was seen as the source of "genuine" healing and knowledge, but its power was feared. Although trees were much appreciated, every man wished that he could have a chainsaw to clear fields quickly and easily. Just because a tree was good for soil fertility did not mean it would be spared the axe during farm creation.

It was clear that new technologies and new crops alone could not solve the problems of soil fertility. A social system that guards the inheritance of the soil for future generations has to be in place. When land and food crops become commodities, they no longer are treated as a patrimony for future generations.

In the past, debt, obligation, and ancestral ties were reckoned and negotiated through the generations and across clans and lineages. Thus, resources essentially belonged to and were managed by the community, represented not only by the present generation but by the past and future generations. To sell land would indicate that these ties are cut and that the past and the future can no longer dictate to the present owner what to do with the resource (see the discussion of intergenerational distribution of benefits in Chapter 11).[10] The elders found, much to their dismay, that food crops were increasingly produced for cash and that the young women no longer had the time to manage their fields well.

Yet the young people needed livelihoods, and many tried to straddle the village and the city in an attempt to fulfil these needs. They saw little in the modern world that respected the reciprocal philosophy espoused by the elders. While living a "double life," they could not maintain soil fertility. In fact, the only informants in our sample who attempted to do so were old and childless. Many young people we interviewed had education and technical training but could not use it because the declining economy offered few jobs. They were (rightly) skeptical about the mythical unity of the past and felt that the elders were using this mythology to justify their dominance over land and marriage.

The social institutions of clan, marriage, and labor exchange were shifting from being client based (controlled by elders) to being land or commodity based (Guyer 1995). This shift certainly started with the spread of the cocoa economy, as men became able to start plantations and attain wealth without necessarily being a client of a powerful elder. Yet the cocoa economy was still at least partially integrated into the reciprocal economy through the circulation of bridewealth, labor exchanges, and the emphasis on household food security. As the cocoa economy collapsed, the last shreds of reciprocity were being torn away.

Most development planners see this trajectory as inevitable and desirable. It is widely believed, for example, that land registration will help resolve soil fertility problems by enabling individuals to invest in their own land (Rudel

1995; Tucker 1999). In Central Africa, this proposition is very dubious, because land is often registered for the purposes of speculation and elite domination. Although Shepherd (1991, 171) supports land registration, she remarks, "When institutions of the strength of the clans and lineages, which have allocated land for hundreds of years, break down, we should not lightly suppose that it will be easy to create ... new village institutions to do their work."

The reciprocal system will not survive in the form that it took during the precolonial or even the cocoa years, but its function must be recognized. State institutions, as they exist in Cameroon at present, will not replace these functions with systems that preserve the resource base because their aim is to maximize revenue. Externally funded conservation organizations have penetrated little into Cameroon society. Intergenerational equity should become a focus for agricultural and forestry development. For example, investment in forest-based industry is a strategy that would spread wealth across generations. Our work led us to believe that the development of forest products such as rattan, nuts, bark, oils, and resins is preferable to the development of intensive agriculture in the forest zone (Ndoye and Russell 1993). Supporting cocoa production in a more sustainable manner is preferable to supporting greater commercialization of food crops because of cocoa's comparative permanence (making more intensive land use possible) and the ecological functions of tree crops.

Most important, it is imperative to look at how land and other resources are linked to the social institutions that govern inheritance and investment. Understanding the extent to which local social units can recognize and respond to resource degradation is important. Individuals and households often are unable to maintain soil fertility under increasing pressure, and larger entities such as the region or the state may have other priorities for the resources (foreign exchange, cheap food for the cities, and so forth). At the time of the cocoa crisis, the lacuna of control, confusion, and degradation of local institutions appeared to be leading to something resembling a "tragedy of the commons."

We also observed how increasing commodification of food crops, exacerbated by the decline of cocoa, led to intergenerational strife. The elder-dominated social institutions that governed land allocation, labor, and marriage conflicted with the younger generation's need to get cash. Cocoa was no longer available as a buffer between the market economy and the reciprocal economy.

The degeneration of soil fertility is both a metaphor for the decline of cooperation and a real outcome of commodification. As land was increasingly used to produce food for the market, there being few other viable alternatives, the soil was degraded. Women were too busy going to market to care for their fields, and men were thinking about raising money for bridewealth or building a house by clearing the forest.

245

The Bënë discourse on soil fertility includes a prescient view of the relationship between natural resources and human behavior. This view, described by Harms (1984) as a "zero sum game," is consistent with the views of many Central African peoples and with a growing view of environmental economists who see clear "limits to growth" imposed by finite natural resources. In this conception, there are no winners without losers. An excess of resources is obtained only through sorcery—often involving the killing or harming of close relatives. People who hoard or show greediness are sorcerers. Jealousy can kill too.

From our study on soil fertility among the Bënë, we deduced certain principles of the imagined reciprocal client–centered worldview espoused by the elders. These principles are closely related to the views of some environmentalists:

- If people do not cooperate, then soil fertility declines and resources are depleted.
- If no structure is in place to bring young people to responsible adulthood, then resources will be depleted.
- Those who take more than their share of resources negatively affect close relatives and neighbors.
- If market production enters significantly into the equation with household food supply, then the household food supply will suffer.
- The division of labor reflects the need to produce and transform different resources in different ways: food crops and forest products, clearing land and planting land, and using land and allocating its use.
- Revenue should be reinvested as social goods back into the community.

The modern beliefs of young people in their struggle for control over their lives contradicted these principles. Agricultural extension agents, foresters, and development groups propagated the following ideas:

- Soil fertility problems are technical in nature and can be solved with new technologies.
- The state can and should manage forests for its own purposes.
- If people have "secure" tenure, as in land registration, they will invest in maintaining soil fertility.
- It is good for individuals to be innovative, to try new techniques, and to be successful in obtaining large revenues.
- Sources of knowledge are centered outside of the community.
- Revenue should be invested in new technology to increase production.

The first set of principles centers on community control of resources, both spatially and temporally. It punishes those who deplete resources and rewards social investment. The risk is the domination of male elders who may not have the best interests of the community at heart. The second set explicitly rewards those who use up greater amounts of resources to generate more revenue. These principles encouraged men and women to enter into the market economy and invest in more intensive production methods. However, the experience of the cocoa crisis made this path look hazardous. The first set defines limits to resource access based on status and division of labor, whereas the second set sees limits in terms of entrepreneurship and ability.

Finally, the first set recognizes that resource management methods (that is, how practices are carried out) are deeply connected to the relationships involved and to productivity. The modern principles posit no such relationship but instead emphasize "correct" or "incorrect" technologies. Village people struggled with the implications of these two sets of principles, finding neither particularly useful. They began to see the death of the first set of principles without finding life in the second.

A more severe crisis looms, however. The richness of local knowledge is linked to use of the land. With increasing land and food sales and the final destruction of the reciprocal economy, the forest becomes more and more a bundle of commodities to be quickly exploited rather than a resource base for a clan or lineage. Children who no longer learn about forest resources (because they are busy with school or grow up in the city) cannot value them. As young people marry at will, strong ties do not bind them to the clan or lineage. Finally, people will be free to sell their land to the highest bidder. Who, then, will ensure that the forest will be there for the next generation? Who will even know that it is gone until it is too late?

The old sorcerers, who live in the trees and bother people who try to "get ahead," are dying out. Many wish them good riddance as they look forward to a lifestyle of mobility, flexibility, and financial rewards. Others feel sadness as they contemplate the shape of a new sorcery that fosters the economic gain of some at the expense of others and, indeed, of the future of the resource base itself.

Acknowledgements

We thank the Rockefeller Foundation, which funded Diane Russell's two-year fellowship at the International Institute of Tropical Agriculture; the Resource and Crop Management Division of the International Institute of Tropical Agriculture (IITA), especially Dunstan Spencer, Karen Dvorâk, and Doyle Baker, who guided the research. (*Note:* These individuals are no longer staff at IITA.) Many thanks also to the Academy for Educational Development, which financed Diane's participation in the National Symposium on Indigenous Knowledge and Contemporary Social Issues and the writing of this chapter, and to Jane Guyer and Carol J. Pierce Colfer, who read drafts and provided extremely helpful comments. Antoine Eyebe and Louis Lekegang, who are now with the Center for International Forestry Research, were key members of our research team.

Above all, we acknowledge the wisdom imparted to us by the men and women in the four Mbalmayo villages. (The village names presented here are not the real names of the villages in our survey.)

We are solely responsible for the content of this chapter.

Endnotes

1. For example, Baka "pygmies" fled the area, the Maka migrated east, and other groups (such as the Ntumu and the Etenga) assimilated.

2. Comparing the Mbalmayo area of Cameroon to the forest area of northeast Congo (formerly Zaire) is instructive. Canoe building, ironworking, meat and fish smoking, and home palm oil processing are very active rural occupations in Congo but rarely practiced in Cameroon (see also Russell 1991).

3. This system enhanced the power of local chiefs to use wives' and clients' labor to maintain food production and supply (Guyer 1978, 597).

4. "Cocoa cropping, at the level of intensity practiced in the Department, is a relatively sustainable enterprise and is ecologically sound.... In the long-term, the farmers' efforts to increase food crop production to replace the cash incomes previously gained from cocoa is likely to threaten the stability of traditional systems" (NRI 1992, 35–6).

5. However, Cameroonian cocoa sells at a premium and is very important to the country's economy, representing about 15% of export earnings (NRI 1992). Consequently, cocoa production will continue to play an important role in the rural economy despite the crisis (Ruf 1995).

6. In fact, people do work for wages; this is a sensitive topic. Bënë are not supposed to work for wages or pay for land, but these transactions occur. Outsiders may be hired for work, but it is often believed that outsiders who stay a long time have been enslaved through sorcery, especially if the outsider is a man who works on the farm of a widow.

7. Geschiere and Nyamnjoh (1998) note that another major issue that provokes witchcraft accusations is the relationship between town-based elites and villagers— land is likely to be at the root of this tension as well.

8. However, it is interesting that the chief of Ekwan remarked, "When the tax collector comes, people do not have cash, when the priest comes, there is no cash, but when Madame A comes, there is cash aplenty to cover her exorbitant fees and demands." In fact, there were sources of cash in the village but, because some of them involved clandestine activities such as distilling whiskey, using chainsaws or guns, and growing forbidden crops, this cash was in the "underground economy" and subject to censure by the elders. Of course, the younger people felt that the elders were just jealous of their cash-earning capabilities. We suspect that elders manipulated younger people in marriage settlements not only with respect to the amount of cash involved in paying bridewealth but also in negotiating acceptable "distance" between clans and lineages of marriage partners and in blessing the unions.

9. The record of introduced technical change in the region is less than impressive. New cocoa varieties are much less resistant to pests and disease than older varieties

without the use of chemical inputs, many introduced food crop varieties are seen to be unappetizing, chemical treatment of plantain seedlings led to widespread disease when the chemicals became unavailable, copper-based fungicides may have caused the disappearance of many treasured mushroom varieties, and plants introduced to restore soil fertility in cocoa plantations were found to harbor snakes.

10. The selling of labor by local people is much frowned upon, but it occurs. In the past, strangers were often employed during the cocoa harvest, but food production (as noted earlier) is generally not remunerative enough for hired labor. Also, the people fear that outside laborers may claim land, especially if a debt is owed them. Thus, few people from the north and west came to Mbalmayo. In one case, where two Chadians were employed to help grow pineapples for the market, the farmer's fields were burned and the Chadians chased out of the village.

CHAPTER ELEVEN

Access to Resources in Forest-Rich and Forest-Poor Contexts

Roberto Porro, Anne Marie Tiani,
Bertin Tchikangwa, Mustofa Agung Sardjono,
Agus Salim, Carol J. Pierce Colfer, and
Mary Ann Brocklesby

In this chapter, we discuss the social principle in the *CIFOR Generic C&I Template* (CIFOR 1999) that deals with human well-being, Principle 3: "Forest management maintains or enhances fair intergenerational access to resources and economic benefits" (Annex 1 in the Introduction). Part of this principle focuses on forest actors' having "a reasonable share in the economic benefits derived from forest use" (Criterion 3.2). Another component requires an on-the-ground assessment of changes in resource accessibility over time. Yet another part deals with people's thoughts and feelings about this question. One of the indicators is, "Local people feel secure about access to resources" (Indicator 3.1.5). A related criterion states that "People link their and their children's future with management of forest resources" (Criterion 3.3).

Access to resources implies some distribution of these resources among stakeholders, and the concept of sustainability inevitably incorporates an intergenerational element. If we want to sustain forests as well as the people within and around them, we must think in terms of the future—perhaps examining the past in our efforts to gain such understanding. We use data collected in various forest-rich and forest-poor contexts to assess two issues: the sharing of benefits among stakeholders and intergenerational equity.

We used results from two pebble-sorting methods to examine the hypothesis that inequitable sharing of benefits (as perceived by local stakeholders) correlates with poor forest quality. Another hypothesis was that people's perceptions of access to resources by their grandparents, themselves, and their grandchildren would reflect forest quality in their area. We anticipated that such an examination might confirm or strengthen the general perception that significant causal links exist between the issues identified in the *CIFOR Generic C&I Template* and sustainability as represented by forest quality.

We selected nine research sites in Indonesia, Cameroon, and Brazil. Five were in forest-rich areas, and four were in forest-poor areas (descriptions of each site are given in the Introduction, Annex 2). We deal first with the question of sharing of benefits among stakeholders and then with intergenerational access to resources.

Sharing of Benefits

We used a pebble-sorting method (described in the Introduction) to assess the sharing of benefits, recognizing that any such assessment is likely to be partial.

The need to tailor the method to different sites precluded the kind of standardization that might be desirable from a statistical perspective. We examined the data using various cluster analysis techniques and selected bar charts and biplots as the best tools for displaying these results (also described in the Introduction).

Forest-Rich Sites

In the five forest-rich sites—Bulungan (Indonesia), Mbongo and Dja Reserve (Cameroon), and Trairão and São João (Brazil)—the stakeholders and benefits were identified by the researchers who worked on that site on the basis of their existing knowledge of the area and/or observations and interviews on this topic. Stakeholder identification also was the topic of a series of methods tests, which contributed to the researchers' selection process (reported in Sections 1 and 2 of this book). Then, local residents allocated these benefits among the stakeholders (Table 11-1).

Bulungan, East Kalimantan, Indonesia. In Bulungan,[1] the government and the logging company—the formal managers of the forest—are seen to dominate in access to cash (Figure 11-1). Local people and formal managers have fairly equal shares of timber. For the rest of the benefits, local people perceive that they have greatest access.

Biplot analysis identified those stakeholders with similar profiles relating to the distribution of benefits. Stakeholders are positioned on these graphs according to their correlation with particular benefits (Figure 11-2). The logging company, whose forest use is concentrated on timber extraction, is located at the bottom of the graph near the arrows for "cash" and "timber."

Table 11-1. *Stakeholders and Benefits in Five Forest-Rich Sites*

| | Indonesia | Cameroon | | Brazil | |
	Bulungan	Mbongo	Dja Reserve	Trairão	São João
Primary stakeholders					
Punan	Natives		Baka (pygmy)	Colonists (female)	*Ribeirinhos* (female)
Lundaye	Nigerians		Nzime	Colonists (male)	*Ribeirinhos* (male)
Abay	Traditional doctors		Kako (in-migrant)	Middlemen	Middlemen
Other ethnic groups	Sawyers		Forest workers	Ranchers	Ranchers
Merchants	Hunters and trappers		Merchants	Loggers	Large-scale fisherfolk
Foreign projects	Cameroon Development Corporation (a planta-tion company)		Logging company	Logging company	Consumers
Logging company	Mt. Cameroon Project		Conservation project	INCRA	Loggers
Government	Logging company		Village development committee	IBAMA	Logging company
	Government officials		Traditional chiefs		INCRA
			Forest service		IBAMA
Forest benefits					
Cash	Cash		Cash	Timber	Timber
Timber	Timber		Timber	Wildlife	Wildlife
Wildlife	Wildlife		Wildlife	Land	Land
Forest foods	Forest foods		Forest foods	Water and fish	Water and fish
Medicinal plants	Medicinal plants		Medicinal plants	NTFPs	NTFPs
NTFPs	NTFPs		Rattan and other NTFPs		

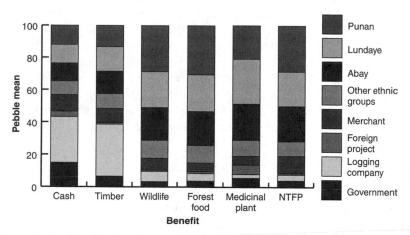

Figure 11-1. *Benefit Sharing among Stakeholders in Bulungan, Indonesia*

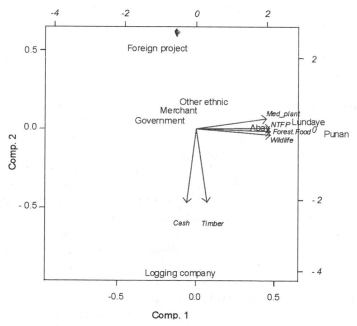

Figure 11-2. *Biplot for Stakeholders from Bulungan, Indonesia*

This location reflects the local perception that the logging company receives a large share of cash and timber but only a small share of other forest benefits. Local peoples (Punan, Abay, and Lundaye) were characterized by their larger share of wildlife, forest foods, medicinal plants, and nontimber forest products (NTFPs) but smaller share of cash and timber (note the cluster on the right side of the graph). The government, the foreign project, merchants, and other ethnic groups were stakeholders with small shares in every benefit. The correlation between timber and cash is positive (as shown by the small angle between their arrows in the biplot), meaning that if a stakeholder has a bigger share of timber, that stakeholder will also have a bigger share of cash. This finding is reasonable given the intensive nature of logging nearby. The correlation among wildlife, forest foods, medicinal plants, and NTFPs is also positive. It reflects the local peoples' strong dependence on those resources for their daily sustenance and livelihood.

Mbongo, Cameroon. In the Mount Cameroon area, the same six benefits as in Kalimantan were identified as contributing to people's livelihoods (timber, cash, wildlife, forest foods, medicinal plants, and NTFPs).[2] In Mbongo, as in Bulungan, the logging company dominates timber, but none of the stakeholders really dominates cash in the clear way seen in Kalimantan (Figure 11-3). Government officials, the logging company, and the Cameroon Development Corporation are perceived to have fairly equal shares of cash. Natives dominate forest foods and NTFPs, whereas traditional doctors seem to have a dominant share of medicinal plants. Hunters and trappers are clearly linked to wildlife in this data set.

Certain activities were likely to be underreported in relation to benefits. The dominant discourse of native rights to manage and control the forest dominates public discussion in this area and often serves to mask the reality of how local inhabitants, irrespective of their ethnic status, balance forest- and agriculture-based activities in pursuit of making a living. Field observations during the study suggested that nonnative Cameroonians and Nigerians had greater—albeit clandestine—access to forest areas than was reported in the pebble-sorting exercise, particularly with regard to forest foods; this phenomenon has been observed in other studies (Rew and others 1996; Brocklesby and Ambrose-Oji 1997). As such, some distinction is necessary between the public domain, where stakeholders might meet to discuss and negotiate forest management issues, and the private domain, where local stakeholders (both natives and nonnatives) pursue their livelihood strategies with significant implications for benefit sharing.

The case of wildlife benefits also highlights the contrast between the public and the private domains. Other studies of the area (Gadsby and Jenkins 1992; Ekwoge and others 1997) demonstrated the economic and social importance of hunting. In Mbongo, most game meat is sold for cash. Depending on the species and amount of game meat extracted, hunters might take their bounty to the local market or more lucrative markets farther afield. It is significant that hunting is acknowledged as a benefit that brings in cash because a great deal of the existing wildlife trade in this area is illegal; the sales take place clandestinely, in informal markets; and hunting in general remains hidden from the public domain.

Natives appear at the far right of the biplot analysis (Figure 11-4), near the arrows for forest foods and NTFPs. These resources are clearly seen as primarily accessible to this group. The proximity of traditional doctors to medicinal plants and hunters to wildlife is also obvious. The one interesting pattern is that wildlife is positively correlated (though not very strongly) with cash, consistent with the pattern described above for the sale of game meat.

Dja Reserve, Cameroon. In the Dja Reserve, 12 stakeholders were considered important in forest benefit sharing (Figure 11-5). The state and logging companies dominate in access to cash and timber, whereas the Kako, Baka, and Nzime (three local ethnic groups) are perceived to receive most of the rest of the benefits. A fair amount of wildlife was attributed to merchants, somewhat indicative of the intensive hunting there.[3]

The biplot pattern for the Dja Reserve (Figure 11-6) is similar to that for Bulungan (Figure 11-2). Timber is positively correlated with money. Forest foods, medicinal plants, wildlife, and NTFPs are positively correlated with each other. The state is the stakeholder most similar to the logging companies, with both dominating in access to cash and timber. Among the local people, the Baka[4] are perceived to have the most access to forest foods, medicinal plants, wildlife, and NTFPs. The Nzime and the Kako follow. In the Dja Reserve, local people's access to resources is definitely linked to the order of their arrival in the area as well as their knowledge of the environment. The Baka have lived in the area for the longest time of all the groups and therefore claim traditional rights to the forest and its resources. Moreover, they have the best knowledge of the forest and thus can make optimum use of the different resources. The Nzime came later, which explains their more limited access to resources (the rights were given to them by the first inhabitants). Finally, the Kako arrived much later. The fact that the Kako came from the savanna explains why they seem to use fewer forest resources; they have limited knowledge about the forest environment.

As in Bulungan, several stakeholders are perceived to not receive great quantities of any forest benefits (local government, the forest service, forest workers, merchants, traditional chiefs, village development committees, and the conservation projects). It is worth noting that in the Dja Reserve, these stakeholders are more or less considered to be intermediate on a continuum, with the government and logging companies on one end, the local populations on the other. One of the hypotheses made by the team in the Dja Reserve was that if these intermediate groups lean closer to the government and logging companies in receiving greater shares of the forest benefits, then

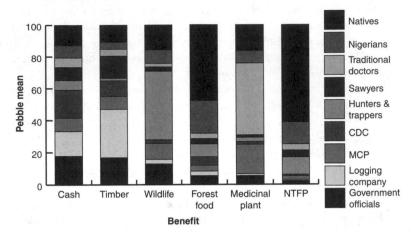

Figure 11-3. *Benefit Sharing among Stakeholders in Mbongo, Cameroon*

Note: CDC = Cameroon Development Corporation; MCP = Mount Cameroon Project.

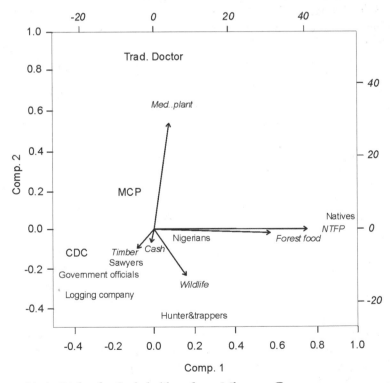

Figure 11-4. *Biplot for Stakeholders from Mbongo, Cameroon*

Notes: CDC = Cameroon Development Corporation; MCP = Mount Cameroon Project.

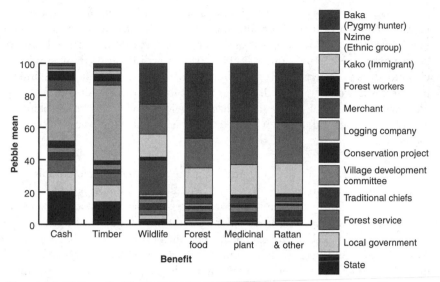

Figure 11-5. *Benefit Sharing among Stakeholders in Dja Reserve, Cameroon*

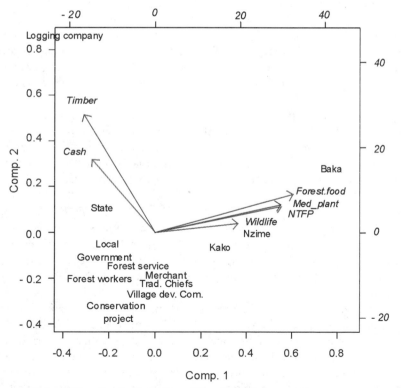

Figure 11-6. *Biplot for Stakeholders from Dja Reserve, Cameroon*

forest management is likely to be inequitable and the system inclined to corruption. Alternatively, if those intermediate groups receive more or less the same benefits as the local populations, then the forest management is likely to be more transparent and equitable. However, it is difficult to draw any conclusions about this hypothesis from our results because the role of the intermediates is not clearly defined. The group of local government officials, forest service, traditional chiefs, and village development committees seems to be rather isolated and does not seem to receive a great share of cash, food, or medicines.

Trairão, Brazil. In Trairão (Figure 11-7), the male and female colonists are perceived to have primary access to land, wildlife, water and fish, and NTFPs. Loggers, the logging company, and ranchers also have moderate shares in these resources. However, timber is dominated by the logging company.[5]

The shares of all benefits except timber are positively correlated (Figure 11-8). The colonists, both male and female, tend to have a large share of every benefit except timber. Loggers and the logging company were characterized as having a large share of timber and some share in the other benefits. It is interesting to note that although there is a clear distinction regarding the benefits actually received by the owners of logging companies and the benefits received by loggers, such a distinction is not captured in the current analysis from this biplot method. In this case, whereas female colonists perceived logging companies and loggers as sharing similar amounts of benefits, male colonists perceived loggers as closer to themselves and the logging company as a distinctly separate dominant group. This response is similar to what we would expect were this method used in a North American logging community, where local people are the loggers. Ranchers get a small share of every benefit (Figure 11-8). The rest of the stakeholders (Instituto Nacional de Colonização e Reforma Agrária [INCRA] and Instituto Brasileiro do Meio Ambiente e dos Recursos Naturais Renováveis [IBAMA]), indicated by a stack of points in the upper left quadrant of the graph, have very small shares.

São João, Brazil. Among stakeholders in São João, the *ribeirinhos*—people who have lived along the rivers of the area for generations—dominate benefits from timber, wildlife, land, and NTFPs (Figure 11-9). Meanwhile, water and fish are perceived to belong to middlemen and large-scale fisherfolk. Female *ribeirinhos* have a relatively more important role in NTFPs than in the other benefits.

As shown in the biplot (Figure 11-10), shares of benefits are correlated with each other, except for water and fish. This means that those stakeholders who have a big share of any one benefit are likely to have a big share of the other benefits, except for water and fish. An unusual pattern here is that the loggers and logging company only have a small share of timber and cash. This perception is due to a high level of community organization that has

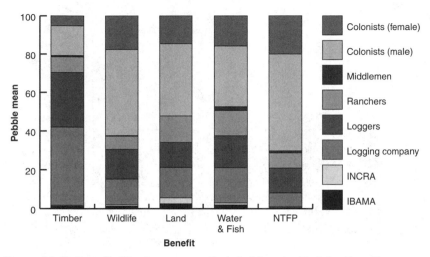

Figure 11-7. *Benefit Sharing among Stakeholders in Trairão, Brazil*

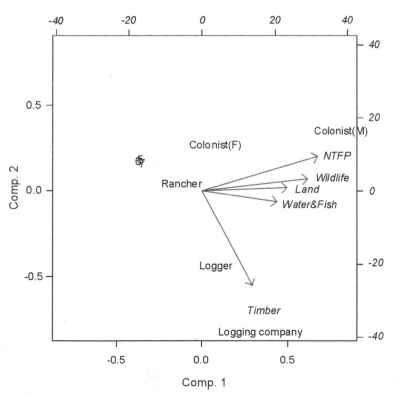

Figure 11-8. *Biplot for Stakeholders from Trairão, Brazil*

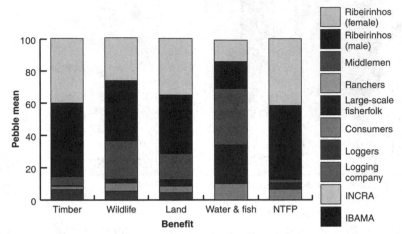

Figure 11-9. *Benefit Sharing among Stakeholders in São João, Brazil*

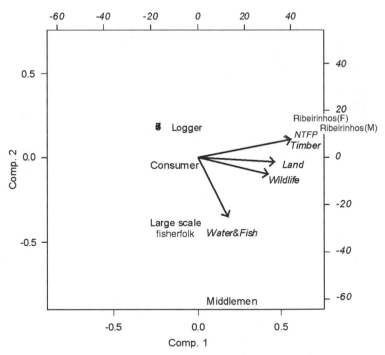

Figure 11-10. *Biplot for Stakeholders from São João, Brazil*

been mobilized to guard against resource depletion. Hence, even though logging companies actively operate in neighboring sites and have tried twice to extract timber from São João, they were promptly expelled by the majority of the local dwellers.

In São João, neither logging companies nor forest dwellers have ever commercialized timber. Local dwellers' expectations, however, are that a community project will be implemented in the future to guarantee their access to benefits from timber. Although local dwellers are not getting any benefits from timber today, this prospect explains the correlations on the benefit distribution graph.

Forest-Poor Sites

Researchers studied four forest-poor sites: Long Segar (Indonesia), Mbalmayo (Cameroon), and Transiriri and Bom Jesus (Brazil). Stakeholders and benefits were identified as for the forest-rich sites (see earlier). Table 11-2 is a list of the benefits as allocated by local residents.

Table 11-2. *Stakeholders and Benefits in Four Forest-Poor Sites*

Indonesia	Cameroon	Brazil	
Long Segar	Mbalmayo	Transiriri	Bom Jesus
Primary stakeholders			
Kenyah	Local people	Colonists (female)	*Ribeirinhos* (female)
Kutai	Immigrants	Colonists (male)	*Ribeirinhos* (male)
Other ethnic	Sawmill workers	Middlemen	Middlemen
groups	Local artisans	Ranchers	Large-scale
Merchants	Logging company	Large-scale	fisherfolk
Foreign projects	State	fisherfolk	Consumers
Logging company		Consumers	Loggers
Government		Loggers	Logging company
		Logging company	
		INCRA	
		IBAMA	
Forest benefits			
Cash	Cash	Timber	Timber
Timber	Timber	Wildlife	Wildlife
Wildlife	Wildlife	Land	Land
Forest foods	Food crops	Water and fish	Water and fish
Medicinal plants	Medicinal plants	NTFPs	NTFPs
NTFPs	Fibers		

Long Segar, East Kalimantan, Indonesia. From Long Segar, seven stake-holders were considered to have a significant role in sharing forest benefits: the Kenyah, the Kutai, other ethnic groups, merchants, a foreign project, a logging company, and the government (Figure 11-11). The benefits identified were essentially the same as in Bulungan, Mbongo, and the Dja Reserve.[6]

The pattern in Long Segar is quite similar to that for Bulungan (Figures 11-1 and 11-2). The logging company dominates in access to timber and cash, whereas the Kenyah and Kutai dominate in access to wildlife, forest foods, medicinal plants, and NTFPs. The resulting biplot analysis (Figure 11-12) is also strikingly similar to the Bulungan results.

Mbalmayo, Cameroon. In Mbalmayo, only six stakeholders—local people, immigrants, sawmill workers, local artisans, the logging company, and the state—were identified as having a direct interest or share in forest benefits. The benefits were similar to those at most other sites: cash, timber, wildlife, food crops, medicinal plants, and fibers. The use of food crops rather than forest foods and fibers rather than NTFPs reflects the forest–poor nature of this locale.[7]

In the pattern of benefit sharing (Figure 11-13), local people share with the state and the logging companies in access to cash and timber. As with other sites, local people dominate in access to more subsistence-oriented products.

The biplot analysis (Figure 11-14) shows that the shares of wildlife, medicinal plants, forest foods, and NTFPs in this area are somewhat positively correlated with shares of cash—unique in our sample of sites. The reason may be the geographical position of the site, close to two towns, Mbalmayo and Yaoundé, the capital of the country. Any forest product can easily be converted into cash. Again, the logging companies and the state are primary recipients of timber because the people think that the state receives an important amount of money from timber sales in the form of duty and taxes. Local people recognize that they receive the most important part of any forest products other than timber. Even so, they do not perceive the sharing of forest benefits as fair, because forest products other than timber are considered to be secondary. Local artisans, immigrants, and sawmill workers are not so clearly linked to any of the benefits as are the other stakeholders.

Transiriri, Brazil. In Transiriri, the same stakeholders and benefits were identified as in Trairão. The logging companies and loggers have the most access to timber. Ranchers are perceived to have most access to land and to water and fish, and colonists also have significant shares of these benefits. Meanwhile, colonists of both sexes have the greatest access to wildlife and NTFPs (Figure 11-15).[8]

Again, the profiles of the stakeholders' shares of benefits seem reasonable and fairly consistent with our findings at other sites, with the exception of

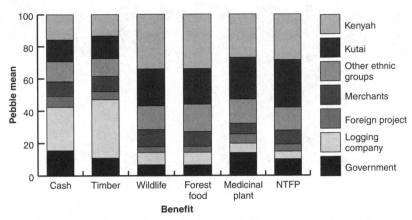

Figure 11-11. *Benefit Sharing among Stakeholders in Long Segar, Indonesia*

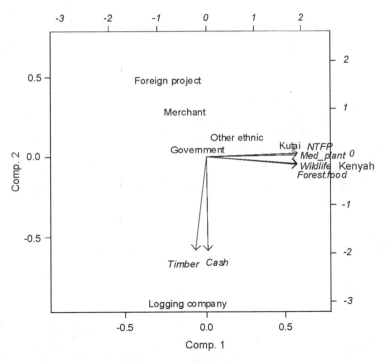

Figure 11-12. *Biplot for Stakeholders from Long Segar, Indonesia*

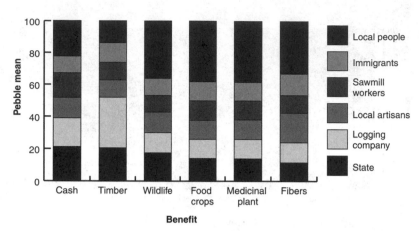

Figure 11-13. *Benefit Sharing among Stakeholders in Mbalmayo, Cameroon*

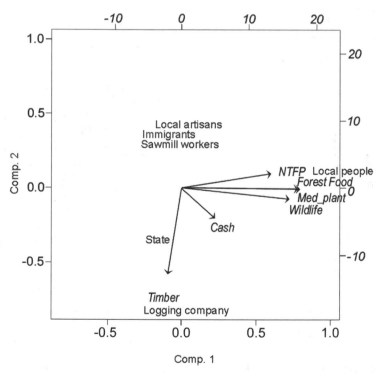

Figure 11-14. *Biplot for Stakeholders from Mbalmayo, Cameroon*

the absence of the large-scale fisherfolk in connection with water and fish in other sites (Figure 11-16). Transiriri is a secondary road off the Transamazon highway, and the village where the test was conducted is 50 kilometers from the Iriri River, where commercial fishing occurs. The case of Transiriri (a site where colonist land tenure is extremely insecure) deserves special attention because this method does not capture the insecurity faced by colonists toward the future. In fact, although during semistructured interviews and informal conversations people in Transiriri constantly referred to the possible expropriation of their lands by the Fundação Nacional do Índio (FUNAI; the Brazilian national authority for Indian affairs) in the near future, the sorting of seeds among plates reflects only the current situation regarding benefit sharing.

Bom Jesus, Brazil. In Bom Jesus, the logging companies and loggers clearly dominated only timber (Figure 11-17). Land access was dominated by middlemen who also had a sizeable share of water and fish. Water and fish seemed to be shared primarily among fisherfolk, middlemen, and the *ribeirinhos*. *Ribeirinhos'* shares were most striking for NTFPs and wildlife.

Biplot analysis (Figure 11-18) confirms that male and female *ribeirinhos* have the most dominant share of NTFPs and wildlife. The logging company and loggers, as usual, dominate in their share of timber. Finally, middlemen, consumers, and fisherfolk together dominate water and fish. The correlation between shares of land and shares of water and fish also is positive. In this case, the attribution of large shares of benefits from timber to logging companies and loggers proves to be an interesting insight of the method. Despite logging companies and loggers no longer (or only marginally) operating in Bom Jesus (and the same applies to Transiriri), male and female *ribeirinhos* attributed to them a large share of benefits exactly because, as one Bom Jesus resident stated, "they are now gone, and nothing is left, even the road is abandoned here. They took all the wealth to outside ... they are making their future and the fools here have nothing to survive."

Cross-Site Comparisons of Benefit Sharing

Generally, in both forest-rich and forest-poor sites, three major groups were identified:

- local people, who have the dominant shares of forest foods, wildlife, water, fish, NTFPs, and medicinal plants;
- logging companies and loggers, who have dominant shares of timber and cash (except in São João); and
- "less significant" stakeholders, who have smaller shares in every benefit.

In the forest-rich sites, stakeholders other than the first two groups often dominate one or more forest benefits. In Mbongo, for instance, traditional

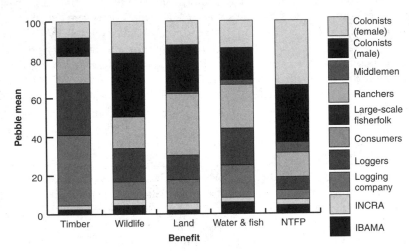

Figure 11-15. *Benefit Sharing among Stakeholders in Transiriri, Brazil*

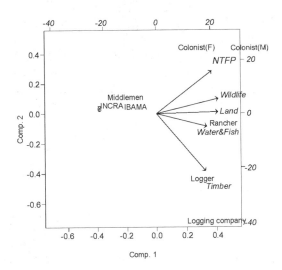

Figure 11-16. *Biplot for Stakeholders from Transiriri, Brazil*

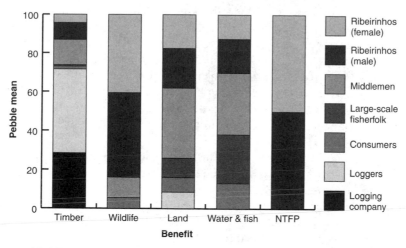

Figure 11-17. *Benefit Sharing among Stakeholders in Bom Jesus, Brazil*

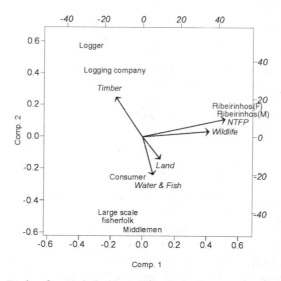

Figure 11-18. *Biplot for Stakeholders from Bom Jesus, Brazil*

doctors are perceived to have a dominant share of medicinal plants, and hunters and trappers to have a dominant share of wildlife. In São João, fisher-folk's and middlemen's shares of water and fish are other good examples.

Attending to gender differences in Brazil—by the use of plates that represented both male and female local dwellers in the pebble-sorting process—produced significant insights into the determinants of gender differences in access to benefits. In comparing the results from Transiriri and São João, we found a seemingly causal link between conscious social organization for conservation of resources and a more balanced access to benefits between men and women. Conversely, the Transiriri community's insecurity about the basic foundations for its integrity (access to forest and land, freedom in allocating labor) seemed to be reflected in perceptions of less equitable access to benefits based on gender.

In concluding this section, we return to our initial hypothesis: that inequitable sharing of benefits (as perceived by local stakeholders) would correlate with poor forest quality. This was not confirmed. Instead, we found a consistent pattern that timber and, usually, cash were seen as taken by timber companies, loggers, and the state in both forest-rich and forest-poor areas and that local people were perceived to have significant shares of other resources in all sites. The differences among sites seem much more closely connected with particular historical chains of events (as Vayda 1996 argues from a more philosophical point of view) and with cultural or regional patterns than with the current quality of local forests.

We cannot conclude from these data and analyses that equitable sharing of benefits is linked to sustainable forest management, even though our other, more qualitative analyses support our perception that such a link does in fact exist. We suspect that one fatal flaw is linking sustainable forest management with current forest quality, which is essentially an artifact of previous management. A second fatal flaw is the reductionist, decontextualizing approach taken in this analysis—an analysis that adheres carefully to the requirements of such science.

Intergenerational Access to Resources

The primary purpose of this study was to assess stakeholders' perceptions of generational access to resources (Principle 3, Criterion 3.3, Indicator 3.1.5 [Annex 1 in the Introduction]) by using a pebble-sorting method in five forest-rich and four forest-poor contexts in Indonesia, Cameroon, and Brazil.

We had to adapt the method for determining intergenerational access to resources in the Brazilian colonist sites, where the local peoples were predominantly recently arrived settlers, whose grandparents (and in some cases, parents) were in no way related to local resources. Some researchers also found it necessary to modify the question. Tchikangwa's team (1998) in

Cameroon, for instance, substituted "the absence or existence/the rarity or frequency of conflicts related to the use of resources" ("l'absence ou l'existence/la rareté ou la fréquence des conflits liés à l'utilisation des ressources") in place of "access to resources" due to problems trying to translate the original concept into local languages. In the Brazilian tests, a similar question was incorporated about respondents' perceptions on how well-being changed over time, across generations. We thought that this additional question increased the explanatory power of the method with regard to the extent that changes in accessibility to forest resources are associated with changes in well-being.

The distribution of local resources among the generations as perceived by respondents in the forest-rich sites reveals a worrying overview of how those people perceive the security of intergenerational access to resources (Figure 11-19). Only in Bulungan do people consider the access of future generations to be secure. In the other forest-rich sites, people perceive decreasing access to resources for future generations (see also the Trinidad case reported in Chapter 7).

Greater allocations for past generations usually indicate both less constrained access and greater supply of resources. In Brazil, the latter factor was often observed to have a major influence on the allocation of seeds to one category or another.

The number of pebbles attributed to "self" tends to be more stable from site to site than those attributed to the other generations. Figure 11-19 shows that, in all sites except Bulungan, people feel their grandparents had access to more resources than they themselves have. The expectation that their children would have still less access to resources was reported in the Dja Reserve, Mbongo, and Trairão. As noted earlier, migration from different areas makes comparisons of the colonist situations complex. Stratification by age group is essential in these cases to minimize bias when averaging allocations made by older and younger stakeholders.

In the forest-poor sites, all the sites except for Long Segar (the other Indonesian site) show a decline in perceptions of access to resources for future generations compared with the past generation (Figure 11-20). Mbalmayo shows a slightly unusual pattern where access was perceived to be greatest for the current generation rather than past and future generations. A small allocation to past generations in Mbalmayo is related to people's perception that in the past there were many resources, but access was restricted due to lack of sophisticated equipment (sawmills and tractors) and infrastructure (roads and bridges).

As we analyzed the data to assess support for the hypothesis that people's perceptions of access to resources by their grandparents, themselves, and their grandchildren would reflect forest quality in their area, we realized that within-country similarities were greater than those based on forest quality. To test for similarities across sites, we used hierarchical cluster analysis, incor-

269

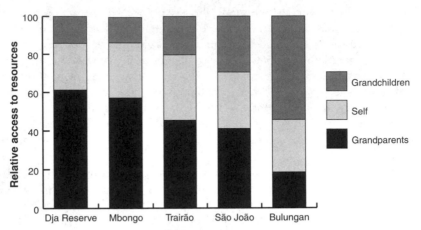

Figure 11-19. *Perception of Intergenerational Access to Resources in Forest-Rich Sites*

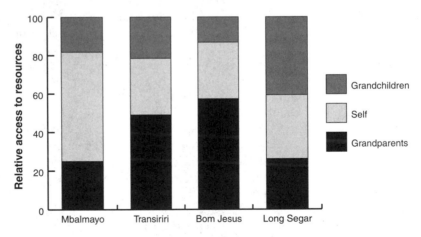

Figure 11-20. *Perception of Intergenerational Access to Resources in Forest-Poor Sites*

porating several linkage methods and distance measures. For determining the optimum number of groups at final partition, we used a formula suggested by Mojena (1977, revised by Milligan and Cooper 1985).[9]

The two Cameroonian sites falling into one cluster and the two Brazilian sites in another cluster reveal similar perceptual patterns relating to security of intergenerational access to resources. People tend to perceive access to resources as declining from past generations to future generations, and those in Cameroon anticipate the most drastic reduction. The Cameroonian perception may well be related to the comparatively extreme antagonism between the generations in that country, where the young often suspect their elders of using sorcery against them. Bulungan, the only forest-rich site from Indonesia, reveals a completely different pattern. These people foresee access to resources for future generations increasing compared with past and current generations (Figure 11-19). Plausible explanations are many, including an inherent optimism in the Bornean psyche, a sense that external actors will bring in modernity with accompanying prosperity, a belief that higher levels of education and technological improvements spell "progress" and wealth, and confidence that their children will be better prepared to defend their resources from powerful intruders.

Among forest-poor sites, cluster analysis shows the same pattern. The most consistent results yield three groups as the optimum. Again, sites from the same country (Brazil) are clustered together, whereas Mbalmayo and Long Segar are clustered individually.

To make any correlation between sites more obvious visually, we use the minimum spanning tree (MST). MST can be thought of as a set of lines that represent the pair-wise similarities that connect all the objects (sites, in our case) by a line of the shortest distance possible (Belbin 1993). The network or tree shown in Figure 11-21 was developed by using this method. MST works like multidimensional scaling to produce a configuration of all objects and then tries to make a circuit (a line that connects all objects) with the restriction that the circuit should have the minimum length possible.

Figure 11-21 makes the within-country similarities quite clear; sites from the same country tend to have closer connections than those from different countries. The one exception was Mbalmayo, which tends to be closer to the Indonesian sites. The differences between countries appear to be greater than the differences based on forest quality.

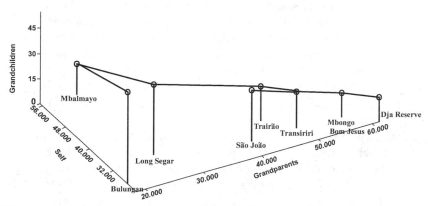

Figure 11-21. *Minimum Spanning Tree (MST) for All Sites*

Cross-Site Comparisons on Intergenerational Access to Resources

Our data suggest that, if we accept local people's perceptions on the basis of this pebble-sorting method, respondents from test sites in Cameroon and Brazil do not expect intergenerational access to resources to be maintained over time, whereas the respondents from test sites in Indonesia do. If resources are not maintained over time, continuing links between people and the forests are likely to be broken and people are likely to feel insecure about their continued access.

Although residents of both Indonesian sites see greater benefits for their grandchildren than for themselves or their grandparents, those in the forest-rich site (Bulungan) see a considerably greater increase for their children than do those in the forest-poor site (Long Segar). Long Segar's pessimism has almost certainly been exacerbated since this 1997 data collection: Long Segar's entire territory burned during the El Niño–induced drought of 1997–1998 (Colfer and others 2000b).

If we compare the two forest-rich Cameroonian sites (Dja Reserve and Mbongo) with the forest-poor Cameroonian site (Mbalmayo), we see a steady decrease in access to resources in the forest-rich sites but a very pronounced spike in access to resources for the current generation in the forest-poor site (see also Chapter 10). Indeed, the amount remaining for the children in the forest-poor site is anticipated to be more than for those from the forest-rich sites. People seem to perceive themselves to be garnering a significant share of forest benefits.

In Brazil, comparing the forest-rich (Trairão) and forest-poor (Transiriri) colonist sites, we find a rather similar, stair-step pattern of resource availability where the parents (grandparents were not included in this test) have the greatest access, the respondents' generation has an intermediate level of access, and the grandchildren receive the least. If we compare the forest-rich (São João) and forest-poor (Bom Jesus) *ribeirinhos* sites, we find that the forest-poor site resembles (with greater intensity) the pattern observed in the colonist sites, whereas the residents of the forest-rich site expect to maintain the level of access to resources they themselves have for their children. This response could be largely related to the perception that São João residents have of their relative success in avoiding the entrance of outsiders (timber companies and large-scale fisherfolk) and preserving natural resources (mainly timber and fish) and to their confidence in institutional arrangements for the purpose of adequately managing these resources in the future.

A side note relates to the method's gender sensitivity in the context of the tests in Brazil. For women in the Transamazon, land inheritance is not automatic, like it is for men; for women it involves such conditions as marriage, separation, or level of achievement in education. Although women are legally entitled to half of the couple's possessions in case of divorce, constraints such as lack of labor in specific phases of the agricultural calendar may impede a woman's use of this access. Therefore, women may have allocated their seeds according to a male-centered view or according to scarcity of the resource, rather than its access.

In summary, although we feel that this method and these analyses have been useful for shedding light on issues relating to the security of intergenerational access to resources, they have not confirmed either of our hypotheses. The unsatisfying nature of these results has provided additional support for the view that contextual, historical, and ethnographic understanding is critical in assessing issues such as the security of intergenerational access to resources.

Endnotes

1. See Chapter 1 on Long Loreh, Bulungan, and Chapters 5, 6, 8, 12, and 16 for comparable forest-based systems. See also Kaskija 1995, Wollenberg and others 1996, Puri 1997, and Wollenberg 1999 on the people of Bulungan.

2. See Brocklesby and Ambrose-Oji 1997 and Brocklesby and others 1997 for details of this particular area; Chapter 9 for a related, forest-rich system in neighboring Gabon; Chapters 2, 3, and 10 for related systems in more forest-poor Cameroonian contexts; Eba'a-Atyi 1998 for a discussion of the timber industry in Cameroon; and Sunderlin and Pokam 1998 on effects of the economic crisis.

3. See Tchikangwa and others 1998 or Chapters 2, 3, and 10 for more forest-poor but related systems in Cameroon; see Chapter 9 for a related forest-rich system in Gabon.

4. The Baka have traditionally been referred to as pygmies in much of the literature.

5. See Porro and Miyasaka Porro 1998 and Chapter 13 for contextual information about all Brazilian sites.

6. See Colfer 1981, 1982, 1983a, 1985a, 1985b, 1991; Colfer with Dudley 1993; Colfer and others 1997a; Sardjono and others 1997; and Chapters 1, 5, 6, 8, 12, and 16 for related systems in other areas of Kalimantan.

7. See Tiani and others 1997 and Chapter 2 for contextual information on this area; Chapters 3 and 10 are also directly relevant.

8. See Porro and Miyasaka Porro 1998 and Chapter 13 for contextual information.

9. The fact that all the methods identify five groups as the optimum solution suggests that clusters are not obvious. By changing the cutoff point to 75% similarity, we get three clusters as the optimum solution, with sites from the same country clustered together.

Rights and Responsibilities to Manage Cooperatively and Equitably

This topic, which we phrase as "acknowledged rights and responsibilities to manage forests cooperatively and equitably," initially emerged as a call for local people to "participate" in forest management (discussed at length in Chapter 12 [Colfer and Wadley]). We largely abandoned the use of the term *participate* when we began to realize the recurrent existence of two or more overlapping management systems operating in the same location. It simply did not make sense to refer to local peoples' "participation in forest management." The phrase includes two loaded assumptions: that local peoples were not doing forest management themselves, and that they should be participating in one of the other external management systems (such as a timber concession).

The relevant literature on participation is extensive. Pretty (in Pimbert and Pretty 1995) developed an excellent typology of categories of participation. Numerous field manuals and "how to" documents guide users in improving the participation of local communities (Carter 1996; Hobley 1996; Wollenberg 1997; Ingles and others 1999). Much of the literature has a serious project focus (for example, Songan 1993; Isham and others 1995; Borrini-Feyerabend with Buchan 1997; Colfer and others 1999d) as opposed to the kind of naturalistic involvement we refer to in this book. Researchers at the International Institute for Environment and Development also have "evolved from the study of participation" (Bass and others 1995) to the study of rights and responsibilities in their work (see also the "4Rs" in forest management [rights, responsibilities, returns/revenues, and relationships] in Dubois 1998, 1999).

A wealth of material, from the 1980s onward, falls under the rubric of farming systems research and development that provides relevant lessons on cooperation between farmers and researchers and could be adapted to the management of natural resources (for example, Shaner and others 1982; Chambers and others 1989; Rocheleau 1999; Collinson 2000). More recently, a rash of work on integrated conservation and development projects also assumes the importance of people's involvement (for example, Wickham 1996; Borrini-Feyerabend with Buchan 1997; Child and others 1997; Nguinguiri 1998; Getz and others 1999); however, some researchers and practitioners have found reason for concern with these approaches (Wells 1997; Murombedzi 1999; Campbell and others 2000a, 2000b).

Studies that emphasize local peoples' own management abound (with numerous contributions by Colfer, Dove, Padoch, Peluso, and Wadley in Borneo alone; Alcorn, Clay, Posey, Schmink, and Snook in Latin America; Diaw, Leach, Nguinguiri, and Katz in Africa; plus compendia by Kunstadter and others 1978; Croll and Parkin 1992; Redford and Padoch 1992; Padoch and Peluso 1996). Obviously, local peoples recognize their own rights and responsibilities to manage forests in their areas. One significant problem is that outside actors are often completely unaware of the complexity and effectiveness of some traditional management systems, as argued in Chapter 12.

The criterion "rights and responsibilities to manage equitably and cooperatively" evolved from an initial concern by local communities about "participation in forest management." We have found in these predominantly developing-world contexts that local voices rarely have legal strength; yet very frequently, those voices are strong—especially in the absence of more formal penetration by the state or other outside actors.

Several rationales explain why researchers perceive this issue of *voice* to be important. The first pertains to the existence of bodies of indigenous knowledge in many forest-dwelling communities. When such knowledge is held by people whose voices are not heard and who have no recognized role in formal forest management, a valuable human resource is wasted. Ample evidence of such indigenous knowledge appears in the anthropological literature and increasingly in other bodies of literature as well. Such knowledge is an important part of forest peoples' cultural repertoire and as such has implications for their own well-being. Global cultural wealth is diminished with the loss of indigenous knowledge.

A second rationale relates to the social capital represented by local institutions and management systems. In tropical forests, complex and enduring management systems typically have functioned well in the past and continue to function well in many areas. Again, not recognizing local rights and responsibilities to manage can be considered a profligate waste of social capital. Such management institutions, as part of local cultures, obviously perform other social functions of importance to local communities, as noted in Chapter 10 (Russell and Tchamou). Cultural diversity, on a par with biodiversity as something worth maintaining, is reduced insofar as local management systems fade away under outside pressures.*

A third rationale is pragmatic. Local people live in local forests and therefore have both a motivation and a capability to perform various management functions that are much more complex for outsiders to perform. Integration of formal (outside) management and local management holds a potential for improved overall management. Agreements about goals and regulations can be reached to make management more consistent, and communication among different stakeholders or interest groups can avert and diffuse potential conflicts.

A fourth and final rationale pertains to local peoples' control over their own destinies. Social scientists recognize that cultures that function well are characterized by both stability and change. Human beings need sufficient stability to raise their young, to maintain a core of cultural values, and to nurture what they hold dear; yet all people seem to want to improve their lives, for themselves and for their children. Some control over the direction and pace of change is important to the mental health of all peoples. For forest dwellers, having both rights and responsibilities to manage local resources is an important aspect of maintaining this balance between stability and change.

In Chapter 12, Colfer and Wadley report research undertaken in forest-rich Danau Sentarum Wildlife Reserve, West Kalimantan, Indonesia (also discussed in Chapters 5, 8, and 16). Their focus is on the results of the initial attempt to study *participation* in forest management. One conclusion is that the concept may reflect a misguided assumption.

In Chapter 13, Brazilian Noemi Miyasaka Porro introduces the various interpretations of *participation* in the Amazon. One of the unique strengths of this contribution is its emphasis on the colonists who come to the rainforest from other environments and have to adapt to the new and alien conditions—some forest rich, some forest poor. Our teams have encountered such people (the equivalent of colonists) in Asia, Latin America, and Africa. Porro's analysis differs from other analyses in this book in its emphasis on narratives from the people (Townsend and others [1995] present the voices of women colonists in Mexico and Colombia).

In Chapter 14, Tchikangwa and others attempt to examine, in a cross-cultural light, rights and responsibilities to manage. The authors examine results from a pebble-sorting method used in Brazil, Indonesia, and Cameroon, examining differences on the basis of national and forest quality dimensions. These results are more significant than are the other attempts at cross-site comparison that we present.

---

*Admittedly in cultural evolutionary processes, ineffective or inefficient systems eventually and naturally die out. But the current context for forest-based peoples and their cultures seems unnaturally antagonistic to local systems, based largely on the extreme inequities in power between forest-based peoples and the groups typically "invading" their areas. The result appears, in our view, to be an acceleration of loss of cultures.

CHAPTER TWELVE

From "Participation" to "Rights and Responsibilities" in Forest Management

Workable Methods and Unworkable Assumptions in West Kalimantan, Indonesia

Carol J. Pierce Colfer and Reed L. Wadley

We report the results of research in West Kalimantan, Indonesia, originally designed to quickly and easily assess the level and nature of participation by local peoples in forest management. During the initial phase of field testing criteria and indicators for sustainable forest management, we had identified that people's participation was an important issue. As we examined our early results, we realized that we could not specify, to our satisfaction, the relevant causal links between sustainable commercial forest management on one hand and the participatory behavior and needs of the people most closely involved on the other.

We found various processes operating in the field. Increased participation provided forest people with a sense of ownership, or a defined "stake," in forest management. It provided other forest managers with useful knowledge about local forest use and management strategies. It provided a means for conflict resolution and empowerment. In other contexts, increased participation seemed to involve increased conflict. From a sustainability point of view, such conflict had positive outcomes in some cases, negative ones in others. The concept of *participation* in commercial forest management seemed to beg further analysis, improved definition, and improved methods to assess it on the ground, reliably and quickly.

This chapter is adapted from Colfer and Wadley 1996.

We selected the area in and around Danau Sentarum Wildlife Reserve (DSWR) in West Kalimantan, Indonesia, to conduct our test, partly because the explicit approach of the Conservation Project there was to work toward co-management of the reserve and significant efforts had been devoted to this goal.

Qualities about the area itself also contributed to its appropriateness. The wildlife reserve was surrounded by timber concessions; two ethnic groups living there had quite different patterns of resource use and, we thought, levels of participation in forest management; and the existence of indigenous, conservation, and production forest management seemed to offer interesting possibilities for comparison.

We report the results of our earliest research on causal links between sustainable forest management and participation. We briefly describe the three methods we found useful and argue that the concept of *participation* as commonly used in natural resource management may rest on an inappropriate assumption: that local peoples should participate in externally defined management systems.

Methods Tested

Our research took place in June and July 1996 and involved two primary researchers (Colfer and Wadley); other team members were recruited locally.[1] We focused on four small communities of 100–200 inhabitants each: Danau Seluang and Nanga Kedebu' (Melayu communities) and Wong Garai and Bemban (Iban communities) (see also Chapters 5 and 8). Another Iban community, Kelayang, was added later, after revising the methods significantly. The Conservation Project had worked extensively with Nanga Kedebu', Bemban, and Kelayang and hardly at all with Danau Seluang and Wong Garai. Nanga Kedebu', located near the Conservation Project Field Center, had involvement in the project since its inception in 1992 (providing guides, research assistants, temporary accommodation, study sites, and so forth). Kelayang had recently been included in an expansion of the reserve boundaries. At the time of the study, both Kelayang and Bemban were involved in the Conservation Project's income-generating efforts (basket weaving and mat weaving).

In selecting appropriate methods, we considered several factors. We felt that a combination of qualitative and quantitative methods would be preferable to either alone. Ultimately, three methods designed to assess participation were found useful: the iterative continuum method, a participatory card-sorting method, and a researcher guide.[2] The methods (described in the Introduction) were translated into Bahasa Indonesia, Melayu, and Iban, as appropriate.

The researcher guide, designed to structure informal interviews, focused on four proposed functions of participation: reducing noncompliance of various stakeholders with regulations by local monitoring; reducing conflict about forest management or converting it to mutual accommodation; contributing indigenous knowledge of forest management; and controlling the speed and direction of social change. The dual purpose of the researcher guide was to allow us to examine the functions in a field context and to remind us of potentially relevant issues.

Results

In this section we present the results, derived from the three methods, that led us to conclude that the term *participation* carries with it some unwarranted assumptions.

Researcher Guide on the Four Functions of Participation

The researcher guide was based on the four functions (listed above) hypothesized to be critical functions of participation in sustainable forest management. In the DSWR, because most management appeared to be in the hands of local peoples, the answers to most questions were, "Yes, they were participating." At the same time they were *not* participating in management that was recognized externally (particularly in commercial timber management—the context for which the questions were initially formulated). This observation or conclusion prompted a thorough review of our assumptions.

In 1994–1995 field tests for the Center for International Forestry Research (CIFOR), we had focused on a formal forest management unit. From the social perspective, we had tried to assess the conditions of the people within that forest management unit. How to involve local peoples in these activities in a mutually beneficial way had been the question in the minds of the field teams. It remains an important question. However, the researcher guide clarified an assumption we found problematic: we had inadvertently assumed the priority of the activities of the timber manager.

In the DSWR, at least four kinds of management were in operation (by the Conservation Project, commercial timber companies, Iban villagers, and Melayu villagers). The question of participation in which system became important. Phrasing the question as we had earlier (how to involve local peoples with the managers of a forest management unit) ignores the many places where local peoples' management has more effect on local forests than does external management. It also reinforces the power of the timber manager or government official vis-à-vis local peoples.

Participatory Card Sorting

With this method, important stakeholders who were ranked included the local community, other communities, the government, the timber companies, the Conservation Project, and traders. The results provide a strong statement about the importance of local communities in the management of the DSWR forests. With "1" representing the most important stakeholder, the average rank for local communities in the DSWR data ranged from 1.00 to 1.93, disaggregated by gender, occupation, or location (Tables 12-1 to 12-10 show the various ways the data sets were analyzed). In all cases, the local community was ranked the most important, and it received the highest average score in frequency of interaction in all but one (Kelayang, Table 12-1). Interestingly, Kelayang is the community that is within the timber concession partially owned by Iban from a nearby area (P.T. Panggau Libau), and participants reported interacting most frequently with the timber company on forest management issues, with the community a close second. This response contrasts markedly with the northwest of DSWR, also within a timber concession (P.T. Militer) but one wholly owned and operated by outsiders. There, the concessionaire is ranked fourth in importance (Table 12-2) and shares second place (11%) with "other communities" and "government" in frequency of interaction, far below the mean for interaction with "community" (49%; Table 12-3).

The Conservation Project rose to third place, both in terms of importance and frequency of interaction (equal with government in the latter instance), in Nanga Kedebu' (Tables 12-4 and 12-5), where the project had been active for four years. This placement was above the sixth place in Danau Seluang (Tables 12-6 and 12-7), a community that had not yet been involved with the project. These findings suggest both less communication with the project than its leaders might wish for Nanga Kedebu' and an accurate depiction of noncommunication between the project and Danau Seluang. As with the Kelayang example, these data show the preeminence of local communities in local forest management.

Most of our card sorting was done in local communities, and one could argue that local communities perceive their own importance to be greater than would other stakeholders in the area. However, the card-sorting data

Table 12-1. *Mean Frequency of Interaction among Six Stakeholders, Assigned by People from the Iban Community of Kelayang*

Stakeholder	Mean frequency of interaction (%)		
	Women	Men	Overall mean
Timber companies	29	27	28
Community	25	24	25
Traders	15	20	18
Other communities	17	17	17
Government	9	8	9
Conservation Project	4	4	4

Table 12-2. *Mean Order of Importance Assigned to Six Stakeholders by People from a Melayu Village, an Iban Village, and a Timber Camp to the Northwest of DSWR*

	Mean order of importance			
Stakeholder	Pulau Duri' (Melayu) (n = 5)	Militer logging camp (n = 4)	Bemban (Iban) (n = 2)	Overall mean
Community	1.95	1.06	3.63	1.93
Other communities	3.40	3.06	3.13	3.23
Government	3.10	4.31	2.13	3.36
Timber companies	4.25	4.13	3.63	4.09
Traders	4.00	4.25	4.13	4.11
Conservation Project	4.30	4.19	4.38	4.27

Notes: Pulau Duri' is a larger Melayu village; Bemban is officially part of Pulau Duri', but its inhabitants are Iban and largely manage their own affairs.

Table 12-3. *Mean Frequency of Interaction among Six Stakeholders, Assigned by People from a Melayu Village, an Iban Village, and a Timber Camp to the Northwest of DSWR*

	Mean frequency of interaction (%)			
Stakeholder	Pulau Duri'	Militer logging camp	Bemban	Overall mean
Community	44	60	43	49
Other communities	12	6	17	11
Government	13	10	10	11
Timber companies	13	12	6	11
Traders	11	6	12	9
Conservation Project	8	6	12	8

Table 12-4. *Mean Order of Importance Assigned to Six Stakeholders by People from the Melayu Community of Nanga Kedebu'*

	Mean order of importance			
Stakeholder	Men (n = 7)	Women (n = 7)	Mixed group (n = 4)	Overall mean
Community	1.00	1.00	1.00	1.00
Other communities	2.96	2.14	2.63	2.57
Conservation Project	3.75	4.43	3.94	3.75
Government	3.64	3.75	3.38	3.93
Traders	4.82	4.50	4.31	4.58
Timber companies	4.82	5.18	5.06	5.01

Table 12-5. *Mean Frequency of Interaction among Six Stakeholders, Assigned by People from the Melayu Community of Nanga Kedebu'*

Stakeholder	Mean frequency of interaction (%)			
	Men (n = 7)	Women (n = 7)	Mixed group (n = 4)	Overall mean
Community	45	46	53	47
Other communities	16	18	16	17
Government	14	12	9	12
Conservation Project	14	11	10	12
Traders	6	6	8	7
Timber companies	5	7	4	6

Table 12-6. *Mean Order of Importance Assigned to Six Stakeholders by People from the Melayu Community of Danau Seluang*

Stakeholder	Mean order of importance			
	Men (n = 5)	Women (n = 7)	Mixed group (n = 4)	Overall mean
Community	1.00	1.00	1.00	1.23
Other communities	3.00	3.00	3.00	2.73
Government	3.00	4.00	4.00	3.53
Traders	4.00	5.00	4.00	4.48
Timber companies	5.00	5.00	4.00	4.52
Conservation Project	5.00	5.00	4.00	4.92

Note: Several Danau Seluang respondents did not rank stakeholders they considered unimportant. In those few cases, we ascribed values, in order of appearance on the form.

Table 12-7. *Mean Frequency of Interaction among Six Stakeholders, Assigned by People from the Melayu Community of Danau Seluang*

Stakeholder	Mean frequency of interaction (%)			
	Men (n = 5)	Women (n = 7)	Mixed group (n = 4)	Overall mean
Community	48	40	45	45
Other communities	20	21	16	19
Government	11	7	10	13
Traders	11	15	11	10
Timber companies	8	10	10	7
Conservation Project	2	7	7	5

from outsiders do not confirm this idea. The P.T. Militer log camp responses in Tables 12-2 and 12-3 and the business and government responses in Tables 12-8 and 12-9 reflect similarly high ranking of communities by outsiders, in terms of both importance and interaction.[3] They provide clear evidence of the near unanimity with which local peoples' management is recognized as preeminent in the area—despite governmental perceptions to the contrary (in the form of timber concessions held by outsiders and the conservation activities under way in the wildlife reserve). Conducting the card-sorting exercise with more distant stakeholders would have been of interest but was outside the scope of our research. Differences in men's and women's ordering for both importance and frequency of interaction (Tables 12-4, 12-6, and 12-10 and Tables 12-1, 12-5, and 12-7, respectively) appear to be minimal.

Table 12-8. *Mean Order of Importance Assigned to Six Stakeholders by People from a Melayu Village, Business, and Government along the Kapuas River*

Stakeholder	Mean order of importance			
	Cincin women (n = 2)[a]	Business (n = 6)	Government (n = 7)	Overall mean
Community	2.38	1.67	1.68	1.77
Government	3.50	2.96	2.54	2.83
Other communities	3.75	3.50	4.29	3.90
Traders	4.50	3.54	4.36	3.90
Timber companies	2.88	4.42	4.04	4.07
Conservation Project	4.13	4.92	4.11	4.55

[a] These two groups comprised a total of six Melayu women.

Table 12-9. *Mean Frequency of Interaction among Six Stakeholders, Assigned by People from a Melayu Village, Business, and Government along the Kapuas River*

Stakeholder	Mean frequency of interaction (%)			
	Cincin women (n = 2)[a]	Business (n = 6)	Government (n = 7)	Overall mean
Community	40	40	31	36
Traders	17	19	10	15
Government	12	12	16	14
Timber companies	21	12	10	12
Other communities	9	12	8	10
Conservation Project	2	4	6	5

[a] These two groups comprised a total of six Melayu women.

Table 12-10. *Mean Order of Importance Assigned to Six Stakeholders by People from the Iban Community of Kelayang*

Stakeholder	Mean order of importance		
	Women (n = 6)	Men (n = 8)	Overall mean
Community	1.54	2.19	1.91
Other communities	2.96	2.28	2.57
Timber companies	3.33	3.09	3.20
Traders	4.33	3.97	4.13
Government	4.33	4.44	4.39
Conservation Project	4.71	4.78	4.75

Note: The assistant conducting this survey neglected to indicate whether the interviews were individual or in groups.

The Iterative Continuum Method

The iterative continuum method (also described in the Introduction) involved the primary researchers making daily entries relating to participation on a form. Both Colfer and Wadley had problems with the initial participation continuum. The day-to-day participation of local peoples in the management of their environment was obvious, and their lack of involvement in timber company management on a day-to-day basis was equally obvious. We therefore reworded the continuum to say "recognized, functioning rights to manage forest resources cooperatively."[4] With this wording, DSWR was placed at "5" on a scale of 0 to 10.

Our 1996 prognosis suggested fewer rights and obligations of local peoples, as national-level penetration increased (although this was not necessarily consistent with local peoples' perceptions).[5] We discuss below how this research identified an important need to rethink our use of the concept of *participation in forest management.*[6]

An Unworkable Assumption, or Participation in What?

In recent years, a sizeable literature has emerged on participation—one that we have welcomed and read with interest. Indeed, we prepared for the West Kalimantan research with this literature in mind. We were convinced of the importance of people's participation in forest management.

An appealing World Bank definition (1996, 75) of *participation* is "a process through which stakeholders influence and share control over development initiatives." Pimbert and Pretty (1995) provide a typology that includes *passive participation* ("where people are 'involved' merely by being told what is to happen") as well as participation in information giving, participation by consultation, participation for material incentives, functional participation, interactive participation, and finally self-mobilization.

The kind of participation we have envisioned, in a sustainably managed forest, seems closest to Pimbert and Pretty's (1995, 26) last two definitions:

- *interactive participation:* "People participat[ing] in joint analysis, which leads to action plans and the formation of new local groups or the strengthening of existing ones. It tends to involve interdisciplinary methodologies that seek multiple perspectives and make use of systematic and structured learning processes. These groups take control over local decisions, and so people have a stake in maintaining structure or practices." (For descriptions of the DSWR Conservation Project's attempt to work cooperatively with local communities and build on their management systems, see Dudley and Colfer 1993; Colfer and others 1996b, 1999b, 2000a; Wickham 1996.)
- *self-mobilization:* "People participat[ing] by taking initiatives independent of external institutions to change systems. Such self-initiated mobilization and collective action may or may not challenge existing inequitable distributions of wealth and power." (See also the discussion of the community of São João in Chapter 11).

Yet as we began conducting the Kalimantan test, we realized that we were operating under an unworkable assumption. We found (and at first unconsciously shared) an assumption that the local peoples would somehow be brought into the management system of the government or of the timber concessionaire (or both; or see also Songan's 1993 analysis of Iban "participation" in a Malaysian agricultural scheme). We illustrate this assumption with two recent quotes. Coakes (1996), for instance,[7] outlines "how [community] stakeholders can be actively involved in the process of forest management." Carter and others (1995) ask whether participation "provide[s] [local communities] with the knowledge they need to take decisions and become managers...?" These questions undoubtedly make sense in many environments, but in West Kalimantan (and, we suspect, other areas as well), the local peoples were actually the stakeholders most actively involved in the process of forest management; they already *were* (under-recognized) managers of the forest.

We also found that much of the literature on participation focuses on projects (for example, Songan 1993; Isham and others 1995; Borrini-Feyerabend with Buchan 1997). The project context is almost by definition limited in time, whereas sustainable forest management assumes and requires a long time horizon. In West Kalimantan, we found it necessary to think about participation in the Conservation Project quite differently from participation in sustainable forest management in general (see Chapter 13 for similar problems). Some communities were quite earnestly involved in income-generating

Preserving Traditional Rights

In 1989, the people of Wong Garai became alarmed when they learned that P.T. Militer had built one kilometer of logging road into their territory on Bukit Pelawan and that surveys were being done for eventual timber cutting. The forest there was mainly old growth that had not been farmed in the past because of the poor soil on the mountain. The people were concerned about preserving the forest for their long-term use. They had seen other longhouse communities in the area whose lands had been logged left with no building materials, no places to hunt and gather forest products, and fouled water sources with declining fish populations.

The Wong Garai leaders notified the district officer and told the local logging manager that they would not allow any cutting within their territory. They also marked trees with red paint along the territory boundary. P.T. Militer offered the headman (*kepala desa*) Rp 7 million (US$3,500 at the approximate 1989 exchange rate of US$1 = Rp 2,000) if he would allow the logging to proceed, but he declined. Fortunately, one of Wong Garai's own was currently serving a term as legislator in the Regency legislature (*Dewan Perwakilan Rakyat*) in Putussibau. With his help and influence, Wong Garai protected its forest from logging. The Regency head (*Bupati*) agreed to support them and issued an order protecting the forest. In the same letter, he also gave official support to what apparently was a traditional Iban practice, prohibiting the farming of mountain peaks to prevent erosion and the drying up of water sources.

This success set a precedent for the community. Wong Garai also has an area of old-growth lowland swamp forest. Some of it had been converted over the past century into swamp rice fields, but a large area still remains. The community has been successful in preserving it from other longhouses that have tried to establish claims on the land (Wadley 1996), and it intends to prevent any logging there, as well.

activities initiated through the project, and it had been quite successful in trying to build on traditional management. But looking at people's participation in sustainable forest management required attention to the activities of timber concessionaires and of shifting cultivators, and we had to ask who was actually doing what that had an impact on the sustainability of forests and on people's ways of life within the forest.

Looking at the forests in the area of Danau Sentarum (both within and outside the wildlife reserve), we see at least three distinct systems of management, each with a completely different approach. That the Melayu fishing people have a complex resource management system is hard to deny (for example, Malvestuto 1989; Bailey and others 1990; Bailey and Zerner 1992;

DSWR Fisheries Management in Action

A *jermal padat* is a large funnel-shaped fishing net used at the mouths of small rivers or along the edges of large ones. At the business end of the net, the mesh is almost as fine as window screening. Huge amounts of fish are caught in these nets (approximately 10% of DSWR's total annual catch, which is estimated at around 10,000 metric tons [Dudley 1996a]). Fisherfolk use 200–300 *jermal padats* in the DSWR area (Dudley 1996b).

In 1992, there was considerable grumbling at DSWR about the *jermal padat,* from fisherfolk as well as officials. All felt that the nets were catching too many fish, both from the standpoint of the fishery's sustainability and from the perspective of equity, because only the wealthy could afford a *jermal padat* (see also Giesen 1987). We were told that use of the *jermal padat* had in fact been declared illegal and was being phased out, but the Fisheries Department was going slowly with implementation so as to minimize losses to current *jermal padat* owners.

In 1996, the same complaints could be heard. Nothing seemed to have changed. Meanwhile, the indigenous fisheries management of local people within the reserve was in the process of being strengthened and supported by the Conservation Project. The project had organized intra- and intervillage meetings, documented existing local regulations, mapped traditional management areas, and was trying to secure letters of authority from the *Kabupaten* (the Regency) to grant fisherfolk more official power in implementing their regulations. These efforts were motivated partly by a recurring management dilemma in the lakes area. Those managing do not have the authority to manage, and those in authority are not only not managing but also, in some cases, interfering with effective management.

In Danau Seluang, we learned of a representative conflict, which demonstrates this situation. Two nearby communities were situated on a com-

Colfer and others 1996b; Dudley 1996a, 1996b; Wickham 1996), as do the Iban swiddeners (Dudley and Colfer 1993; Wadley 1996; Wadley and others 1997, in preparation; Colfer and others 2000a). Both ethnic groups have areas with defined borders, sets of regulations, sanctions, and mechanisms for applying those sanctions when their regulations are not followed. Melayu management focuses primarily on fisheries, though they also manage their rattan and wood supplies (see also Peters 1993, 1994). Iban management is centered on forest resources as part of a complex system of agroforestry that involves, in part, the cycling of long-fallow forest for swiddening, the cultivation of forest fruits, and the preservation of both large and small tree stands for multiple use (Wadley not dated; Wadley and others 1997). The operation of these systems—though imperfect—is visible on a daily basis.

mon lake. One of the communities (Tayak) had regulations that they enforced against the use of the *jermal padat* in their territory; the other (Dayong) allowed its use. Tayak complained to the authorities, citing prohibition on use of the nets from the governor, the regent, the county officials, as well as a treaty signed by all DSWR communities and many surrounding it in 1994, promising not to use poison, electric fishing, or the *jermal padat* in the lakes area.

A meeting was held in Tayak in June, attended by head fishermen from five fishing communities; heads of five villages; county officials; and representatives of fisheries, the military, and the police. At this meeting, although everyone agreed that the nets were undesirable from the perspectives of fisheries and equity and were prohibited by various supposed managers, no conclusion could be reached. Why? Because the people of Dayong had a written permit from the Fisheries Department, for which they had paid Rp 30,000 (at which time the exchange rate was roughly US$1 = Rp 2,300), to use the *jermal padat*.

When the county commissioner was asked why no conclusion could be reached, he said that the government didn't have a legal leg to stand on, because another government department had given permission, and the *jermal padat* owner had paid for a license. When the Fisheries Department personnel were asked the same question, they looked uneasy, saying that this was an intercommunity boundary dispute, over which the Fisheries Department had no authority.

Local people, trying to manage local resources in a responsible fashion, cannot consistently count on the support of the institutions formally charged with management. At the same time, formal government management is ineffective because of inadequate staff, low staff qualifications and motivation, underbudgeting, and the myriad other problems that plague bureaucracies in the developing world.

A second system of management derives from the commercial timber industry in concert with the Indonesian government. Timber concessions were given to several companies to harvest timber in areas all around the reserve. These companies (or their predecessors) have been operating for roughly two decades. Again, they have defined borders, regulations, sanctions (theoretically), and mechanisms for dealing with infractions. All of these management components, however, differ significantly from those of local peoples and, with the exception of harvesting and log transport, are much less obvious than is community management.

In 1992, a third system of management emerged with the beginning of the Conservation Project in the lakes area. This project tried to manage the resources cooperatively with local peoples. It focused on strengthening existing management by local peoples as a means of attaining the sustainable management and conservation of a unique biome. DSWR has a third set of borders, with somewhat different regulations, sanctions, and mechanisms for conflict resolution. One could in fact identify still other systems of management with intended effects in this area (such as that claimed by the Fisheries Department).

Who, then, is managing the forest? Results from our study suggest that in the DSWR area, local peoples are almost unanimously perceived to be the most important actors in forest management. As noted before, not only villagers from both ethnic groups and various villages but also government officials, traders, and timber company employees share the perception that local peoples are the most important stakeholders in forest management and report that they interact most with local peoples on such issues.

The perceived importance of these people in forest management probably would decline as distance from the site increased (such that officials in Pontianak—and even more so in Jakarta—would consider local peoples to be less significant). Yet day-to-day management obviously occurs at the local level, which can be defined as a *community* or as a *forest management unit,* depending on one's orientation.

In fact, the traditional local system is the only management that seems fully operational. Although the concessionaires have regulations they are bound by contract to follow, many of these regulations appear to exist on paper, with little or no on-the-ground monitoring. When asked, for instance, what regulations the concessionaires had to follow, one of two local forest agents listed three kinds of tax the companies are required to pay. When Colfer clarified her intent by asking about regulations such as the minimum size of a tree to be cut or road-building requirements, the forestry agent looked blank. After thinking a minute, he acknowledged that such regulations did exist and were monitored by the Forestry Department. However, it took considerably more thinking to decide that it must be the Forestry Department in Pontianak that did this. Similarly, the Conservation Project, in its laudable attempt to work cooperatively with local peoples, does not enforce the regulations of the Perlindungan Hutan dan Pelestarian Alam (PHPA; the forest conservation and nature protection agency). Indeed, if it were to do so, it would have to evict several thousand current residents!

What are the implications of this kind of situation for "people's participation in forest management"? These people are involved in forest management already. The four functions that participation in forest management were hypothesized to fulfil are indeed operative, but within their own system—not as their activities pertain to the other systems. One might more reasonably ponder how the timber concessionaire could more meaningfully "participate" in the local peoples' management, if sustainable forest management is our goal.

Before proceeding to a discussion of some concepts that we feel have a bearing on how we look at local peoples' participation in what might be alien systems of management, we briefly touch on some subsidiary findings pertaining to conflict (another issue considered important in all CIFOR's field tests). The initial tendency of these teams was to consider conflict as a negative occurrence, with potentially adverse effects on sustainable forest management. We joined them in seeing the extremes of war as a natural progression in levels of conflict that ultimately would destroy both the forest and the lives of the people in it.

However, in West Kalimantan, disputes[8] appeared to have salutary effects on the well-being of local peoples. In the past, the inherent conflicts among the three major management systems operating in and around DSWR have been avoidable (consistent with Indonesians' common cultural preference for overt harmony) for the following reasons:

- the area in question was sparsely populated, and the concessionaires could cut a sufficient volume of timber in comparatively distant areas;
- the availability of other areas to the concessionaires allowed them to accede to community wishes in many cases of disagreement;
- the activities of outsiders included some minor benefits to the local peoples (such as improved access, a small amount of employment, increased exposure to the outside world); and
- an ethic of hospitality in Borneo (which perhaps derives from the historical sparseness of population) welcomes newcomers (see also Peluso's "ethic of access" 1994).

However, recently, the available forest has shrunk. The timber companies have moved into areas the local peoples wanted to protect for their own (and their children's) use. So far, the government and the concessionaires have not found it expedient to disregard community sentiment completely—perhaps in partial local recognition of the costs of management that fall to local peoples while many of the benefits from the forest accrue to outsiders. But the concessionaires and the Forestry Department, particularly at high levels in the bureaucracy, have considered the concessionaires' claims to be preeminent. This situation holds considerable potential for serious disputes that could have detrimental effects on local peoples. At the time of this 1996 research, the basic laws regarding traditional rights to land were subject to varying interpretations (see Colfer with Dudley 1993, 75–80). The timber companies consistently had better access than local peoples to information, financial resources, and the more powerful regional and national decision-makers. Now, the situation is far more confusing because Suharto's regime made way for B. J. Habibie's, which has now been replaced by Abdurrahman Wahid and Megawati Sukarnoputri. What this will mean for forest management in general is anyone's guess.

An Explanation for Our Unworkable Assumptions about Participation

The crux of the problem, in our view, can be summed up in a concept originally put forth by Brigitte Jordan regarding, strangely enough, childbirth (see Jordan with Davis-Floyd 1993; Jordan 1997; Davis-Floyd and Sargent 1997):[9] *authoritative knowledge*. Jordan (1991) also has applied this concept in the context of U.S. air traffic control. She (Jordan 1997) explains

> For any particular domain [such as forest management] several knowledge systems exist, some of which, by consensus, come to carry more weight than others, either because they explain the state of the world better for the purposes at hand (efficacy) or because they are associated with a stronger power base (structural superiority).[10] ... A consequence of the legitimation [sic] of one kind of knowing as authoritative is the devaluation, often the dismissal, of all other kinds of knowing. Those who espouse alternative knowledge systems then tend to be seen as backward, ignorant, or naive trouble makers. Whatever they might think they have to say about the issues up for negotiation is judged irrelevant, unfounded, and not to the point.... The constitution of authoritative knowledge is an ongoing social process that both builds and reflects power relationships within a community of practice.... It does this in such a way that all participants come to see the current social order as a natural order, that is, the way things (obviously) are. The devaluation of nonauthoritative knowledge systems is a general mechanism by which hierarchical knowledge structures are generated and displayed.

She also writes, *"The power of authoritative knowledge is not that it is correct but that it counts"* (Jordan's italics) and "By authoritative knowledge, I specifically do *not* mean the knowledge of people in authority positions.... Authoritative knowledge is an interactionally grounded notion."

We suggest that indigenous knowledge and management systems of local peoples in DSWR and in many other rural forested areas of the world are ignored—indeed, not even perceived—because such knowledge and systems are not considered by the more powerful to constitute *authoritative knowledge* (see also Chambers 1994; Long and Villareal 1994; Salas 1994; or more philosophically, the writings of Foucault [1980], particularly on the link between knowledge and power). The global treatment of traditional vis-à-vis "modern" forest management is a comparable situation (see Fairhead and Leach 1994/95; Leach and Mearns 1996). The existence of and general compliance with locally defined regulations, which mark the traditional management system in Danau Sentarum, for instance, stand in marked contrast to the apparent irrelevance of modern Forest Department regulations in the surrounding area.[11]

Noncompliance with regulations by officially sanctioned (that is, modern) forest managers is a chronic problem in tropical forestry. The global evidence for the sustainability of many traditional systems consistently mounts (for example, Clay 1988; Fortmann and Bruce 1988; Redford and Padoch 1992; Colfer and others 1997a) alongside the increasing evidence for nonsustainable practices and policies within the modern sector (for example, Head and Heinzman 1990; Ascher 1993; Barbier and others 1994 on Indonesia specifically).

The irony is that continued access to the resources by stakeholders unconcerned with sustainability issues effectively undercuts the strength of the traditional systems (see, for example, Peluso 1994). When other stakeholders are free to harvest local resources at will, outside the control of either the indigenous or the modern system, convincing local peoples to moderate their own use becomes much more difficult. Traditional regulations lose their force. Why try to conserve a local resource when unsustainable harvesting by others cannot be controlled?

The suggestion here is definitely not that traditional systems are the "be-all and end-all" of sustainable management. They quite obviously have their failings as well, particularly in situations where change is occurring rapidly. However, sufficient evidence indicates that people's participation in forest management is important. We must consider the possibility of turning our current assumptions on their heads.

The concept of authoritative knowledge in this context is useful in explaining why the management of forests continues to be perceived as rightfully in the hands of the modern sector—in the face of mounting evidence of biological, human, and economic unsustainability therein (see also Leach and Mearns 1996, for comparable African examples). The prestige of science, modernity, and the West is wrapped up in "modern forest management." This perception can be seen in operation from the highest level, where developing countries compete to show their compliance with the

---

When Others Don't Follow the Rules

The people of Danau Seluang told of an incident that took place in 1982–1983. An Iban community living upriver from Danau Seluang used *tuba*, an illegal fish poison, in a river that drained into the nearby lake. Eleven times the people of Danau Seluang caught them and reported the incident to the Fisheries Department. Eleven times, nothing was done.

After the twelfth time, the people of Danau Seluang decided to use *tuba* themselves, as a form of protest. Although 250 individuals from the community were taken to jail, they succeeded in making their point and were soon released. The use of *tuba* by the Iban was not a problem for some time after that. (The Iban view of this incident may, of course, differ.)

wishes and cultural preferences of their previous "colonial masters" to the village level, where a forest ranger's personal prestige depends on the local peoples' acceptance of modern forestry (Umans 1995; Colfer 1983b for this issue more generally). Ultimately, recognition of the legitimacy of indigenous management may be a prerequisite for sustainable forest management in areas such as West Kalimantan.

The ideas of social and cultural capital have been discussed increasingly in recent years. Ostrom (1994) proposes social capital as relevant in considering the relative importance of indigenous vis-à-vis modern systems. Following Coleman (1988) and Putnam (1993), she defines *social capital* as "the shared knowledge, understanding and patterns of interaction that a group of individuals brings to any productive activity.... [It] is created when individuals learn to trust one another so that they are able to make credible commitments and rely on generalized forms of reciprocity rather than on narrow sequences of specific *quid pro quo* relationships" (Ostrom 1994). Pretty and Ward (1999) include relations of trust; reciprocity and exchanges; common rules; norms and sanctions; and connections, networks, and groups as elements of social capital. Berkes and Folke (1994; see also Barbier and others 1994) discuss a similar concept, *cultural capital.*

These concepts in fact reflect the existence of human resources that can and, we argue, should be used in resource management. To date in the DSWR, despite the existence of a governmental "will" to manage the forest effectively, the means to do so has been lacking (because of problems such as inadequate staff numbers, poor training, small budgets, and logistical problems that interfere with the supervision and monitoring of concessionaires).[12] Local social and cultural capital provide the means to some extent already. Recognition and strengthening of that capital may be the most valuable role for outsiders (such as government and private industry) in moving toward more sustainable forest management.

Jordan finds some contexts (such as rural Mexican births and air traffic control) in which input from various actors comes together in improved analyses of situations and solutions to problems that emerge. She concludes, "In some groups, differing kinds of knowledge come into conflict, in others, they become a resource for constructing a joint way of seeing the world, a way of defining what shall count as authoritative knowledge" (Jordan 1997). Such a goal—constructing a joint way of seeing the world—is being called for by an increasing number of researchers (for example, Filer with Sekhran 1998; Nguinguiri 1999; Buck and others in press).

In international forestry, and certainly in the DSWR area, dramatic changes are occurring. The traditional systems that were sustainable under previous conditions are under great stress; and in many cases, the modern approaches are devastating to tropical forests and the species within them. Neither approach alone is or will be adequate in the changing circumstances. Nor is it likely that a "final solution" is possible.[13]

Forest management, intimately connected with the needs and wishes of human beings, must inevitably deal with change. Beginning to search for a dynamic, joint way of seeing or knowing the forest is probably the only hope for encouraging more sustainable forest management in general.

Theoretical possibilities for sustainable forest management in the area abound:

- The Indonesian government could begin to firmly enforce its forestry regulations (pertaining to both timber companies and local communities). In this case, the concept of *participation* with which we began this test would be appropriate, because such a policy would probably begin with one-way, directive, top-down communication.
- The indigenous system could be acknowledged, legalized, and supported by the government—*devolution,* involving significant gains in access to and responsibility for resources by local peoples, and significant losses for powerful stakeholders such as timber companies, could be implemented.[14]

Or, more possibly,

- A process of negotiation could be undertaken that would involve recognition of the existing costs and benefits to each stakeholder and collaborative planning for the future.[15] This could involve either (a) a division of benefits, responsibilities, or areas or (b) a series of compromises among subsistence, commercial, and conservation interests in the area.

To evaluate the sustainability of forest management in a particular environment, one of our first steps must be to identify who is really managing that forest. Only then can we develop ways to integrate that management with the interests of other stakeholders (Lewis 1989). In the DSWR, participation of the timber concessionaires in the traditional management of local peoples would definitely be a more sustainable option than persuading the local peoples to follow in the footsteps of the timber concessionaires.

Conclusions

In this chapter, we describe a set of methods used to assess participation in forest management in the DSWR in West Kalimantan, Indonesia, along with our results. We found the card-sorting method and the iterative continuum method useful for understanding the roles of different stakeholders in forest management. The inapplicability of the researcher guide particularly highlighted unwarranted assumptions related to the concept of *participation in forest management* as commonly used.

The conclusion that local peoples were the primary managers of the forest in the area prompted us to rethink our assumptions about people's participation in forest management. We realized that we had assumed the preeminence of timber concessionaires in forest management—consistent with national laws—with local peoples potentially participating in that modern system. However, we found that traditional forest management appeared far more sustainable and operational than did management by the timber concessionaire.

We then proposed use of Jordan's concept of *authoritative knowledge* to account for the curious dismissal of indigenous systems in contexts such as this, where local management is so obvious and so widely acknowledged locally. The modern approach to forest management is, at the moment, generally considered authoritative; the traditional system is not—regardless of evidence to the contrary. This approach must change in the interest of fostering more sustainable management. The existence of a functioning indigenous management system constitutes a form of social or cultural capital that can serve an important function if recognized and integrated with other management systems.

On the basis of the problems identified with the concept of *participation in forest management,* we concluded in 1996 that, in the DSWR, it made sense to focus on local peoples' "rights and responsibilities to manage the forest cooperatively"[16] (see also Drijver 1992; the "4Rs" [rights, responsibilities, returns/revenues, and relationships] framework of Dubois 1998, 1999). Our current, more global phrasing of the issue is reflected in Principle 4: "Concerned stakeholders have acknowledged rights and means to manage forests cooperatively and equitably" (CIFOR 1999). The definition of who has which rights and obligations may need to be assessed for each forest management unit and defined locally.

We continue to work on the issues initially called "people's participation in forest management." Several additional issues may be important.

- Determining how "good" for forests and people the various local management systems are: For instance, the conversion of East Kalimantan's natural forests to industrial timber plantations—a very modern change—is well under way; yet this land use is obviously not compatible with maintaining the previous biodiversity or the many functions the natural forests have provided for local inhabitants.
- Identifying what values are held by different stakeholders (some being more conducive to sustainable forest management than others): For instance, in Brazil, *índios* are hunter-gatherers who have a very extensive, nonintrusive management system that does not produce a significant surplus; *ribeirinhos* are river-dwelling peoples who have a less extensive, more intrusive, and more productive shifting cultivation system; government and timber company officials and *colonos* (settlers or colonists) would prefer to cut the forest down, either to secure the timber or to convert it to pasture and agricultural lands, thus producing something completely new. The implications of such different approaches for participation in forest management must be examined further (see Chapter 13).

- Assessing how productive of forest products each system is and can be: This question is complicated by the fact that assessors, depending on their orientation, tend to evaluate productivity on the basis of one or two products, whereas natural forests are likely to have many products.[17] The usefulness of various forest products also depends on the needs and knowledge of the various users.

In summary, we are convinced that cooperation and mutual accommodation will be necessary among the various stakeholders connected with natural forests if they are to survive and be managed sustainably for all concerned. We hope this discussion can move us forward both in developing useful tools for assessing the social dimensions of sustainable forest management and in bringing about conditions that contribute to human well-being.

Acknowledgements

Several people other than ourselves played important roles in this research, including Ravi Prabhu (project leader, Center for International Forestry Research [CIFOR]); Sahardi (Agency for the Conservation of Natural Resources, Danau Sentarum Wildlife Reserve [DSWR], and resident of Nanga Kedebu'); Ida Marlia (periodic resident of Cincin, Lawa, and Bukit Rancong, DSWR); Gideon Ilong, Yustina Lenjai, and Dana Atam (residents of Wong Garai); Lasah, Tamin, and Ajo' (residents of Kelayang); and Emily Harwell (then a Yale University doctoral student working in Kelayang; now a private consultant). The cooperation and assistance of so many stakeholders in and around DSWR is most gratefully appreciated.

Thanks also to Dr. Eva Wollenberg and Nicolette Burford de Oliveira for constructive comments on early drafts of this chapter. The financial and administrative support of CIFOR, the U.S. Agency for International Development, Wetlands International–Indonesia Program, Indonesia's Agency for the Conservation of Forests and Protection of Nature, and West Kalimantan's Natural Resources Office are gratefully acknowledged, as well as the informal cooperation of members of the U.K. Department for International Development team at DSWR. We also are grateful to James Mayer and Olivier Dubois for their constructive comments as reviewers.

The help of the people in the lakes area has been the most critical. We thank them for their patience and good humor during our research.

Endnotes

1. Sahardi from Kantor Sumber Daya Alam (KSDA, Kalimantan's natural resources office) and Nanga Kedebu'; Ida Marlia from Cincin and Lawah; Gideon Ilong, Yustina Lenjai, and Dana Atam from Wong Garai; Lasah, Tamin, and Ajo'

from Kelayang; and Emily Harwell, at that time a doctoral candidate from Yale University's School of Forestry and Environmental Studies.

2. We rejected a fourth method, a communication network analysis instrument.

3. The exact number of respondents cannot be given because of our use of groups in some cases, but at least 44 individual villagers and at least 17 outsiders were queried. We also queried several traders whose insider/outsider status was ambiguous.

4. This wording has evolved to the current "concerned stakeholders have acknowledged rights and responsibilities to manage forests cooperatively and equitably" in the *CIFOR Generic C&I Template* (CIFOR 1999).

5. Significant changes in forest policy and field realities are under way because Indonesia has left the Suharto and Habibie regimes behind. What the future holds remains unclear, but some encouraging signs indicate that local peoples' rights and responsibilities may be increasingly recognized.

6. During the course of the project on criteria and indicators, Colfer encountered strong resistance from at least two European social scientists (Bo Ohlsson and Jan Kressin) to the term *participation*. Both felt it was a patronizing concept, but they were not explicit about why. The conclusions of this field research support their views, which Colfer initially resisted.

7. No special criticism is intended for these authors; they are simply illustrative of an assumption that is very widely held, and probably warranted, in many contexts. Our point is simply that this assumption is not warranted in all cases.

8. As distinct from *conflicts,* which of course always abound. See Chandrasekharan 1997 for a thorough and fascinating series of papers and discussions on this issue.

9. In the American, high-tech birth context, she found that certain knowledge counts (*authoritative knowledge*), and other knowledge does not. She used examples of childbirth in Mexico and the United States, for instance, to demonstrate the difference. In the United States, authoritative knowledge about childbirth is (or was at that time) held exclusively by medical personnel. The birthing mother's knowledge (of her bodily functions, for instance) is ignored and trivialized. In Mexico, a birth attendant works closely with the birthing mother, acknowledging both the mother's and the birth attendant's authoritative knowledge about what is occurring. Jordan provides evidence of this, from both participant observation and videotapes of births in both countries.

10. *Structural superiority* is closely connected to the *power deficit* we have proposed as a dimension in determining who should "count" in sustainable forest management at the forest management unit level (Colfer and others 1999c).

11. Dudley (1996b) has taken the eight design principles for collective management of a common property resource compiled by Ostrom (1990) and evaluated their fit with conditions in the Danau Sentarum Wildlife Reserve area (see also Hobley and Shah 1996 for an examination of many of the same issues). Dudley's conclusions focus on fisheries, but they illustrate the extent of indigenous natural resource management in the reserve.

12. The concessionaires themselves have not had the motivation to manage sustainably, partly because of short, 20-year concession contracts with the government. Barr (in press) makes a convincing analysis that questions the viability of sustainable logging in Indonesia at this time.

13. See also Fox's (1996) comment about community resource management, which he says "is a process; it will never be finished; there is no single solution; and conflict cannot be escaped, it is part of the process." Roe's (1994) analyses of "policy narratives" also suggest that this is true.

14. Given the current race for "decentralization" in Indonesia in late 2000, we would like to clarify that we refer to a transfer of authority to communities, nongovernmental organizations, and other elements of civil society more than the current transfer to lower level government officials.

15. See also Behan's (1988) note that "'Participative' public involvement ... is a democratic, personalized, dynamic, interactive process of bargaining, negotiation, mediation, and give-and-take among and between the constituents and managers alike.... Public involvement becomes a continuous process, no longer a series of discrete events."

16. One reviewer commented that "cooperatively" here might better be replaced by "jointly."

17. For an example of research on this variety of products in the Amazon, see Shanley 1999 and Shanley and others 1996, 1998.

CHAPTER THIRTEEN

Rights and Means to Manage Cooperatively and Equitably

Forest Management among Brazilian Transamazon Colonists

Noemi Miyasaka Porro

In this chapter, I examine the rights and means to manage forest resources from a perspective that accounts for the effective participation of forest dwellers in forest management initiatives aimed at sustainable forest management (SFM). I examine social situations observed in the Brazilian Amazon in July and August 1998.[1] The research team spent 45 days in the field, testing 13 methods of data collection (several of which were designed to assess local dwellers' participation in forest management) in sites considered "forest rich" (Trairão) and "forest poor" (Transiriri) as defined in this book (see site descriptions in Annex 2 of the Introduction). In the process of carrying out participant observation, testing methods, and interviewing colonists along the Transamazon highway, the importance of effective participation as a foundation for achieving rights and means for any cooperative action toward SFM became clear. I also learned that different interpretations of what *participation* means should be considered from the very beginning of management initiatives. My objective in this chapter is to discuss the reasons for and implications of these findings.

The importance of *participation* to rights and means for cooperative management have been discussed in works from the Center for International Forestry Research (CIFOR; Colfer and others 1997c; see also Chapter 12 of this book) and led CIFOR to draft Principle 4: "Concerned stakeholders have acknowledged rights and means to manage forests cooperatively and equitably" to contribute to the achievement of SFM (Colfer and others

1999a). According to this principle, in an ideal situation, true participation is achieved when the rights of men and women are recognized and used for gaining the means to manage forests cooperatively.

Although widely used, the term *participation* can carry with it an undesirable assumption that local people take part in someone else's system of forest management (Chapter 12). Therefore, for every initiative, the conditions and consequences related to Principle 4 demand a clear understanding of what we mean by *participation*. While testing the methods, I observed and noted different interpretations of the term, from governmental and nongovernmental agents in the capital and in central towns to rural leaders and residents in the villages. Such different interpretations affect how men and women perceive and construct their rights and means to manage the forest, and how outside agents conduct their management projects—whether or not they are called "participatory."

In the current state of forestry affairs in the Transamazon, equal participation by all stakeholders definitely is not a given and does not have a consensual meaning. Rather, participation is being defined and practiced differently by various forest actors, resulting in inequitable management. In the Transamazon, the lack of success of several initiatives and projects has been linked to the inaccurate assumption that all actors had the same understanding about participation and rights.

First, I assess participation among Transamazon colonists by examining narratives about social situations in which we can observe distinct interpretations of *participation* in defining rights and means to cooperative management. Although I recognize that some issues demand quantitative, statistical methods, the questions addressed in this research—which focus on perceptions of rights and means—are better answered through participant observation and ethnographic interviews. The narratives presented below illustrate concepts extracted from observed case studies and interviews, examined in their contexts, and obtained through ethnographic field methods. The validity of such qualitative findings and the accuracy of interpretation are based on in-depth, long-term experience of the ethnographer in the field (I was trained in agriculture and anthropology, spent nine years in the Amazon as a practitioner, and have carried out anthropological research in the region in the past five years). Qualitative data are supplemented by results obtained by using the card-sorting method and the pebble-sorting method focused on rights and means to manage forests (both methods are described in the Introduction of this book).

Then, I relate the concept of *participation* to CIFOR's Principle 4 regarding rights and means as part of human well-being assessment. I conclude the chapter with a discussion regarding the legal concepts of *acquired rights* and *attributed rights* and means to manage the forest cooperatively in the Transamazon. I argue that one of the conditions for SFM in the colonization schemes in the Transamazon is to incorporate the colonists' understanding of *participation* and to insert that perspective into the political agenda.

Assessing Participation among Transamazon Colonists

The Transamazon highway (BR-230) was built in the 1970s as part of development policies related to the so-called Brazilian Miracle. According to government propaganda, the highway would "give to men without land, land without men."[2] West of the town of Altamira, colonization brought in farmers and peasants[3] from different Brazilian states and confined the indigenous Arara people to reservations. In the area that today comprises the municipality of Uruará, colonization followed the classic fishbone pattern, in which hundreds of kilometers of secondary dirt roads (*travessões*) extended from the Transamazon highway. The colonization project was designed to develop the Amazon region along the highway, but it both had disastrous results for the indigenous people and created a unique social group of farmers and peasant immigrants, so-called colonists (*colonos*).[4] Displaced by development programs and land concentration in their regions of origin, colonists came to obtain their own plots of land and thereby escape subjugation by landlords elsewhere. Margarete, a 41-year-old colonist, explains:

> My dad told me that they lived in [the eastern state of] Ceará. His father worked for a landlord there, but it was like slavery, because it was work only for food and they were not allowed to go anywhere, let alone to leave that place. Then, one by one—my uncles, my grandfather, aunts—one by one they escaped from the landlord's hand. It was on Sundays, market days, or at night during busy times throughout the years. The only one who was left behind was an uncle, but it was because he himself and his wife did not want to leave. I think he had already adopted a slave mentality; he stayed there until he died. But then [the other members of the family] moved to other landlords' places as they did not have anywhere else to go. Finally my dad heard about the Transamazon. He heard that the government was giving land to the poor, and that there were even houses for all, ready to move in.

However, as the interviewees stated, adaptation was much harder than they had anticipated. Over the past 25 years, colonists along the highway have faced numerous obstacles presented by an unfamiliar environment and abandonment by the government. They had no experience working in the rainforest and were isolated from basic public services such as education and health care. Farmers from the south, used to a more constant link to the market, had to cope with the limited and unpredictable access to markets in Amazonia. Migrant peasants from the northeast also faced formidable challenges because in the Amazon—at least in this initial phase—they lacked the social networks and knowledge of ecosystems that had been the basis of their peasant system of agro-extractive production.

Because most of the colonists had a more agriculture- than forest-oriented tradition than the *ribeirinhos* (traditional riverine inhabitants) and the ethnically distinct *índios*, they viewed forests as an abundant resource for slash-and-burn cultivation; this perspective led to different levels of forest degradation.[5] In their efforts to survive both the difficult initial phase and subsequent long years under conditions of economic and social marginalization, the colonists were unable to thoroughly understand the new ecosystem and establish a coherent and sustainable pattern of resource use. Moreover, once becoming part of the Transamazon colonization program, colonists struggling to be successful adopted certain prescriptions of extension models developed by the state in efforts to protect themselves from being marginalized and displaced yet again. In the 1970s, the government promoted the production of annual crops for subsistence, but this approach resulted in inadequate income and tended to degrade soils. During the 1980s, many colonists followed state tutelage that favored perennial crops (cocoa, black pepper, and coffee), which commanded higher prices, had less pronounced environmental impacts, and thus ostensibly could promise a better livelihood. However, by the late 1980s, declining prices and plant disease had reduced the attractiveness of cultivating perennial crops. More recently, cattle raising has become the symbol of success, despite its questionable social and environmental effects.

Because colonists' agricultural fields require clearing forests each year and because there are no signs of sustainable practices related to timber and game extraction, some observers think that colonists are far from having a conservation ethic and that their participation in forest management should be directed by external forest managers. A key objective of Brazil's proposed development model is the increased integration of colonists into the market, which can be expected to increase forest clearing and the depletion of forest resources. One may conclude that the continuing colonists' practice of opening small gaps within the forests each year is due more to the lack of financial ability to clear larger areas than to a "spiritual link" to the forest or a conservation ethic. Indeed, while in the field, I observed no evidence that they have an attachment to the forest due to a historical past or cultural tradition toward forest conservation.

However, this behavior should not relegate colonists' participation in forest management to mere labor providers. I encountered strong evidence that their saga throughout the country consolidated their reluctance to be subordinated to landlords and their determination to have a piece of land of their own. Their history of struggle, I would argue, entitles them to rights and means to manage the resources on their hard-earned land and, furthermore, to rights and means to do so in a sustainable way.

Moreover, shifting cultivation can be more or less destructive in many situations that involve colonists. Cultural backgrounds and economic means define their tactics in their pursuit of success. The more capitalized colonists and those with more experience in market transactions use cheap labor provided by poorer colonists who have land but no income from commercial production. The latter need to sell labor or forest products to buy clothes, kerosene, salt, and other industrial products while planting their own subsistence fields (*roças*). A *roça* is a gap cleared within the forest where peasants practice slash-and-burn cultivation in a rotational system. For Transamazon colonists, the forest is the natural resource that makes the *roças* possible, because it provides fertile soils and does not demand much labor for weeding (Miyasaka Porro 1997).

Poorer colonists created a unique culture of survival in the new social and natural environment of the Transamazon. In many *travessões*, this marginalized population constituted new peasant social groups as they practiced a unique mode of production mostly based on a subsistence economy. Although they eventually did buy and sell in markets, their economic decisions were not necessarily based on profit. Rather, their final goal was to maintain the material and social lives of their families, on their own land, without subjugation by landlords. Such motivations differ markedly from those of most indigenous forest dwellers (see the Cameroonian and Indonesian cases discussed in Chapters 3, 4, 5, and 10, for instance).

The colonists' *roças* mean more than short-term economic gain. Rather, they seem to carry a political meaning associated with independent access to land for subsistence. Through local people's statements, we discovered that *roças* represent freedom from exploitative landlords and a means of subsistence within an antagonistic dominant society and an uncaring government. This feeling is illustrated by Socorro, a 16-year-old colonist:

> My dad lived in so many places, in so many states. Always working for others. In each and every place that he worked, he left poorer than when he had arrived, whereas the landlord got richer than before. It is only here that we have our own land, free of the landlord. [My father] first needed to work for others to gain the means for us to settle on our own plot. My aunt is employed in the city in Minas Gerais; she works for a boss, but she respects my dad because he has wandered so much, and now he is the owner of his own things.

Colonists in the Transamazon acknowledged the decline in forest area, and whereas they recognized outsiders' and their own impacts on forests, they also pointed out their lack of alternative means of making a living. The fact that they continued to clear the forest to plant their *roças* did not imply that they did not also count on the forest when assessing their well-being. Conservationists and external forest managers aiming for sustainability tend to see colonists in the Transamazon as unknowing agents of deforestation, especially compared with indigenous peoples or peasants who live in less pressured areas, whose systems of production are reported to be more ecologically sound.[6]

I contend that deforestation caused by colonists engaged in a peasant economy is not a matter of an inadequate system of production but of inadequate policies that constrain the system. As suggested by Chambers (1988) and Goodman and Redclift (1991), poverty—not the poor—is the real cause of deforestation by forest dwellers. The socially and environmentally deleterious results of policies that drive development in the Brazilian Amazon confirm this argument (see also Browder 1988; Hecht and Cockburn 1989; Schmink and Wood 1992).

Despite being relatively recent immigrants to the area, colonists have an interest and the potential to maintain a close, long-term relationship with the environment. This interest strengthens their claims to rights and means to manage the forest. Their background is not related to tropical forested areas, and they acknowledge this fact when referring to their entry into the frightening "big, closed forest full of jaguars" that initially was so strange to them. However, they constantly remind us of their oppressed background that compelled them to struggle for this place and gave meaning to their attachment to it. These considerations suggest a potentially strong role for colonists in the discussion of responsibilities and their participation in the search for alternatives to deforestation.

To achieve effective participation leading to SFM, we needed to better understand the meanings of roça, forest, and logging to the Transamazon colonists. For the colonists involved in a peasant economy, the roça is not only a field or a physical place to engage in subsistence activities. It also carries cultural meanings that sustain their social group and has become an instrument of political negotiation with other stakeholders. A colonist who has a roça that ensures his or her livelihood has greater bargaining power. Therefore, the roça also has political significance if we consider politics as the negotiation of social status and relations based on power relations between the different sectors of society. The roça is what colonists can rely on to avoid being trapped in the oppressive relations of labor exploitation, which once expelled them from other places. However, if the roça is indeed politically relevant for the colonists but also appears to work against forest conservation in the current agrarian context, then what is the importance of the forests and forest management to Transamazon colonists?

The roça is central to the colonists' material and cultural life. As such, the forest is viewed as the essential renewable source of extractive products and especially of biomass (see Anderson and others 1991) that make the roça possible. The forest is seen as a kind of checking account. On the other hand, timber is perceived almost as a by-product, or the equivalent of a savings account. In both the forest-rich and forest-poor sites that we studied, peasants view logging mainly as a means to access markets. Logging companies provide transportation, road construction, and infrastructure maintenance that make it feasible for the colonists to bring roça products to market. Logging companies also are sources of cash for colonists who have trees to sell, but the income obtained is small compared with that of agricultural production. Colonists welcome loggers because they build skid trails to access trees;

these trails also serve to reach colonists' *roças* and allow them to transport the harvest. Although the colonists recognize the deleterious effect of logging on the ecosystem, the combination of beneficial services and lack of alternatives makes logging a relevant factor in achieving subsistence, because the government does not effectively support their survival. In forest-poor sites along the Transamazon, colonists complain that, because the loggers are gone, there is no longer transportation, road maintenance, cash for occasional jobs, or timber sales to support them until the next harvest.

In summary, forest conservation is important to colonists because it provides resources that allow people to plant *roças*. Logging, in the current context, is important because it provides some means to commercialize part of their production. How they balance forest clearing, trading with loggers, and dealing with conservation practices (that is, how they manage the forest) is based on their perceptions of well-being and defines the sustainability of the forest. We need to understand this balance to understand how they manage the forest on their own and to assess their participation in forest management initiatives.

An example from my fieldwork demonstrates the importance of the *roça* related to participation in forest management as expressed by local people in the Transiriri *travessão*. Dona Dalila, a 53-year-old colonist, explains why her household was not clearing as much forest as before. Through her narrative, I tried to understand how she defined her participation in practices that affected forest management.

> The same drudgery I have suffered since I was a girl, I still suffer today. There is no such thing as, "Ah, today I am not going to work...." No. Everyday, everyday, it is grinding rice, it is taking the meals to the workers in the *roça,* it is treating the pigs, the chicken, it is cleaning the house, and it was struggling with all those children. Since I was 7 years old, this is my life. Then one day, I said, "Look, Manelo [her husband], if you want to work, you work. I just cannot handle this anymore. From this day on I am not working on *roças* anymore." Before, I worked on *roças* from the beginning to the end. It was planting, it was harvesting, and it was either with that big belly or it was just after the delivery.

Later, while we were removing the grains from the ears of corn, she told me that she had decided to stop working with him in their *roça* because he had started to sleep with another woman "as if she were his wife." After that, working together in their *roça* had no meaning because they "were not really together anymore." Her refusal to work on the *roça* was a symbol of a broken alliance.

By including these narratives, I am not attempting to generalize, even though the situations of other colonists may be similar. They clarify the diverse social situations that surround the *roças* and, consequently, participation in forest management. Learning about these relations of love, hate, and mutual dependence—each involving gender, economic, and cultural issues—I began to rethink my questions: If the *roça* is what drives people's relationships with the forest, then what does participation mean in this context? If the *roça* is part of their style of forest management, then are people going to bring all these involved social relationships into formal forest management to really participate? Relationships involved in the *roça* may be not based on the logic of a capitalist market economy; are formal agents of forest management aware of their complexity?

I imagined the questions I would ask to potential participants in a formal forest management project:

- Would you participate in a forest management program?
- Would you participate in a conservation project?
- Would you participate in a community group?
- Would you participate in a *roça*?
- Would you participate in a household?
- Would you participate in a marriage?

Then I realized that after the third question, the word *participate* did not fit well. People do not participate in the *roça;* they just live their *roças*, their households, their marriages. If someone else—other than the people themselves—were to direct or make decisions about the *roça,* then it would not be a *roça* anymore. *Participation,* as defined by most formal development agents and forest managers, implies not only taking part but also letting others (who may not share the same understanding about *roças*) take part and, often, letting other more powerful stakeholders take directive roles. To consider this kind of *participation* part of the *roça* would necessarily change the meaning of "*roça.*" In my understanding, for colonists in the Transamazon, the *roça* had become a symbol of hard-earned freedom from oppressive landlords and a symbol of an alliance between a man and a woman in building a household. Such a significant institution requires careful consideration prior to opening up its management to others.

Understanding that the *roça* represents much more than just an economic activity or asset to colonists, we may realize that, just as outsider development agents have a hard time sharing decisionmaking power with local people, colonists have a hard time being convinced to share decisionmaking that affects their *roças*. The prospect of bringing a third party inside a sphere as intimate and complex as an alliance between a man and a woman in a household, as hard-fought a material and political position as independent control of production, demands serious thought—especially if this third party is a development agent or a forest manager who frequently is identified as an agent of a historically uncaring government.

This selectiveness in involving other parties in subsistence activities linked to colonists' forest management and the importance of having colonists themselves driving these processes is seen in the results of social methods we tested in the Transamazon (and in another site in the riverine area [*várzeas*], in the municipality of Porto de Moz). Our research team conducted card-sorting and pebble-sorting methods on rights and means to manage cooperatively (for more information about these methods, see the Introduction; Chapter 14; Porro and Miyasaka Porro 1998). These methods seek to assess the respective involvement of local stakeholders in forest management; identify the division of management responsibility, the capability among stakeholders, and people's agreement about it; and determine the amount of interaction among stakeholders. They also are useful for identifying gender inequalities regarding current forest management, because we disaggregated the colonist categories by gender.

In the card-sorting method, interviewees were asked to consider four forest management activities and then put cards in order according to the importance of each stakeholder. (The list of stakeholders had been elicited previously, through individual and collective interviews.) Results are presented in Figures 13-1 to 13-5. The vertical scale represents the ordering position for each stakeholder, where 1 is the most important stakeholder.[7] Results are shown of the assessments made by males and females of the importance of the different stakeholders listed on the horizontal axis, in four locations. Colonists, for instance, consistently received the lowest score (that is, the most important role) in the views of both male and female respondents.

In the pebble-sorting method on rights and means to manage, interviewees were asked to distribute pebbles among plates assigned to the stakeholders defined earlier (see Chapter 14 for a comparative analysis that includes Brazil). After confirming which stakeholders the interviewee thought were most relevant, the assigned task was to "distribute the pebbles among these stakeholders according to the rights and means they have regarding [management functions]." Surprisingly, they always assigned first place and a greater number of pebbles to themselves, the colonists.[8]

Results of this method are presented in Tables 13-1 to 13-4. When asked who were the most important people making decisions about the main *roça* and forest use activities, interviewees in both forest-poor and forest-rich sites assigned first place to the colonists. Both men and women considered colonists to be more important than other more powerful stakeholders, such as loggers, logging companies, middlemen, cattle ranchers, the Instituto Nacional de Colonização e Reforma Agrária (INCRA; the Brazilian governmental institute for colonization and agrarian reform), and the Instituto Brasileiro de Meio Ambiente e dos Recursos Naturais Renováveis (IBAMA, the Brazilian governmental institute for the environment and renewable resources).

Colonists certainly are aware of the differentials in power and conditions for effective decisionmaking among the different stakeholders. Data regarding this awareness were collected using the "Who Counts Matrix" (Colfer and others 1999c), focus group methods (Porro and Miyasaka Porro 1998; Colfer

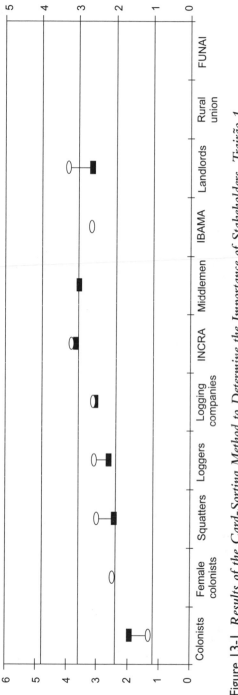

Figure 13-1. *Results of the Card-Sorting Method to Determine the Importance of Stakeholders, Trairão 1*

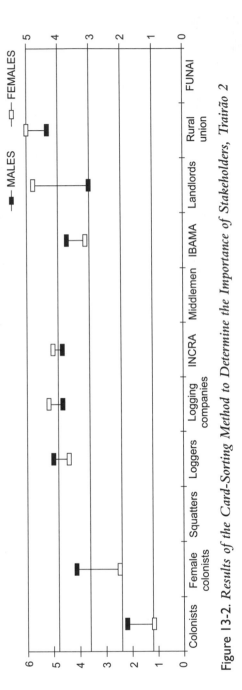

Figure 13-2. *Results of the Card-Sorting Method to Determine the Importance of Stakeholders, Trairão 2*

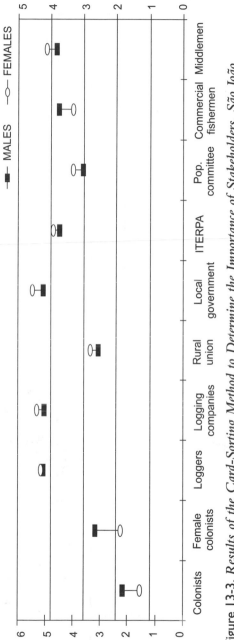

Figure 13-3. *Results of the Card-Sorting Method to Determine the Importance of Stakeholders, São João*

Note: ITERPA = Instituto de Terra do Pará.

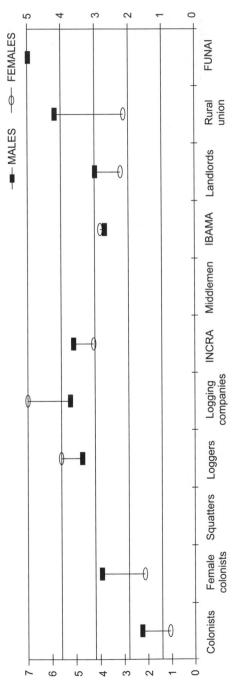

Figure 13-4. Results of the Card-Sorting Method to Determine the Importance of Stakeholders, Transiriri

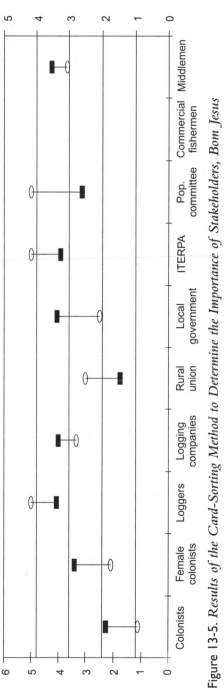

Figure 13-5. *Results of the Card-Sorting Method to Determine the Importance of Stakeholders, Bom Jesus*

and others 1999a), and ethnographic interviews. In the next section, I discuss why—although aware of these differentials—the colonists persisted in identifying themselves as the most important people to make decisions regarding *roças* and forests. Why do they consider themselves to have rights and means to manage the forests, despite their deficit in power and material conditions? Does this perception influence the way peasants view participation?

I argue that this perception affects participation in forest management initiatives proposed by outsiders. Although peasants were well aware of the power and significance of other stakeholders and recognized that participation in processes that affect their *roças* and the forests they use might involve other parties, in the sphere of their own forest management defined at the household level, they considered themselves to be the drivers of the process. During fieldwork, I learned that colonists have strong feelings about this position. Interwoven in their narratives was acknowledgement of their own attained rights.

Rights and Means to Co-Manage Forests Equitably[9]

Thus far, I have discussed why *participation in forest management* in the Transamazon should take into consideration the perceptions of the local forest dwellers and recognize their rights as drivers of this process. This approach requires taking into account rights attained by practice as well as rights attributed by formal laws. According to Brazilian law, the rights to direct a forest management initiative are contingent on approval by IBAMA according to a previously established protocol entitled Plano de Manejo Florestal Sustentável (Sustainable Forest Management Plan, Portaria #48/95, July 10, 1995). After approval, logging companies, through their foresters, may apply for aid in implementing the project, such as state or federal credit and subsidies.

Throughout the Amazon region, most of the so-called forest management projects begin as proposals submitted by logging companies to IBAMA, the government agency responsible for regulating forest management initiatives. The submitted proposal must include an evaluation of socioeconomic impacts and suggestions for minimizing the negative impacts of the project. However, none of the projects that we visited during our fieldwork in the Transamazon effectively addressed human well-being or participation by local dwellers. Indeed, according to the logging company officers we interviewed, loggers hire services or have a forester or similar professional employed to write the required plans. In our research areas, none of the local residents or people in the rural union we interviewed ever had met such professionals. There was no evidence that social and ecological aspects had been addressed as components of effective forest management.

Despite this situation, once a project is approved, the logging company is lawfully registered at IBAMA and has the right to extract a quota (in cubic meters of timber) per species, per project. Each time a company truck with logs or lumber passes through an inspection post, governmental officials deduct the amount loaded on the truck from the total quota of that given firm. However, according to our informants (both in the field and in the IBAMA office in the capital), officials have no means to effectively check whether the loaded timber actually came from the site assigned in the approved project. Nevertheless, because loggers hold the license and project approval, they are entitled to the rights conceded by the regulating laws.

In this section, I distinguish between *rights sanctioned by laws* and *rights attained by practice*. Currently, there are two main schools of thought dealing with Brazilian legislation and advocating different views for such a distinction. Along the lines of *positive rights*,[10] law is understood as intrinsically linked to the state, and *laws and rights* are used almost interchangeably because, under this rubric, rights exist only through laws. Rights emanate from laws, which are expressed in texts and manifested through the state's institutions. For the advocates of positive rights, a law regulates the social chaos that would be generated in the absence of the state, assuming that norms or rules defining the rights of each individual do not exist without the state. From this perspective, if there is a need to create rules for forest management, one should propose a law, and the law would modify social practices toward such resources, establishing attributed rights. In this case, the stakeholders with greater access to legislation in the state would determine such laws and attributed rights.

The second view offered[11] is related to increasingly popular movements among jurists and scholars who constitute a new school of juridical thought. Movements belonging to this school deal with concepts such as alternative rights and insurgent rights. From this perspective, laws are completely distinct from rights. Rights may be, but are not necessarily, an attribution of the state. Rights are viewed as the expression of power relations in a given society, which may or may not be sanctioned by the state through a law. Aguiar (1994) proposes a view in which the state is neither the only nor the primary source of rights. He illustrates this view with examples from some Amazonian indigenous societies in which people not only live without a state but also practice "millennial juridical norms" that aim to permanently avoid the formation of the state. Aguiar does not argue in favor of a prestate society but for a view in which rights are above the state and the law. Consequently, laws are never created and applied in a vacuum of rights; they are both a tool and a process for consolidating rights generated by relationships among social groups that hold different amounts of power and by the relationships between them and the environment.

Adopting a historical approach, we find that Brazilian legislation has hardly been a tool of equal access and use for all the sectors of society. This situation is even more true for colonists, because their own social origin is a product of exclusionary laws. Although marginalized, this social group was able to survive by means of a peasant mode of production. Throughout history, this peasant mode of production established a set of norms and rights to regulate forest management, constructing what I call *attained rights*. These rights are especially important for those colonists less integrated into the market economy, who have less capital, and therefore are destined to receive the worst land and conditions of access by the colonization program.

This situation of social, economic, and geographical marginalization generated an organization of labor and management of resources according to attained rights, independent of current laws. In this historical context, it is not surprising that colonists who have been marginalized perceive themselves to be the most important people to decide about forest management and with greater rights and means to manage the forest (as shown by the results of the pebble-sorting method on rights and means to manage the forest). However, in forest-poor Transiriri, men stated that they need to share these rights with the ranchers who live nearby (Table 13-1). They establish temporary and occasional alliances with ranchers, politically negotiating their conditions, to better face the constraints of their livelihoods. Women, on the other hand, assigned a governmental agency as responsible for protecting boundaries and resolving conflicts (Table 13-2).

For the few colonists who were recently objects of governmental attention (forest-rich site, Trairão[12]), peasants share the rights and means to manage the forest with the government (INCRA) (Tables 13-3 and 13-4). Because they benefited from INCRA in being considered a settlement project (despite some colonists' having been there for more than 15 years), the interviewees considered that INCRA had recognized their attained rights. The people of forest-poor Long Segar, Indonesia (described in Annex 2 in the Introduction), had a similar settlement history.

However, recognizing attained rights—including colonists' rights to direct forest management initiatives—is not enough to ensure social sustainability for the colonists or for the ecosystem. Although the established norms and attained rights of access to (and use of) resources hold valuable knowledge about the environment, they are not guarantees of conservation or of the sustainable development of people. Effective participation in forest management initiatives aimed at sustainability should include ongoing comprehensive assessment of economic, environmental, and social factors (Prabhu and others 1996). These assessments should be available to the people who direct SFM initiatives when attempting to institutionalize the rights and means to manage cooperatively.

Conclusions

Brazilian legislation has not always considered forest dwellers as holders of any attained rights or responsibilities. The Forestry Code of 1967 (Senado

Table 13-1. Importance of Stakeholders as Assigned by Male Colonists (n = 30) in Forest-Poor Transiriri

Management function	Colonists	Squatters	Forest managers		Loggers	Logging company	Ranchers	Total
			INCRA	IBAMA				
Land use planning	30	15	25				30	100
Planning timber extraction	31		3	10	27	60	31	100
Planning to set fire				38	21		21	100
Hunting planning	39			19			21	100
Protecting boundaries	45	8					46	99
Conflict resolution	60		6		10	12	12	100
Avg (no. of pebbles)	34	4	6	11	10	12	23	100

Table 13-2. Importance of Stakeholders as Assigned by Female Colonists (n = 10) in Forest-Poor Transiriri

Management function	Colonists	Squatters	Forest managers		Loggers	Logging company	Ranchers	Total
			INCRA	IBAMA				
Land use planning	100							100
Planning timber extraction	100							100
Planning to set fire	100							100
Hunting planning	100							100
Protecting boundaries			100					100
Conflict resolution			100					100
Avg (no. of pebbles)	67	0	33	0	0	0	0	100

Table 13-3. *Importance of Stakeholders as Assigned by Male Colonists (n = 23) in Forest-Rich Trairão*

Management function	Forest managers							Total
	Colonists	Rural union	INCRA	IBAMA	Loggers	Logging company	Ranchers	
Land use planning	30	5	30	24	7	3	3	102
Protecting boundaries	18	16	50	16				100
Conflict resolution	18	23	51	8				100
Avg (no. of pebbles)	22	15	44	16	2	1	1	101

Table 13-4. *Importance of Stakeholders as Assigned by Female Colonists (n = 12) in Forest-Rich Trairão*

Management function	Forest managers							Total
	Colonists	Rural union	INCRA	IBAMA	Loggers	Logging company	Ranchers	
Land use planning	55		44					99
Protecting boundaries	23	4	52	9	4	8	2	102
Conflict resolution	32	14	31	19	5	3		101
Avg (no. of pebbles)	37	6	42	9	3	3	1	101

Federal 1996), for example, actually prohibited the unregulated exploitation of primary forests in the Amazon basin. According to this law, these forests could be used only under the observation of technical norms and management, excluding the practices and attained rights of the peasant, riverine, and other local communities. However, this rule—sanctioned by the state as having the force of law—could not keep up with economic policies implemented by the same state. Having stimulated development efforts through the colonization program for the Transamazon, the state was not able to supervise the compliance with technical norms and management. Therefore, colonists were left at the margin of the law, struggling with rights obtained or taken within the context of power relations between them and dominant sectors of the society, such as merchants, loggers, and landlords.

In this context, the fact that colonists managed to survive using newly constructed social norms regarding the use of natural resources for consumption and marketing, even when the market economy was unfavorable to their products, demonstrates that attained rights may rule above ineffective constituted laws. However, such a cultural construction has been increasingly challenged, especially since the early 1970s, when the state began to impose its model of development in the Amazon more clearly and permanently. New regulations were created to rule this social space as if it were devoid of pre-existing rights. Therefore, we observe a conflict between attained rights and rights provided by laws. A classic example is *direito de posse* (land tenure rights), discussed at length in Schmink and Wood (1992).

However, in the past two decades, the juridical scene has changed due to political mobilizations that demand greater popular participation. Although the Brazilian legislation in general still excludes marginalized sectors of the society by not recognizing many attained rights, much has also changed in terms of environmental legislation as forest dwellers have developed increasing political power in this arena. Progress has been made in the sense that some laws are open enough to be accessed and used by forest managers and forest dwellers in defense of their rights. Recognizing both the inevitability of value judgements in assessing sustainability issues and building on our early definition of well-being as including justice, I conclude that colonists' struggles against exploitative landlords have entitled them to a strong political stake in the Transamazonian forest management arena. They have, in this view, attained rights to effectively participate in SFM.

Forest management among colonists who practice a peasant economy is intimately linked to the organization of the *roça* and the use of forest resources. Therefore, in determining the place of their attained rights in the legal system, it is imperative that we understand their interpretation of *participation*. Only then can formal forest management agents begin to participate in the colonists' projects and contribute to truly participatory SFM, in which there is no place for landlords. Dona, a 59-year-old colonist born in the state of Bahia and living in Trairão, provides a fitting conclusion to this chapter:

[In Bahia], we always worked on the landlords' land. When we finally managed to get land, it was this very one here. I was raised without a father; when my mother had eight children, he died. It was me and my older sister [who had to work]; the others were all too little. We were raised on other people's land....

Then I got married, still all the time working on the landlords' land. Then [her husband] bought a little piece of land, but it did not go well, we had to sell it.... All the time living and working on other people's land.... Then one day my brother came here to look around. He said, "In Altamira [a city in the Transamazon] there is land where the poor can settle themselves." We said, "Let's go."

[Her husband] came first; he stayed one year and a half, and I was [in Bahia]. All the women, working like donkeys, working for the others, planting cocoa, eucalyptus for the landlords, the firms. Then, we came ... forests, big forests, and jaguars....

One day a rich old lady showed up, saying she was the landlady, and she wanted to take the land from us. She talked and talked, but we kept working. She said everything was hers.... We were struggling in these forests, suffering all sorts of needs and she wanted us to work for her. We were working, planting our *roças,* and she wanted to take everything from us. She said those things, and I said, "I may leave this place, but I am not going to work for you."

Now, she is gone, and we are still here.... Ah, after living everything that I have lived, another landlord? Never more.

Endnotes

1. This fieldwork was carried out in collaboration with Roberto Porro as part of an ongoing CIFOR research project. For a description of the methods and their rationales, see Colfer and others 1999a; for the testing procedures and evaluation, see Porro and Miyasaka Porro 1998.

2. This famous phrase disregarded that "men without land"—mostly descendants of enslaved Africans, uprooted migrants, and detribalized indigenes—were expelled from their original places by development programs that favored land concentration, such as soy cultivation or cattle ranching. In addition to ignoring the women, this modern colonization slogan disregards that the "land without men" was inhabited by indigenous societies long before Portuguese colonization.

3. For the purposes of this chapter, I consider *farmers* as those who work on agricultural activities through family labor, whose economic decisions are mostly based on a market economy. By *peasants* I refer to those who work on agricultural activities that rely on family and community labor but whose economic logic is not driven by the market economy. Although they use some capitalist concepts, their final goal is not necessarily profit. This kind of economy has been conceptualized as a peasant economy in Wolf 1967, Scott 1976, Mintz 1985, Chayanov 1986, Roseberry 1989, and Shanin 1990. (See also Deere 1990 and Kearney 1996.)

4. *Colonist* is an operational category created by the governmental agents of colonization to identify the beneficiaries of colonization projects. It became also a category of representation used by farmers and peasants to affirm their rights over colonization projects' benefits (title, rural credit, education, and so forth). Within this category, we find people of different historical and cultural backgrounds who practice different kinds of economy and systems of production.

5. According to Brazilian environmental laws, a landowner can cut forests on his or her property in the Transamazon as long as 50% of its area remains covered by forest.

6. For more on this topic, see Myers 1980, 1991, 1993; Fearnside 1993. Works on forest sustainability among colonists in Brazil are scarce compared with those among indigenous or folk groups. See Posey and Balee 1989 for several examples on management of natural resources by these social groups.

7. It is important to remember that we were testing the methods and making adjustments to them. In Trairão 1 (Figure 13-1), for example, we still had not disaggregated the colonist category by gender when interviewing. Also, some categories were not ordered because in that given site, the interviewees did not consider these categories as stakeholders. Squatters and merchants did not emerge as relevant stakeholders in Trairão 2.

8. This finding is consistent with Colfer and Wadley's findings in West Kalimantan (see Chapter 12).

9. *Editors' note:* This wording has been changed in the final *CIFOR Generic C&I Template* to "rights and means to manage forests cooperatively and equitably" partially in response to the issues identified by Porro and Miyasaka Porro.

10. For a discussion of *positive rights* relating to the environment, see Tostes 1994 (on the Brazilian environmental legislation system) and Silva 1994 (on constitutional environmental rights) (both in Portuguese).

11. To examine this line of reasoning, I selected Aguiar 1994—not only because of its content but also because it is a representative product of a political momentum in that governmental institution.

12. Trairão is the name of a dirt road; Trairão 1 and Trairão 2 are sections of this road where we observed distinct social situations. Both are forest-rich areas, whereas the Transiriri road is forest poor.

CHAPTER FOURTEEN

Rights to Manage Cooperatively and Equitably in Forest-Rich and Forest-Poor Contexts

Bertin Tchikangwa, Mary Ann Brocklesby,
Anne Marie Tiani, Mustofa Agung Sardjono,
Roberto Porro, Agus Salim, and Carol J. Pierce Colfer

"Concerned stakeholders have acknowledged rights and means to manage forests cooperatively and equitably" (Principle 4, *Generic C&I Template*, CIFOR 1999). This idea was considered to be an integral part of sustainable forest management (SFM) by all the research teams that participated in our field tests (including tests in South America, Asia, and Africa; developed and developing countries, or, the North and the South). It emerged partially in response to a common pattern whereby people living in forests had little or no rights or power to affect formal management while they often had functioning and unacknowledged management systems. The more powerful stakeholders (such as government and timber company officials) typically did not perceive local peoples to have either rights or abilities to manage forests in their areas.

This unanimous conclusion involved several factors (discussed more fully in the introduction to Section 5):

- the existence of bodies of indigenous knowledge, potentially useful in improved forest management, in many forest-dwelling communities;
- local social capital in the form of complex and enduring systems of management;

This chapter, like Chapter 11, draws on preliminary analyses reported in Colfer and others 1998b.

- the availability of local peoples who live in local forests and therefore have both a motivation and a capability to perform various management functions; and
- local peoples' rights to and need for some control over their destinies, including maintaining a viable balance between stability and change.

All of these factors also contribute to global cultural diversity—something that many people consider as important as biodiversity—as a risk-aversion mechanism for the human species and as an inherently valuable part of the human heritage.

In this chapter, we examine the views of selected local stakeholders about their roles and the roles of others in the area to ascertain whether we could identify patterned differences in these perceptions that correlate with forest quality. Our hypothesis was that the *link* between SFM (as represented by existing good-quality forest) and the acknowledgement of the rights of concerned stakeholders to manage the forest (as represented by our results) might be clarified or confirmed. In retrospect, we recognize that using current, good forest quality as a proxy for SFM was not warranted. However, hindsight is notoriously better than foresight, and in seeking generalizable links between SFM and people's involvement in management in the reductionist approach taken in this study, we needed a fairly straightforward mechanism for differentiating contexts. The existence of good-quality forests in an area seemed to suggest that management had at least at some point been good (a view we still hold—the catch is the phrase "at some point," because time is needed to degrade good forests, if only a few years).

One would expect that in forests where the rights and functions of all stakeholders are taken into account in resource management, the chance that they are sustainably managed is greater. Indeed, the intervention of outsiders (for example, logging companies, migrants, and commercial hunters) in a forest area often limits the space and resources available to local peoples, who then may be forced to start using the resources in an unsustainable way.

We selected eight of the nine research sites analyzed in Chapter 11, in three countries: Indonesia, Cameroon, and Brazil. Within each country, sites were selected on the basis of the forest-rich/forest-poor continuum. Five sites, Bulungan (Indonesia), Mbongo and the Dja Reserve (Cameroon), and Trairão and São João (Brazil) were considered forest-rich sites. Data from only three sites—Long Segar (Indonesia), Mbalmayo (Cameroon), and Transiriri (Brazil)—were in a form we could compare for the forest-poor sites.

Methods and Results

We used a pebble-sorting method (described in the Introduction) for assessing local stakeholders' perceptions about the division of management rights and responsibilities among significant stakeholders.

Interviewing interest groups other than the local communities and indigenous groups, as we did in West Kalimantan, Indonesia (Chapter 12), seems important for assessing rights and responsibilities in forest management. Unfortunately, time constraints precluded obtaining such quantitative data from other relevant stakeholders (such as logging company personnel and government officials) at these sites. The comparison of the perceptions of people in other social categories with those of local dwellers will be pursued in further stages of our research.

In some Brazilian situations, respondents associated the concepts *responsible* or *important* only with what they perceived to be positive forest management outcomes. As a result, they excluded stakeholders that were in effect performing important, though negative, forest management functions.

In the research, we differentiated between rights and responsibilities, and respondents allocated pebbles for both topics. However, in the analysis, we found that the differences between responses on these two issues were so minimal that we report only results on "rights" here.

In our analyses we have sought distribution patterns of management responsibilities, focusing on possible differences between the forest-rich and forest-poor sites. Although the stakeholder categories varied from site to site, we have arranged them on the bar charts in such a way that there is a progression from external to local (for example, from the government to local authorities, logging companies, local businesses, and on to workers and local peoples). The numbers across the bottom of the charts represent the following functions of forest management identified here in Bulungan and most other sites:

1. Defining/protecting boundaries
2. Developing/applying rules and regulations
3. Monitoring compliance
4. Conflict resolution
5. Providing leadership
6. Assessing fines/sanctions

The functions were modified somewhat for the Brazilian sites. Biplot analysis is also provided (see Introduction for methodological discussion).

We begin with the forest-rich sites, followed by the forest-poor sites (each of which is described in Annex 2 in the Introduction). We conclude with a comparison of the results obtained from the two kinds of sites.

Forest-Rich Sites

Bulungan, East Kalimantan, Indonesia. The Bulungan test site is remote, inhabited primarily by indigenous Dayak who practice swidden agriculture. Logging has been under way for a decade or more, and development-related activity has increased just east of the surveyed area. The six management functions identified above were used.[1]

Initial assessment suggests that all stakeholders have fairly equal rights to manage the forest. The relevant local stakeholders included the government, the logging company, and three local ethnic groups (Abay, Lundaye, and Punan). The cumulative rights of the three local ethnic groups are represented by respondents as greater than either the government's or the logging company's rights (Figure 14-1).[2] The Punan, a group of hunter-gatherers, are considered locally to have somewhat more rights than the other ethnic groups. This situation is interesting because the Punan, like the Baka (a pygmy group in the Dja Reserve, Cameroon), are widely held in low esteem by outsiders. This marginalization in both places is longstanding and relates to their forest-based, relatively mobile lifestyle combined with outside perceptions that they are "primitive."

In Figure 14-1, the government's perceived rights to *define/protect boundaries* (function 1) and *develop/apply rules and regulations* (function 2) are somewhat greater than its rights relating to the other functions. The strong perceived role of local groups to *assess fines/sanctions* (function 6) is also of interest. It probably reflects the real lack of enforcement capabilities of the central government in the area and the existence of a system of traditional sanctions.

In the biplot analysis, the stakeholders are positioned according to their correlation with particular rights (Figure 14-2). There is a strong positive correlation among the various management functions (indicated by the small angles between the arrows), which means that stakeholders who have more rights relating to one function tend also to have more rights relating to the adjacent functions. The one exception is *assessing fines/sanctions*. Traditional local sanctions appear to play a significant role compared with sanctions potentially imposed by other stakeholders. The government is positioned in the middle of the various rights, quite far along the direction the arrows are pointing, indicating that respondents perceived the government to have very strong rights to perform those management functions. The three local ethnic groups (Punan, Abay, and Lundaye) are positioned behind the arrow's direction, indicating fewer overall rights in this respect. Their close proximity suggests similar patterns of rights; their positioning closer to the two functions *assessing fines/sanctions* and, to a lesser extent, *monitoring compliance* suggests that their rights in these two spheres are stronger than their rights relating to the other four management functions. The biplot analysis explains more than 90% of the data structure with these two axes. The significant point on the graph, representing the graphed position of each stakeholder and management function, lies at the beginning of the text (for example, in Figure 14-2, the point where the "P" in Punan begins marks the point that these stakeholders occupy in the graph vis-à-vis other stakeholders and the management functions).

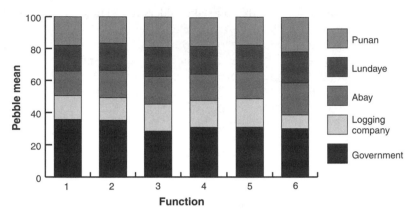

Figure 14-1. *Rights to Manage among Stakeholders in Bulungan, Indonesia*
Note: Functions 1–6 for this site are listed in Methods and Results.

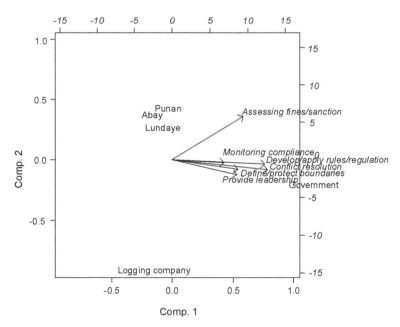

Figure 14-2. *Biplot for Data from Bulungan, Indonesia*

Dja Reserve, Cameroon. The Dja Reserve is a conservation area in francophone central Cameroon, considered comparatively remote, locally. Local peoples, in-migrants, logging companies, and conservation activities compete for local forests that remain in good condition to date. The same set of functions was identified for the Dja Reserve as for Bulungan.[3]

The local ethnic groups (Baka, Kako, and Nzime) are seen to dominate in *defining/protecting boundaries* (function 1) and *developing/applying rules and regulations* (function 2); the Nzime's role is particularly strong—the exact reverse of the Bulungan situation. The government is seen to have the most rights for the rest of the functions, with the most extreme being *providing leadership* (function 5) (Figure 14-3). The elders are significantly involved in *conflict resolution* (function 4) and *assessing fines/sanctions* (function 6). This finding supports the view that, following customary regulations, the local populations are in the best position to define and limit the boundaries of their territory. Tchikangwa considers this component vital to enabling local populations to directly protect their resources from outsiders. In the Dja Reserve, all other functions (such as *monitoring compliance* [function 3], *assessing fines/sanctions* [function 6], and *providing leadership* [function 5]) are left to the government, partly because the local communities do not have a strong sociopolitical organization to fulfil these functions, even in their own territory. On the other hand, local populations acknowledge that the forest belongs to the government and thus the responsibility for protection remains with the government.

A biplot analysis (Figure 14-4) yields two clusters of functions. The first consists of *conflict resolution* and *assessing fines/sanctions*. The second clusters the remaining functions (*defining/protecting boundaries, developing/applying rules and regulations, monitoring compliance,* and *providing leadership*). The elders are seen as the most important stakeholders in the first cluster, as shown by its location closer to the top and center of the diagram (closest to the arrow tips for *conflict resolution* and *assessing fines and sanctions*). The government is the most important stakeholder in the latter cluster, positioned as it is centrally among the other functions and to the right side of the diagram near the arrow tips. The local ethnic groups (lower left) are clustered nearest the arrow tips that represent functions over which they have control, *developing/ applying rules and regulations* and *defining/protecting boundaries* (as discussed in the previous paragraph).

Mbongo Village, Mount Cameroon, Cameroon. Mbongo village is in a fairly accessible, anglophone part of northwest Cameroon where in-migrants from Nigeria complicate ethnic relations and forest management. Logging and conservation activities compete with local needs, as in the Dja Reserve, and plantation agriculture represents another important feature. The relevant functions for this site were the same as for the previous locations.

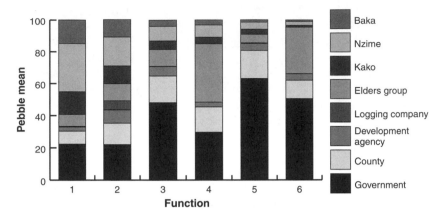

Figure 14-3. *Rights to Manage among Stakeholders in the Dja Reserve, Cameroon*

Note: Functions 1–6 for this site are listed in Methods and Results.

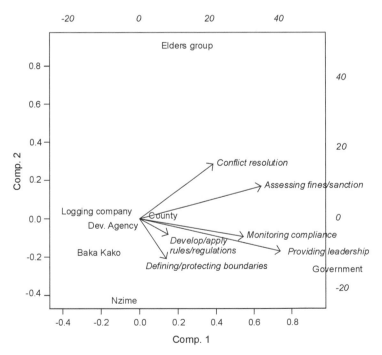

Figure 14-4. *Biplot for Data from the Dja Reserve, Cameroon*

Five stakeholders were identified as playing important roles in managing this forest, but our respondents perceived only three as significant.[4] The government and local ethnic groups share fairly equal rights, and the Mount Cameroon Project has a small share. This pattern is clearest in *providing leadership* (function 5) and *monitoring compliance* (function 3) (Figure 14-5). *Rights* are divided between the government and local ethnic groups, which is clear in the biplot analysis (Figure 14-6). The government develops/applies rules and regulations, as in Bulungan, whereas the local ethnic groups provide leadership, in sharp contrast to the other forest-rich Cameroonian site, the Dja Reserve. The positions of the Mount Cameroon Project (MCP; conservation), the logging company, and the Cameroon Development Corporation (CDC; plantation) have minimal rights to manage, according to these data.

In Mbongo, management was directly associated with ownership and control. *Responsibilities* were interpreted as responsibilities toward the rightful owners of the forest and then, by extension, to the forest itself. Public discussion of different kinds of management was regarded, particularly by native respondents, as in some way conferring legitimacy to nonnative claims (for example, by logging companies) to a stake in deciding the present or future use of native land—a notion contrary to accepted cultural norms. This strong sense of ownership over not only the forest but also the discourses connected with it helps explain the distinct demarcation of rights between the government and local ethnic groups.

Second, in the local language the term *to use* was interchangeable with the term *to value,* and in both senses was used to mean "management." Where this detail gained significance was in the very clear distinction local ethnic groups made between the forest and the land it stood on. The land was deemed valuable, and their inalienable right to the land gave them the right to manage. In this sense, *rights* was interpreted as the right to dispose of forest resources in the way local ethnic groups deemed fit. This interpretation was highlighted by the local interpretation of a *forest reserve.* In stark contrast to the official meaning of state-owned land, formally registered and legally under government control with restricted access and rights for other stakeholders, local ethnic groups strongly felt they had equal rights to manage. *Reserve* for them meant reserved, with the help of government, so as to be available for future use by themselves in ways that they controlled.

Trairão, Pará, Brazil. Trairão is a colonist area along a secondary road (on the Transamazon highway) in the municipality of Uruará, where settlers hold plots that range from 10 to 50 hectares. Logging and livestock are important components of the system. We identified seven stakeholders, four of whom have significant rights in managing the forests (Figure 14-7).[5] In contrast with the previous three sites, here we only consider three functions to describe the division of rights to manage the forest among stakeholders:

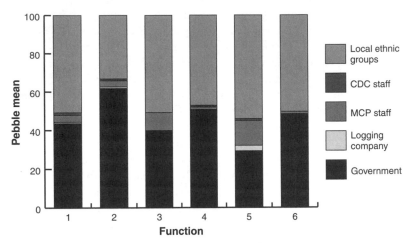

Figure 14-5. *Rights to Manage the Forest in Mbongo, Cameroon*

Note: Functions 1–6 for this site are listed in Methods and Results.

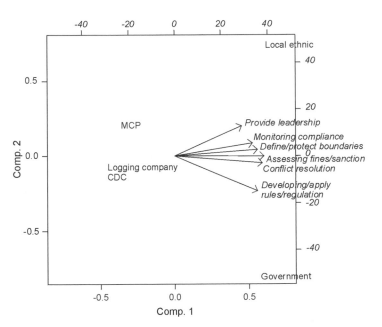

Figure 14-6. *Biplot for Data from Mbongo, Cameroon*

1. Defining/protecting boundaries
2. Land use planning
3. Conflict resolution

The Instituto Nacional de Colonização e Reforma Agrária (INCRA, Brazilian national institute for agrarian reform and colonization) is seen to have the most rights to perform these three functions. The colonists, the rural union, and the Instituto Brasileiro do Meio Ambiente e dos Recursos Naturais Renováveis (IBAMA, Brazilian institute for the environment and natural resources) also are seen to have some rights. Porro remains skeptical about the replicability of these results if the same investigation occurred at a different time. INCRA acquired greater overall importance upon providing access to rural credit for colonist families a few months before our study. However, respondents did not consider a large timber company that had temporarily ceased operations in the area to be significant.

Biplot analysis (Figure 14-8) confirms our impression that the colonists have a comparatively large role in *land use planning*, whereas the other two functions are seen to be dominated by a government agency, INCRA. The rural union is seen to have a minor role in *conflict resolution* and *defining/ protecting* boundaries; loggers, ranchers, and the logging company are portrayed to have almost no such rights.

São João, Pará, Brazil. São João is a site of *ribeirinhos* (long settled, river-dwelling people) in the municipality of Porto de Moz along the Xingu River, where loggers were expelled twice. Fishing and animal husbandry have supplemented agriculture in local livelihood strategies. Of the seven stakeholders identified in São João, four emerged as having significant rights in managing the forest.[6] Six management functions, different from the initial set, were identified for describing the division of rights among stakeholders:

1. Defining/protecting boundaries
2. Land use planning
3. Timber extraction planning
4. Proposing projects to community
5. Conflict resolution
6. Determining fishing location

The identification of relevant stakeholders in this site reinforces conclusions in Chapter 11 about the importance of community organization for SFM. Here, the level of organization is the criterion used to distinguish among the social categories. Roughly, we see that both organized and unorganized fisherfolk have the most rights, and the Natural Resources Committee (another stakeholder related to the process of community organization) has a small share (Figure 14-9). The rural union also plays an important role in some functions, such as *defining/protecting boundaries* (function 1), *land use*

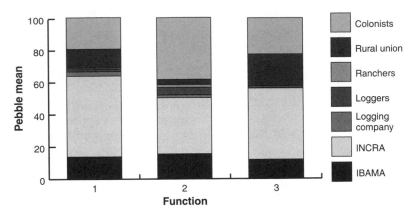

Figure 14-7. *Rights to Manage among Stakeholders in Trairão, Brazil*

Note: Functions 1–3 for this site are listed in Methods and Results.

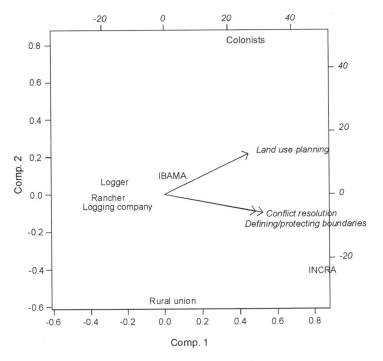

Figure 14-8. *Biplot for Data from Trairão, Brazil*

planning (function 2), *proposing projects* (function 4), and *conflict resolution* (function 5).

The biplot in Figure 14-10 shows that the right to *plan timber extraction* is related to *determining fishing locations*. These rights are strongly held by unorganized fisherfolk and organized fisherfolk, respectively. On the other hand, the rights to *resolve conflicts* are highly correlated with *defining/protecting boundaries, land use planning*, and *proposing projects to the community*. The rural union is comparatively important in this bundle of rights, as is the Natural Resources Committee, but to a lesser degree.

Forest-Poor Sites

Long Segar, East Kalimantan, Indonesia. Long Segar is a community of technically resettled[7] Dayak who moved to their present location in the 1960s and 1970s from a more remote area. Since that time, the area has been subjected to logging, industrial timber plantations, and transmigration influxes as well as severe wildfires in 1982–1983 and again in 1997–1998. We identified the same functions as in Bulungan and most other locales.[8] All stakeholders have significant rights, and the government has somewhat more rights than the others (Figure 14-11).

The biplot analysis (Figure 14-12) shows the government to be neatly situated in the heart of management rights. The local ethnic groups, Kenyah and Kutai, have more rights than does the logging company in *assessing fines/sanctions* and *providing leadership*. On the other hand, the logging company has more rights than local ethnic groups in *monitoring compliance* and in *conflict resolution*.

Mbalmayo, Cameroon. Mbalmayo is a large town very near to Yaoundé, Cameroon's capital. The area is ethnically mixed and subject to intense development pressures; forests in the area are mere fragments. In Mbalmayo, we used the management functions used for Bulungan and most other locales.[9] The government is seen to dominate all management functions (more than in Long Segar); the local peoples, the logging company, and artisans are seen as having rights in decreasing order (Figure 14-13).

The biplot analysis (Figure 14-14) confirms this pattern, without providing much additional information. The placement of local peoples closer to *defining boundaries* suggests that they are seen to have a somewhat greater role in that function than in the others, but the government clearly is seen to dominate most management functions. Locally, all of these functions are perceived to be interwoven in daily life. Tchikangwa suggests that the stronger role of government in forest-poor sites such as Mbalmayo may be related to the fact that these areas have been opened to logging already. By doing so, the government has strengthened its claims on property in the area, weakening the rights of local peoples.

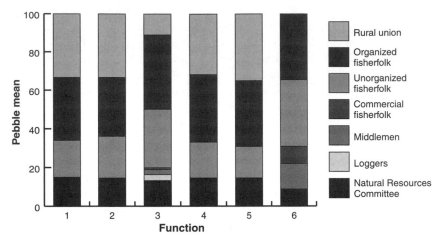

Figure 14-9. *Rights to Manage the Forest in São João, Brazil*

Note: Functions 1–6 for this site are listed in Methods and Results.

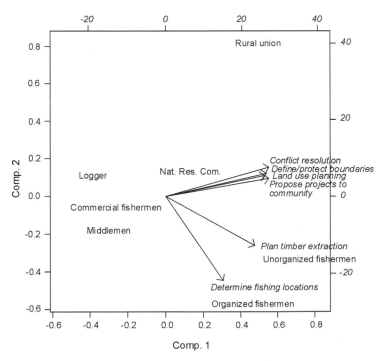

Figure 14-10. *Biplot for Data from São João, Brazil*

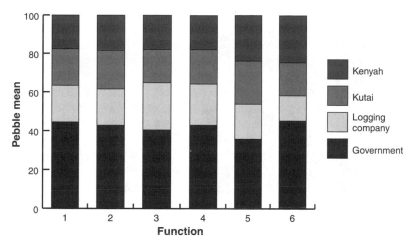

Figure 14-11. *Rights to Manage the Forest in Long Segar, Indonesia*

Note: Functions 1–6 for this site are listed in Methods and Results.

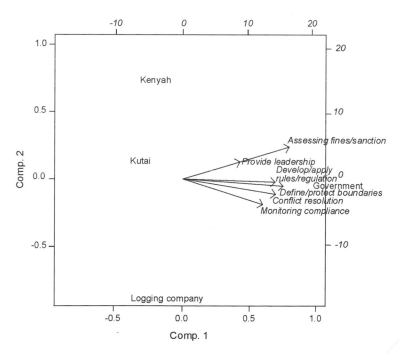

Figure 14-12. *Biplot for Data from Long Segar, Indonesia*

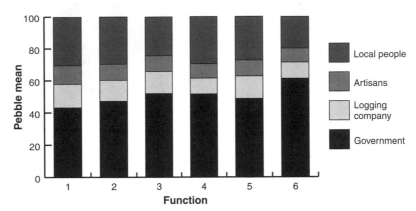

Figure 14-13. *Rights to Manage the Forest in Mbalmayo, Cameroon*
Note: Functions 1–6 for this site are listed in Methods and Results.

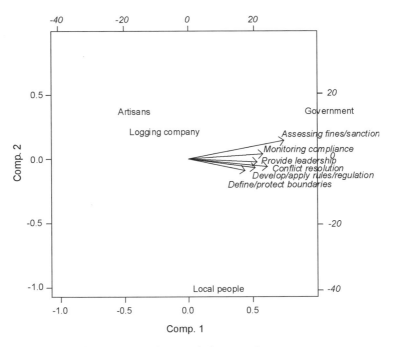

Figure 14-14. *Biplot for Data from Mbalmayo, Cameroon*

The most recent national forestry legislation (Law no. 94/01, January 20, 1994) stipulates, among other things, that

- the forestry concession agreement is made between the forest company and the state. (However, local populations are not consulted when logging agreements are made.)
- the local "commune" (an administrative unit above the village level) receives the sale price from forest products and taxes calculated annually based on their communal area. (Yet, local peoples are not convinced that the "communes" work for their interests, and the people consequently do not concern themselves particularly with the stipulated tax receipts.)
- local peoples must receive the sale price for products taken from the forests of which they are the owners. (But on one hand, the villagers are only owners of the trees found on their fields and in their fallows. On the other, the loggers do not see this last point as a duty toward local populations; rather, they see commodities, with the notion of a sale price being converted into "compensation" for which neither the quality nor the price is fixed.)

The relationships between local peoples and the logging companies are generally conflict-ridden. Typically, if these conflicts over resources are not resolved at the family level or at the community level in a customary tribunal, the state has the last word. Within communities, rules exist that precisely define the conditions and modalities of access to resources, particularly in the case of succession, new acquisitions, gifts, sale, or legacy.

Transiriri, Pará, Brazil. This colonist area in the municipality of Uruará was officially settled in 1970 and 1971, and additional settlement continued into the 1980s. Timber was extracted for more than 15 years; now the area is subject to serious controversy relating to the land rights of members of the indigenous Arara indian tribe. In Transiriri,[10] we identified the following six functions:

1. Defining/protecting boundaries
2. Land use planning
3. Timber extraction planning
4. Hunting planning
5. Planning to set fire
6. Conflict resolution

The colonists are seen as having the most rights related to these functions (Figure 14-15). INCRA, IBAMA, and ranchers also have some rights. The logging company is seen to dominate in matters pertaining to *timber extraction planning* (function 3), and squatters have fewer rights in only *defining/protecting boundaries* (function 1) and *land use planning* (function 2). Transiriri exemplifies situations in which the local perception of the terms *relevant* or

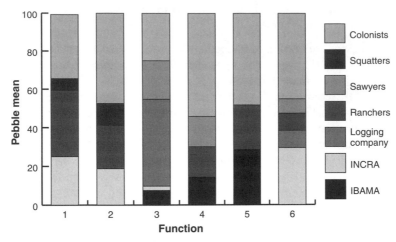

Figure 14-15. *Rights to Manage the Forest in Transiriri, Brazil*
Note: Functions 1–6 for this site are listed in Methods and Results.

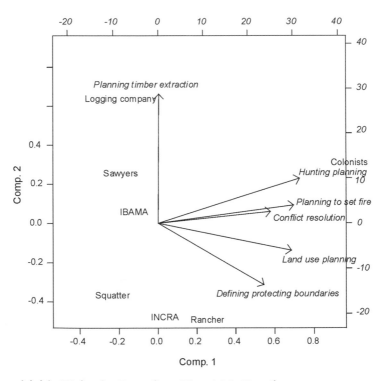

Figure 14-16. *Biplot for Data from Transiriri, Brazil*

important, as associated with positive forest management outcomes, rules out stakeholders that in fact have an important role. Transiriri respondents perceived the Fundação Nacional do Índio (FUNAI, the federal agency in charge of indigenous policy) as not relevant for forest management because the agency is viewed as the main threat to colonists' tenure security and access to resources.

The biplot (Figure 14-16) once again strengthens the conclusion that *timber extraction planning* was absolutely dominated by the logging company. *Hunting planning, planning to set fire,* and *conflict resolution* are highly correlated, and colonists are seen to have a central role. *Defining/protecting boundaries* has a different pattern; colonists, ranchers, and INCRA share most rights.

Forest-Rich/Forest-Poor Comparisons

Qualitative examination of the data does not provide a clear pattern differentiating forest-rich and forest-poor sites. However, several points do stand out: the differences and similarities in stakeholders across sites, the necessity to alter the management functions significantly in the Brazilian cases, a greater correlation between management functions in Indonesia and the Mbalmayo (Cameroon) site, and an increase in the perceived role of government in the forest-poor sites in Cameroon and Indonesia.

In a previous analysis (Colfer and others 1998b), we compared two sites with rich forest quality and two sites with poor forest quality. The forest-rich sites were Bulungan (Indonesia) and the Dja Reserve (Cameroon). Forest-poor sites were Long Segar (Indonesia) and Mbalmayo (Cameroon). In the following analysis, we have added Mount Cameroon (another forest-rich site) to our analysis.

One of our research questions is whether the allocation of rights to manage the forest is similar among stakeholders from forests of the same quality (that is, forest rich or forest poor) in different countries. We used cluster analysis (Everitt 1993) to group stakeholders with similar rights-related characteristics. The algorithm used simply separates groups of stakeholders with different profiles and groups those with similar profiles. There are difficulties in determining how many groups we need to get the optimal separation. The algorithm used in this paper was proposed by Milligan and Cooper (1985) and helped us to determine the final number of groups by picking "the best cut-off point" (the point at which the similarities clearly delineate a useful number of clusters).

For the forest-rich sites, the best cutoff point was 70% (see Figure 14-17); it produced three clusters. The first cluster contained the Cameroonian government from the Dja Reserve data set (*govt(1)*) and local ethnic groups from Mbongo (*local eth. grp(3)*), the second cluster contained only the Indonesian government from the Bulungan site (*govt(2)*), and the third cluster contained the rest of the stakeholders (including *govt(3)* from Mount Came-roon).

The difficult-to-read Figure 14-18, the related biplot analysis, shows that the Dja Reserve's *govt(1)* had an outlier value on functions 5 and 6 (*providing leadership or organization* and *assessing fines and sanctions*, respectively), much higher than the other stakeholders. The state, *govt(2)*, in Bulungan is seen to have greater rights than average to exercise every management function. Local ethnic groups and the logging company in Bulungan lie crowded around the point of origin but still in the same direction as the variable arrows for all rights and responsibilities. This biplot shows that they still have greater rights than do those in the Dja Reserve, who lie mostly in the direction opposite from the variable arrows. The Nzime ethnic group is an exception; they have fairly clear rights to *define/protect their boundaries* and *develop/apply rules/regulations*.

For the two sites with poor forest quality (Long Segar and Mbalmayo), the clustering process yielded two groups with a cutoff point of 70%. The first cluster contained two government stakeholders (*govt(1)* and *govt(2)*), and the second cluster contained the rest (Figures 14-19 and 14-20). Because the two governments were grouped in one cluster, it is evident that their differences were not perceived to be as clear as in the forest-rich case.

Conclusions

We began this research hoping to establish whether SFM was directly and causally linked with the condition stipulated by our Principle 4: "Concerned stakeholders have acknowledged rights and means to manage forests cooperatively and equitably."

On a site-by-site basis, we collected and presented evidence that local communities on each site see themselves as having some share of management rights, differing in strength and emphasis. This finding in itself is interesting, given the frequent marginalization of forest people in respect of their formal political position. The need to revise the management functions significantly in the Brazilian case is also intriguing and warrants further attention.

In the comparison between forest-rich and forest-poor areas, our study suggests that government involvement (vis-à-vis the involvement of other stakeholders in forest management) is seen as stronger in forest-poor areas than in forest-rich areas. One factor in this pattern probably relates to the congruence between roads, development-related infrastructure, and governmental access (and thus presence) in the forest-poor areas versus the comparative lack thereof in the forest-rich areas. Also, local peoples tend to be somewhat more involved in forest management in the forest-rich sites than they are in the forest-poor sites (Figures 14-18 and 14-20). Again, the

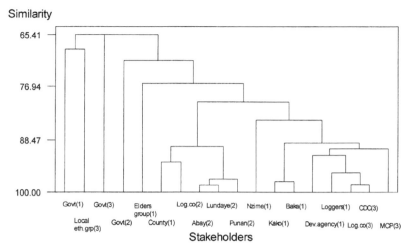

Figure 14-17. *Clustering Stakeholders from Forest-Rich Sites across Countries Based on Rights to Manage the Forest*

Note: Stakeholders for this site are discussed in Methods and Results.

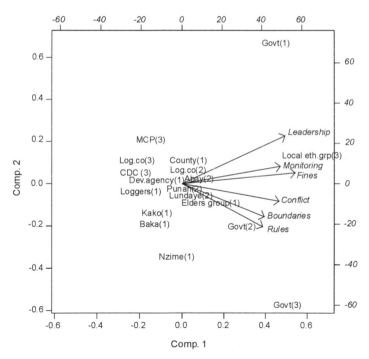

Figure 14-18. *Biplot for Comparing Forest-Rich Sites across Countries Based on Rights to Manage the Forest*

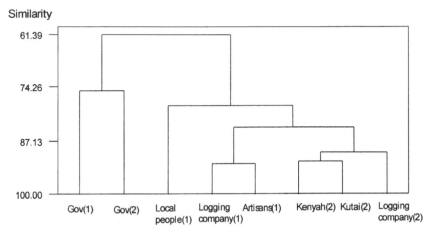

Figure 14-19. *Clustering Stakeholders from Forest-Poor Sites across Countries Based on Rights to Manage the Forest*

Note: Stakeholders for this site are discussed in Methods and Results.

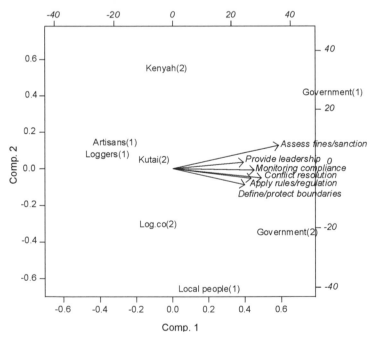

Figure 14-20. *Biplot for Comparing Forest-Poor Sites across Countries Based on Rights to Manage the Forest*

comparative difficulty of governmental access may leave management in local hands in the typically more remote, forest-rich areas.

Although our hypothesis—that local peoples' rights to manage would be stronger in forest-rich areas—is confirmed, we cannot discount our fatal flaw in using current forest quality as a proxy for SFM. Good forests can exist, in the short run, under poor management. SFM is probably possible in degraded environments (see the discussion of degradation in Chapter 15). In fact, we found no cases of forests that we could truly consider to be sustainably managed in our research sites—whether forest rich or forest poor. It is also true that although local peoples' rights to manage in forest-rich sites seem somewhat stronger than in forest-poor sites, they are not stronger than, for instance, government's perceived rights to manage in either quality of forest. However, the stronger involvement of governments in forest-poor sites and, alternatively, the stronger involvement of local communities in forest-rich sites constitute interesting and suggestive evidence that we may be on the right track in considering that local peoples' involvement is important for good forest management.

Endnotes

1. For ethnographic detail about this area, see Puri 1997, Kaskija 1995, Wollenberg and others 1996, Wollenberg 1999; see also Chapters 1, 4, 5, 8, 12, and 16 for similar systems.

2. This finding is consistent with our findings using a participatory card-sorting method in West Kalimantan, Indonesia (Chapter 12; see Colfer and others 1997c).

3. See Tchikangwa and others 1998, for less forest-rich contexts in the same culture area; Chapters 2, 3, and 10.

4. For more on Mount Cameroon, see Brocklesby and others 1997, Brocklesby and Ambrose-Oji 1997; for other Cameroonian contexts, see Chapters 2, 3, and 10.

5. For additional ethnographic detail, see Porro and Miyasaka Porro 1998 and Chapter 13.

6. For pertinent ethnographic context, see Porro and Miyasaka Porro 1998 and Chapter 13.

7. They were "technically resettled" because in fact most of them moved to Long Segar on their own; once they were there, they were declared a "government resettlement village" (in 1972). A similar experience is described in Chapter 13.

8. For ethnographic descriptions of the Long Segar context, see Colfer with Dudley 1993, Colfer and others 1997a; Chapters 1, 4, 5, 8, 12, and 16 also provide information about related systems.

9. For ethnographic context, see Tiani and others 1997 and Chapter 2; for related systems, see Chapters 3 and 10.

10. See Porro and Miyasaka Porro 1998 and Chapter 13.

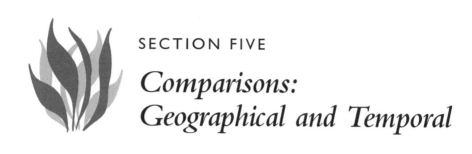

SECTION FIVE

Comparisons:
Geographical and Temporal

In this section, we take a different approach to comparisons from those represented in Chapter 6 on a conservation ethic, Chapter 11 on access to resources, and Chapter 14 on rights to manage forests. In Chapter 15, Tainter analyzes sustainability issues, particularly pertaining to people, in the United States; and then in Chapter 16, Dennis and others temporally compare different management strategies in three communities in West Kalimantan, Indonesia.

The difference between Tainter's analysis and those of other contributors to this volume is reminiscent of the blanket analogy discussed in the Introduction. There we noted that as we move from one context to another, the heights of the hills (or C&I) vary. In Brazil, the workers' rights hill emerged as a critical one; in Côte d'Ivoire, during one of our C&I tests, the health hill (both physical and mental) rose to a new height; in Kalimantan, the cultural integrity hill grew. These differences are partly related to the interests of the researchers but also reflect legitimate differences "on the ground." We need to develop new models to replace the inadequate, though useful, hierarchical model that focuses on principles, criteria, and indicators.

Tainter argues that sustainability is inherently value laden—a view we share. But he also prefers to make a sharper separation between human well-being and sustainability in general than the rest of the authors in this book. Where we have considered human well-being to be an integral part of sustainable forest management, he argues that this concept muddies the waters unnecessarily and makes the concept difficult to defend. His discussion is thought-provoking.

The principles, criteria, and indicators that Tainter discusses as suitable for the United States also are of interest, both for their own sake and for their differences with the criteria and indicators that evolved in developing-world studies. One question that remains of course is, to what extent are the differences simply the usual topological variations—the "lumps" that distort our "blanket"? We would argue that issues such as security of intergenerational access to resources and the rights and responsibilities to manage cooperatively and equitably remain important in developed countries; those countries may simply have "higher scores" on those criteria. Developed countries may do less well on issues such as the health of local cultures (cultural integrity, cultural diversity) through processes such as globalization, urbanization, and other homogenizing forces.

In Chapter 16, Dennis and others report one attempt at placing three sites in West Kalimantan, Indonesia, on the forest-rich/forest-poor continuum (see also Dennis and others 1999). We had hoped to obtain similar data for all our sites, but this goal proved too ambitious.* The Danau Sentarum Wildlife Reserve (DSWR) in general is at the forest-rich end. The historical perspective, though research intensive, provides a much better sense of sustainability trends than can be obtained by single-slice views of a forest landscape (the problematic forest-rich/forest-poor distinction used in this book). The authors also provide a contextualizing, landscape-level analysis of the area, a useful complement to the social science emphases in the previous related chapters.

The variation among the three forest-rich DSWR sites examined in detail suggests significant roles for insecurity of tenure and conflict in those areas where forest management appears less sustainable. The site with the least forest cover change of the three is also in a habitat that is less suitable for agriculture than the others, having almost no dry land on which to plant crops. The apparent sustainability of the swidden cultivation system is also of wider interest, lending credence to ethnographic reports that describe nearby swidden systems as complex and comparatively sustainable (for example, Dove 1985; Colfer and others 1997a; Wadley 1997).

---

*As this book goes to press, Dennis and others have just completed a similar analysis of the forest cover around Long Segar, East Kalimantan, one of our forest-poor sites (Colfer and others 2000b).

CHAPTER FIFTEEN

Sustainable Rural Communities

General Principles and
North American Indicators

Joseph A. Tainter

Sustainability inherently involves values and other intangible social qualities. This characteristic leaves efforts to promote sustainability vulnerable to misuse, political misunderstanding, and extensions of meaning beyond the original purpose. The irony of the most successful mammalian species fretting over its future suggests that sustainability will always be a dilemma. It is a human dilemma, because ecosystems cannot care whether they leak nutrients, lose species, or become less diverse. Many people do care whether ecosystems degrade, yet to sustain nature we build systems of management and jurisprudence so cumbersome and costly that the management effort itself may not be sustainable (Tainter 1997). As researchers seek concrete indicators of sustainability, many argue that the concept is inherently vague, relative, and value laden. It is not surprising that sustainability evokes such a mixture of veneration, opposition, and confusion.

When considering a statement containing the word *sustainability*, one should always ask, of what, for whom, for how long, and at what cost? Few writers on sustainability have thought to raise such questions, and fewer still have tried to answer them. Yet we must insist on these questions, because the term *sustainable* is rapidly filling the niche once occupied in popular discourse by *ecological*. In the absence of clear definitions and limiting conditions, sustainability is a carrier of social, political, personal, and even commercial meanings that we project on it. Someday, no doubt, we will be offered "sustainably produced" toothpaste, just as it has been profitable to market goods as "natural" or "ecological."

In the social realm, sustainability has meant various conditions that are thought to be good: better wages and work conditions, improved health, recycling, spiritual well-being. Sustainability reflects the manifold meanings that we project on it; it also strains under the many purposes we ask of it. As long as this condition persists, politicians will be able to endorse *sustainability* as a concept while denying that it calls for concrete action, and will define the terms of the sustainability debate as consumption and employment versus sacrifice and unemployment. We need an understanding of sustainability that is both more concrete and more nuanced.

In this chapter, I have three objectives: to sketch an understanding of sustainability based on research that combines hierarchy, systems, and complexity theory; to discuss the development of the criteria and indicators (C&I) of social sustainability; and, based on the first two points, to present a set of C&I of the sustainability of small rural communities in the United States and Canada.

A Concept of Sustainability

Certain dilemmas illustrate why sustainability is such a conundrum. Along the Rio Grande where it runs through central New Mexico, for example, the native cottonwood forests are being displaced by saltcedar, an introduced species. Many people who are aware that the cottonwood forest is dying support efforts to sustain it. The reasons are both historic and aesthetic; cottonwood is the dominant native tree of the riverine forest, and it grows to majestic heights with leaves that flutter in even light breezes and provide shade from the intense summer sun. Cottonwood formerly regenerated when spring snowmelt would flood the gallery forest. Today, runoff is controlled, and overbank flooding is a thing of the past. Individual cottonwoods must be planted as poles, and over an area of several thousand hectares, this method is economically infeasible. Ironically, although a saltcedar forest is more sustainable under today's conditions, we try to suppress it; although cottonwood is less sustainable under today's conditions, we try to promote it, largely for aesthetic reasons.

Degradation is one opposite of *sustainability*. Sander van der Leeuw (1998) and his colleagues have studied degradation across parts of Europe and the Mediterranean Basin. In the Vera Basin of Spain, degradation manifests itself in a way that is commonly understood—as erosion. Yet in Epirus, in the northwest of Greece, degradation appears not as the decline of vegetation but quite the opposite—increases in shrub and tree cover that choke a formerly open landscape. Centuries-old pastoralism, which helped villages to be sustainably self-sufficient, is now impossible. To urban residents the landscape now appears "natural," but to Epirotes it has been degraded. As van der Leeuw points out, *degradation* is a social construct; it has no absolute biophysical references. In Epirus, the spread of shrub and tree cover has reduced the supply of groundwater and the flow of springs. As mountain vegetation thrives, the health of other vegetation lower down declines. There has been erosion in Epirus, yet when soil was washed from the mountains, it formed rich deposits in valleys that sustained agriculture for millennia (Bailey and others 1998; Bailiff and others 1998; Green and others 1998; van der Leeuw 1998). Mountaintop soils degraded while agroecosystems thrived. Recall the questions: sustain what, and for whom? In the realm of sustainability and degradation, there are winners and losers. The only constant is that these terms mean whatever people want them to mean under specific circumstances.

Clearly, sustainability is more than biophysical processes. The difficulty in defining sustainability is precisely that it is a matter of values—which vary among individuals, groups, and societies—and those values change over time. This "time variability" has profound implications, particularly when planning what to sustain for the long term. For at least 5,000 years, until the nineteenth century, the ideal landscape was agricultural. A landscape of peasant cultivators produced food for the cities, taxes for the state, and sons for the army. Early in American history, such a landscape was considered the basis of Jeffersonian democracy. It is still the ideal landscape in much of the world. In North America, though, a landscape of small farmers is largely a quaint remembrance, valued more for nostalgia than for political economy. North America today (at least in the United States and Canada) is substantially urban, and many urban residents value land that is managed to appear natural (in their conception) rather than managed to produce commodities. Someday that value, too, may change. In the case of forestry, we may wonder why we struggle to sustain forests that take centuries to mature, when centuries from now no one may care.

Although many advocates of sustainability feel that the things they seek to protect have intrinsic value, this is clearly not so. The things we wish to sustain have only the values that we assign to them, which are transient, variable, and mutable (for example, Tainter and Lucas 1983; Allen and Hoekstra 1994; see also Feyerabend 1962). It is necessary to understand this concept to reduce both disciplinary miscommunications and the political invective that often accompanies sustainability debates. Deciding what to sustain and how to accomplish it are matters for negotiation and consensus.

The least useful conceptions of sustainability are narrowly focused within specific disciplines, such as conservation biology. The practitioners of such disciplines assume that the phenomenon to be sustained has intrinsic values that should be evident to all. Single-discipline concepts or definitions are almost guaranteed to generate opposition from people with competing interests and to result in political stalemate. More sophisticated definitions, on the other hand, respond to the value-laden nature of sustainability by being gen-

eral. The one most widely cited was offered in 1987 by Gro Harlem Brundtland, then Prime Minister of Norway: "Sustainable development is development that meets the needs of the present without compromising the ability
of future generations to meet their own needs" (World Commission on
Environment and Development 1987, 43). This widely used definition has
spawned a miniature industry of writers suggesting clarifications, modifications, or extensions (see Introduction to this book). Writing from the perspective of economics, for example, Pearce and others (1994) modify and
condense Brundtland's definition to mean "nondeclining human well-
being," because *well-being* bears a meaning that economists prefer to *needs*.
Colfer and others (1995, 2) modify Brundtland's definition to address sustainable forestry: "Sustainable forest management aims to meet the needs of
the present without compromising the ability of future generations to meet
their own needs." Two conditions indicate sustainability for this definition:
ecosystem integrity is ensured/maintained, and well-being of people is
maintained or enhanced (Colfer and others 1995). Prabhu and others (1996)
incorporate these indicators by defining sustainable forestry as "a set of
objectives, activities, and outcomes consistent with maintaining or improving
the forest's ecological integrity and contributing to people's well-being now
and in the future."

In these definitions and others like them, the terms *needs* and *well-being*
can be interpreted to extend beyond concern for the material requirements
of life and to cover the intangibles that many people in industrial societies
value, such as ecological processes, endangered species, and uncut forests.
Such matters lie at the core of many sustainability debates, so the definitions
could be improved by making this interpretation explicit (but see Prabhu and
others 1996). Although these definitions do not state that sustainability is a
value judgement, the Center for International Forestry Research (CIFOR)
recognizes that this is so (Colfer and others 1999c). If such thoughtful writers
as these do not state explicitly that sustainability is value laden, though, less-
thoughtful participants in sustainability debates will not realize that it is. Stu-
dents of sustainability will continue to miscommunicate, like ships passing in
the night.

With my colleagues Allen and Hoekstra, I am developing a different
approach to sustainability (Allen and others 1999, not dated). It builds on a
foundation of hierarchy, systems, and complexity theory (Allen and Starr
1982; Tainter 1988, 1995, 1996; Allen and Hoekstra 1992; Ahl and Allen
1996). Five lessons that emerge from this work guide the development of
indicators for rural communities of the United States and Canada, summarized here (Allen and others not dated):

- *Sustainability is achieved by managing for productive contexts rather than for out-*
  puts. Managing to maintain the outputs of productive systems amounts
  metaphorically to sticking a finger in a leaking dike. Leaks spring inevita-

bly (in the form of fluctuations in output) and must constantly be plugged. The problems generating the leaks are never addressed, so the costs of management can never be controlled. This approach is how we typically have practiced agriculture, forestry, and many other forms of management. It is a style in which the integrity of the productive system may never be addressed.

- *Manage systems by managing their contexts.* In a hierarchy, any system is controlled, one level up, by its context (Allen and Starr 1982; Allen and Hoekstra 1992). Management efforts are best focused on the contexts that regulate systems (landscapes perhaps consisting of overly dense forests or monocropped fields) rather than directly on the immediate concern (such as pest outbreaks in forests or agricultural fields). Pests may be most effectively controlled over the long term not by applying pesticides (which equates to another finger in the leaking dike) but by manipulating their reproductive or environmental contexts.

- *Identify the missing components of dysfunctional systems, and supply only those.* This is common-sense management. One might apply fire, for example, to a plant community in which biomass is not decomposing as rapidly as desired. It would be inefficient management to provide more than that. The emphasis of our approach is to call for management that is both more focused and more knowledge-intensive than is now often the case. To know what ecosystems lack requires broad-based research and monitoring of many processes.

- *Use ecological processes to subsidize management.* In a world where fiscal constraints will be a factor for the foreseeable future, management can accomplish most by using ecological subsidies whenever possible. In the case of the Rio Grande gallery forest, for example, reforestation can be accomplished manually (and expensively) by planting cottonwood poles. On the other hand, spring flooding (now controlled by upstream dams) causes detritus to decay more rapidly and allows a fresh generation of cottonwood seedlings to establish themselves where there is open canopy. At the University of New Mexico, researchers are experimenting with riparian management and cottonwood reforestation using seasonal flooding (Crawford and others 1999). The management objective of forest regeneration (and the public's objective of sustaining the cottonwood forest) can be achieved by using subsidies that are free and available whether we use them or not: winter precipitation and gravity.

- *Understand complexity and costliness in problem solving.* As management challenges grow in scale and complexity—a seemingly inexorable phenomenon—management tends to grow more complex in response. Complexity is often successful in problem solving, but it also can be cumbersome and costly. In southeast Alaska, for example, conflict between timber production and Native American subsistence hunting generates litigation and regulation so costly that the net value of subsistence hunting has declined

(Tainter 1997). Historically, problem-solving systems that develop in this way collapse, are terminated, or come to depend on subsidies (Tainter 1988, 1995, 1996).

These five points lead to the following definition for *sustainability:* "maintaining, or fostering the development of, the systemic contexts that produce the goods, services, and amenities that people need or value, at an acceptable cost, for as long as such things are needed or valued." This definition addresses the concerns raised in this chapter. It is avowedly anthropocentric and admits concern for sustaining intangible facets of the environment that we need (undegraded ecological processes) or value (such as pleasing landscapes and uncut forests). It reflects the concern to focus on productive systems rather than outputs.

The questions raised earlier (sustainability of what, for whom, for how long, and at what cost?) must be addressed to implement the definition. Perhaps most important, by focusing explicitly on values, it forces those concerned about sustainability to approach it in the context of the wishes or goals of a community or group. These goals may be ecosystem integrity, biological diversity, cultural diversity, recreational experiences, employment, the continuity of one's community, or anything else. The important thing is that the sustainability goal must be established and clearly articulated. Failing to do so guarantees confusion and conflict. Articulating the sustainability goal is the first step toward clarity of purpose, communication, negotiation, consensus, and implementation. This definition may be less euphonious than that of Prime Minister Brundtland, but its advantage is that it can guide sustainability actions.

Evaluating Social C&I

A sustainability goal is only the first step; one has to know when it has been achieved. Thus, much thought and effort have gone into designing C&I of sustainability. C&I pertaining to ecosystems are vexing enough to develop; those pertaining to societies are even more so. Biological scientists may confront complexity in the ecological sphere as daunting as that in the social, but they are privileged by a great advantage: ecosystems can't talk back. Humans, in contrast, give meaning to their world, and the meanings they assign are as diverse as humanity itself. A social approach to sustainability involves the intersection of a community's or group's perceptions, values, and goals; the sustainability analysts' perceptions, values, and goals; and the biophysical characteristics of the environment.

In this section, I discuss two topics. The first is whether social C&I for industrialized nations must differ from those developed elsewhere, or whether universal C&I are feasible. The second is whether certain social

C&I genuinely measure sustainability. Both topics are important for developing C&I for North America.

Are Universal C&I Feasible?

In CIFOR's experience, testing C&I across various locations (Côte d'Ivoire, Indonesia, and Brazil) yields variable consistency. C&I related to ecology showed 72% commonality, whereas social C&I reached only 34%. In between were forest management and policy at 60% and 57%, respectively (Prabhu and others 1996). It appears from these numbers that C&I consistency varies on a gradient of human involvement (Figure 15-1). At the top of the gradient is applied ecological science. Its degree of human involvement is largely restricted to the specialists themselves. Ecologists are surely as disputatious as any other scientists, but disagreements about theory and application are mostly held within this limited group. Their subject matter cannot object to what ecologists do, and it takes time (up to centuries in some cases) to learn whether the science requires revision. With the least human involvement, ecological C&I display the highest homogeneity.

Forest management ranks second in commonality. Forestry is an established field, with respected authorities and a long tradition of knowledge. Foresters are trained professionals who, to the extent possible, apply what they have been taught. These conditions generate comparatively high homogeneity. At the same time, forest management is very much a matter of human judgement. Foresters may disagree on which treatment to apply to a

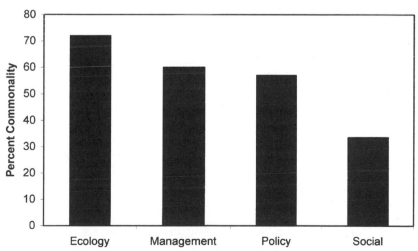

Figure 15-1. *Commonality of C&I across CIFOR Tests*
Source: Data from Prabhu and others (1996).

specific situation. Foresters usually report to higher authorities, who may ultimately be politicians with goals other than biophysical sustainability. Forestry can never achieve the consensus of ecology.

It is no surprise that policy ranks only third on this gradient, although one might not have anticipated that the agreement on policy would be as high as it is (57%). Policy generates enormous disputes. People use policy to express their values and goals, which as emphasized are transient, variable, and mutable. Policy C&I will never achieve consensus except at the level of weak generality that politicians find so appealing.

Last, of course, are the social C&I. They manage barely one-third commonality in tropical forest countries. Throw in C&I for developed nations in temperate zones, and commonality would drop still further (Woodley and others 2000; see later). Once we understand that sustainability is a value judgement, these results are unremarkable. What could be more variable than perceptions of human well-being? It is what all people seek and what all societies claim to provide. Conflicting perceptions of well-being have led to wars. Can we expect them to converge harmoniously in C&I of sustainability?

These numbers raise unavoidable warnings about the feasibility of developing universal (or even nested universal) C&I. Whether to work toward universal C&I is in part a question of whether they should be contextualized. To try to universalize C&I is implicitly to argue that they should be decontextualized—removed from or abstracted beyond the cultural matrix in which they originate, or imposed from the outside. Benefits and costs associated with each approach are worthwhile to understand.

Contextualized C&I are inherently variable. They are difficult to aggregate for higher-level reporting and may offer little opportunity for cross-cultural synthesis. Accordingly, administrators always will dislike them. It is difficult to implement contextualized C&I in the setting of international agreements—apart from the problem of noncomparability, one can never know with certainty that a neighbor isn't cheating. This problem is known in game theory as "the prisoner's dilemma" (Has my partner confessed? Should I remain quiet, or confess for leniency?). It presents a persistent temptation to assume that other parties to an agreement cheat (exemplified in the regular accusations of cheating among members of the Organization of Petroleum Exporting Countries [OPEC]) and so to cheat oneself. On the other hand, only contextualized C&I can reflect the sustainability values and goals of local communities and groups and thus are ideally developed in cooperation with them. Local people are more likely to support contextualized C&I than decontextualized ones.

Decontextualized C&I gladden the administrative heart. They allow the creation of statistics and charts that look like they describe actual conditions. They facilitate international agreements, because on the surface each nation's report appears comparable to the others. Global C&I potentially lessen the "prisoner's dilemma," because cheating is difficult if all partners report openly and identically. Yet in the United States—where agencies such as the U.S. Forest Service contend with rumors that they are conspiring with international bodies to take away local independence—imposing C&I from the top down is not a matter for the faint of heart. People who believe that the United Nations and the United States wish to deprive them of their land are unlikely to accept indicators of sustainable land use derived through an international forum. More broadly, sustainability goals are inherently local, variable, and changing. To remove them from their local contexts is to render them meaningless—along with the aggregated statistics reported to international bodies.

It is pertinent to note the experience of the team evaluating C&I for southern Idaho. This team rejected many of the indicators and verifiers developed for tropical countries, such as the following:

- sufficient breaks in the working day to prevent accidents, and holidays are at least the legal minimum (Brazil);
- transportation for staff is comfortable (according to local norms) and is well-maintained for safety (Brazil); and
- the existence of a "slave labor" relationship (Brazil) (Prabhu and others 1996, 186–7).

Such matters are simply not pertinent in industrialized North America (Woodley and others 2000). If asked to report such matters, forest managers in the United States or Canada would feel that their time was wasted. Thus, many C&I developed for tropical forest communities do not seem suitable for broader international application (Prabhu and others 1999).

Have Social C&I Measured Sustainability?

Conceptualizing sustainability in terms of human well-being potentially opens Pandora's box. It can allow a plethora of C&I to emerge that are intended to advance well-being but may appear to some observers to have little to do with sustainability. Consider, for example, the following:

- Historically, culturally, and ecologically important characteristics of forests are distinguished and appropriately appraised (Indonesia).
- Respect for and protection of cultural and religious sites of special significance has priority over any use (Brazil).
- Health care and safety are at least the legal minimum (Brazil).
- Training corresponds to necessary skills required for the type of work (Brazil).
- People's incomes have increased in real terms since the company's arrival (Indonesia).
- Use of roads and other appropriate infrastructure is guaranteed to local populations (Brazil).

- Workers' representatives have access to workers' living and working sites (Brazil) (Prabhu and others 1996, 109–12).

Each of these points (and others like them) represents an admirable goal, intended to better human well-being, but they must be scrutinized closely for precisely this reason. Do they, for example, advance goals that are worthy but not clearly related to sustainability? Do they aim to improve well-being rather than to sustain it? Do they advance social goals that allow politicians to dismiss sustainability as poorly defined and allow people with conservative social views to reject it? If so, such indicators may harm sustainability efforts in political contexts far removed from the community in question.

The two arguments underlying use of such indicators (regrettably, often left implicit) would be that human well-being is a prerequisite to sustainable forest management (Prabhu and others 1999) and that human well-being in tropical forest communities studied by or through CIFOR would be rated as poor (personal communication from Carol Colfer, 1999). If we consider that humans today are the context of the world's ecosystems (Allen and others 1999), these arguments have merit. Human actions—intentional or unintentional, individual or collective—control sustainability and rates of degradation for both the biophysical and social worlds. This is the case for all scales of analysis, from microbial processes through species, plant communities, ecosystems, landscapes, and the biosphere. Human tropical forest communities that are impoverished, uneducated, in poor health, and exploited by outsiders and that do not control their own destinies are unlikely contexts to ensure their own sustainability or that of their forests (Colfer and others 1995). Why then do such indicators occasionally seem questionable?

One reason is that there cannot be clear rules to distinguish between social improvements required for sustainability and those needed for other reasons (see also Prabhu and others 1999). The permeability of the distinction between social improvements that are desirable for sustainability and those that are merely desirable leaves sustainability more vulnerable to political attack than would otherwise be the case. Sustainability can be dismissed, by those so inclined, as conventional social activism. On a broader level, if sustainability becomes the context to remedy all social ills of the tropical forests, then it is a concept that will have lost its meaning.

Another—and primary—problem related to these C&I is that they are presented in a vacuum. That is, they are presented without regard to either a sustainability goal or a sustainability assessment. Historical analysis is central to determining a sustainability trend (for example, Tainter 1995; Colfer and others 1997b). If a historical sustainability assessment should show that communities in Indonesia, Brazil, or elsewhere are unsustainable, then a program of improving well-being is called for (see also Prabhu and others 1996). Depending on the missing components of these systems, such a program might focus on improving health, wages, or working conditions. In that context, C&I

such as these become necessary. In the absence of explicit sustainability goals and assessments, however, they are easy targets for critics of sustainability.

As an example, the health of tropical forest actors (Colfer and others 1995) is often included in C&I. Health is a topic that illustrates why it is important to be intellectually rigorous about sustainability, because it can potentially cover all of human life. Time is the ultimate test of sustainability. People who have lived in forested landscapes for generations, centuries, or millennia will likely be exposed to illnesses that originate in such environments. The forest actors' long-term persistence in a forest environment would indicate that such maladies are not related to sustainability; the people and their communities have proven themselves sustainable notwithstanding endemic illnesses. However, maladies arising from, or exacerbated by, such things as recent population growth or new forest uses may be related to sustainability and thus potentially are included in C&I. Of course, it is desirable to alleviate long-endemic illnesses, but however good our intentions, we should not label that desire a goal of sustainability. In sustainability analyses, terms such as *health* and *well-being* must be deconstructed—broken down into those components that truly underpin sustainability and those that do not.

Sustaining Rural American Communities

Given the foregoing arguments, C&I of social sustainability should be designed to meet several standards. They should

- specify clearly what is to be sustained,
- emphasize productive systems rather than outputs,
- focus on the context of the human community,
- seek to identify what unsustainable communities lack,
- minimize the costs of monitoring, and
- limit the inclusion of social goals that are not clearly related to sustainability.

The C&I that I present for rural North American communities (see Annex 15-1) were formulated to address these points. They are refined slightly from a set developed in June 1997 (with comments from J. Kay, T. Hoekstra, and T.F.H. Allen) in the context of planning for the Boise, Idaho, test of C&I (Hoekstra and others 2000). They were not actually used in that test (Woodley and others 2000) and so remain unevaluated in the field.

These C&I are meant to monitor the well-being of small American communities that depend on producing or processing commodities such as forest products. Currently, they are the least sustainable communities in the interior western parts of the United States and Canada. Many of them are experiencing net emigration, loss of local retail businesses, greater travel distance to goods and services, and other processes that suggest declining well-being. Not all of these developments, of course, can be traced to local land use. The development of large chain stores would have harmed smaller businesses regardless of other trends. Thus, there is a need to determine, in each community, whether individual indicators are related to local land use and management.

These C&I are contextualized for small communities within the area of a forest management unit. Medium-sized and larger communities depend more on manufacturing, retailing, and services and have more diversified economic bases. Their sustainability depends on different factors; monitoring social indicators at the regional, national, and international levels is best done by institutions at those levels.

These C&I assume that the people still residing in such communities wish to sustain the contexts that give them the opportunity to do so. This assumption must be tested for each community. I give explanatory notes for selected indicators where the reasoning behind them might not otherwise be clear. Data for several of these indicators are readily obtained from censuses and local administrative records. Many of these records are available electronically. This ease of use provides a dual advantage in that data on the indicators can be collected economically by local officials, and there is little room for the collector to insert value judgements or extraneous social goals.

Considered as a set, these indicators are a first attempt to monitor social sustainability in a way that focuses primarily on contexts, shifts the emphasis away from goals that may be socially valuable but are not clearly related to sustainability, minimizes the cost of monitoring, and increases objectivity. No doubt these indicators can be refined and improved upon; some perhaps should be discarded. The challenge is to do this in a way that retains or enhances the necessary qualities.

Concluding Remarks

Sustainability will never achieve the mechanistic certainty of some fields. No simple formulas or graphs will allow people to specify what they wish to sustain, tell them how to sustain it, or indicate that they have been successful. We never will know that we have become sustainable, or even that we have the best possible set of C&I. The pursuit of sustainability always will be ongoing. Changes in values, technology, and the environment guarantee that the effort to be sustainable will be a never-ending exercise in negotiation, implementation, evaluation, and revision. The strengths and weaknesses of efforts to develop C&I for tropical and temperate forest communities may be understood in that context.

Acknowledgements

I am pleased to express my appreciation to Carol J. Pierce Colfer for the opportunity to prepare this chapter; to Carol and Tom Hoekstra for comments on an earlier draft; and to Tom, Tim Allen, and James Kay for comments on the first draft of the criteria and indicators.

Annex 15-1: Principles, Criteria, and Indicators for the Sustainability of Rural North American Communities

Principle: The contexts that support the sustainability of small rural communities are maintained.

Criterion 1: The social and economic structures that promote sustainability are maintained.

Indicator 1.1: Does the rank-size distribution of communities in the area of the forest management unit (FMU) indicate stable, increasing, or decreasing stability of small communities?

> *Note:* Rank-size distribution reflects the degree to which economic production and benefits are dispersed. Underdeveloped economies typically have one enormous city, with a great drop-off to the next-sized place, suggesting a disproportionate concentration of economic production in a single place. The population of Bamako, Mali, for example, is about one million people, whereas that of Mali's next-sized city is about 40,000. Better-developed economies have a gentler slope from large locations to small locations, indicating a dispersion of economic opportunity.

Indicator 1.2: Is the distribution of monetary incomes per capita across communities in the area of the FMU stable or becoming more or less equitable?

Indicator 1.3: Is the distribution of monetary incomes per capita within communities in the area of the FMU stable or becoming more or less equitable?

Indicator 1.4: Is the population age distribution among smaller communities within the FMU stable, becoming older, or becoming younger?

> *Note:* The point here is to assess change in proportions of age classes, such as retirees, those in prime wage-earning years, and children. If a local population is aging, for example, it would be worthwhile to investigate further to determine whether it is because retirees with pensions are moving in (which many communities would consider desirable) or because wage earners are moving out.

Indicator 1.5: Is the number of locally owned businesses stable, increasing, or decreasing?

Indicator 1.6: Is the number of businesses in small communities stable, increasing, or decreasing?

Indicator 1.7: Is the level of economic activity among businesses in small communities stable, increasing, or decreasing?

Indicator 1.8: Is economic diversity in the area of the FMU stable, increasing, or decreasing?

Indicator 1.9: Is the travel distance to basic goods and services stable, increasing, or decreasing for members of small communities?

> *Note:* Throughout the rural United States, small communities are losing local businesses as chain stores are built in regional centers. This trend decreases local community sustainability, because there are fewer ways to earn a living. This indicator is suggested only for consideration because it may not be related to land use.

Indicator 1.10: Is public infrastructure (for example, roads, bridges, and municipal buildings) being maintained, extended, or allowed to deteriorate?

Indicator 1.11: Is the suite of government structures and services within the area of the FMU stable, expanding, or contracting?

Indicator 1.12: Does the local government have a land-use planning mechanism intended to ensure social and ecological stability?

Indicator 1.13: Do nongovernmental organizations provide a suite of services to the population of the FMU that is stable, expanding, or contracting?

Criterion 2: The social and economic processes that promote sustainability are maintained.

Indicator 2.1: Are monetary incomes within the area of the FMU stable, increasing, or decreasing?

Indicator 2.2: Is access to capital within the area of the FMU stable, increasing, or decreasing?

Indicator 2.3: Is the net capital flow into or out of the area of the FMU?

Indicator 2.4: Is the proportion of children choosing to remain in smaller communities stable, increasing, or decreasing?

Indicator 2.5: Are human subsidies from outside the area of the FMU (for example, from the federal government) stable, increasing, or decreasing?

> *Note:* In a time when all institutions are experiencing fiscal constraints and will do so for the foreseeable future, communities that are less dependent on external subsidies will be more sustainable.

Indicator 2.6: Is the ratio of property value to local income per capita stable, increasing, or decreasing?

Indicator 2.7: Is the ratio of property taxes to local income per capita stable, increasing, or decreasing?

Indicator 2.8: Is the community sense of identity stable, improving, or declining?

> *Note:* This indicator was developed for the original set in June 1997 (Hoekstra and others 2000), but I include it here with reservations. It should be used only if a sustainability goal specifies maintaining a sense of community identity and if a way can be developed to measure that.

Indicator 2.9: Is the flow of information between local communities and their contexts beyond the FMU stable, increasing, or decreasing?

Note: In a time when local communities are affected profoundly by a global economy and flow of information, it is imperative that localities understand the contextual forces that influence their sustainability and their place in the global system (Tainter 1999).

Indicator 2.10: Is the responsiveness of the external institutional context of the FMU stable, increasing, or decreasing?

Note: This indicator (which clearly needs some verifiers) is intended to determine whether institutions at higher levels (for example, state, federal, and international) are responding effectively to local needs. Very often, such institutions respond most effectively to information communicated horizontally (from and about similar entities) and poorly to information communicated vertically (Tainter 1999; Crumley 2000; McIntosh and others 2000).

Criterion 3: The interaction between social and ecological systems promotes community sustainability.

Indicator 3.1: Is the contribution of the FMU to regional, national, and global indicators of social and ecological sustainability stable, increasing, or decreasing?

Indicator 3.2: Does the local land–use planning mechanism operate cooperatively with the FMU planning mechanism?

Indicator 3.3: Is the proportion of economic activity derived from local land use stable, increasing, or decreasing?

Note: Local land use is one of the most stable guarantors of local sustainability, but it may degrade ecological processes. Whether it does must be determined in each case.

Indicator 3.4: Is access to forest resources for basic needs (for example, grazing, timber, firewood, and subsistence foods) stable, increasing, or decreasing for people within the area of the FMU?

Indicator 3.5: Is the responsiveness of forest management and use to the values and preferences of local communities stable, increasing, or decreasing?

Indicator 3.6: Is the participation of local people in deciding the management direction of the FMU stable, increasing, or decreasing?

Indicator 3.7: Is the dependence of forest management on local labor and expertise stable, increasing, or decreasing?

Indicator 3.8: Is consensual dispute resolution (in relation to land use and management) stable, increasing, or decreasing?

Note: The reasoning behind this indicator is that disputes that cannot be resolved consensually generate litigation, regulation, and political posturing that are costly and often require external subsidies (Tainter 1997).

CHAPTER SIXTEEN

Forest Cover Change Analysis as a Proxy

Sustainability Assessment Using Remote Sensing and GIS in West Kalimantan, Indonesia

Rona A. Dennis, Carol J. Pierce Colfer, and Atie Puntodewo

In other parts of this book, the Center for International Forestry Research (CIFOR) research on developing criteria and indicators (C&I) for sustainable forest management has been discussed, as has interest in better understanding the links between those C&I and sustainability. To understand such links, of course, we need reliable methods to determine the biophysical sustainability of management in the forests. An ideal assessment of sustainability would require the ability to predict whether the forests will be maintained into the future (Prabhu and others 1998).[1]

In the unfortunate absence of such prescience, we look at current and past conditions. In this way, we at least determine the direction of change. We are satisfied that showing the quantitative change in forest cover in time and space can make a valuable contribution to the assessment of sustainability levels by providing more concrete and visual evidence to indicate the direction of change. However, complex factors contribute to sustainability (including varying definitions of the concept itself), and the kind of analysis based on remote sensing that we provide in this chapter cannot provide the complete picture.

We used a time series of remote sensing data from 1973, 1990, and 1994 to assess the sustainability of forest use in three villages of the Danau Sentarum Wildlife Reserve (DSWR) in West Kalimantan, Indonesia. This time period is characterized by a shift among three separate, recognized management systems: indigenous management, timber concessionaire management, and management as a wildlife reserve.

The actors in the DSWR area are multiple.[2] For at least two centuries, the Iban and Melayu have been alternately warring and peacefully coexisting (Colfer and others 1997b; Wadley 1999b). The Iban are Dayak swidden cultivators who have strong spiritual ties with the forest, which in turn forms the basis for their subsistence (rice cultivation, NTFP collection, hunting, and so forth). The Melayu depend most fundamentally on fishing, supplemented by agriculture in areas where water levels permit. In the 1970s, timber companies arrived in the area, with official permission from the Indonesian government to harvest timber and a formal mandate to manage the local forests. By the 1990s, the DSWR was established. The Melayu area serves as the core of the reserve, surrounded by traditional Iban areas on drier land. Since that time, the number of inhabitants of the DSWR has increased, most dramatically among the Melayu population. The opportunity to examine these changes in light of changes in forest cover was seen as potentially fruitful.

Remote sensing seemed the most obvious technique available to quantitatively compare forest cover in both spatial and temporal dimensions (Tucker and others 1984; Nelson and Holben 1986; Sader and others 1990; Malingreau 1991; Skole and Tucker 1993; Curran and others 1995). Many examples of research where the disciplines of landscape ecology, human ecology and remote sensing have come together are now available (Moran 1990; NASA 1990; Brondizio and others 1994; Liverman and others 1998). Apart from remote sensing, we wanted to place the forest cover change maps within the context of forest uses: traditional, forest concession, and wildlife reserve. We therefore chose to analyze the changes using a geographical information system (GIS), which provides the functionality to integrate many spatially referenced data. The meeting of remote sensing, human and landscape ecology, and GIS has previously been explored (Behrens 1990). However, ethnographic data—such as indigenous resource management systems—have been integrated in fewer examples (Brondizio and others 1994).

We first describe the general location of the research site, to which several of the analyses in this book pertain. We then show maps of each village's territory, indicating sole use zones, shared zones, and areas of conflict with neighboring communities. Finally, we present and discuss the maps for 1973, 1990, and 1994, showing the change in forest cover.

Study Area

Institutional Context

Our focus is on three village territories in and around the DSWR in West Kalimantan, Indonesia (Figure 16-1). The ecological uniqueness of the Danau

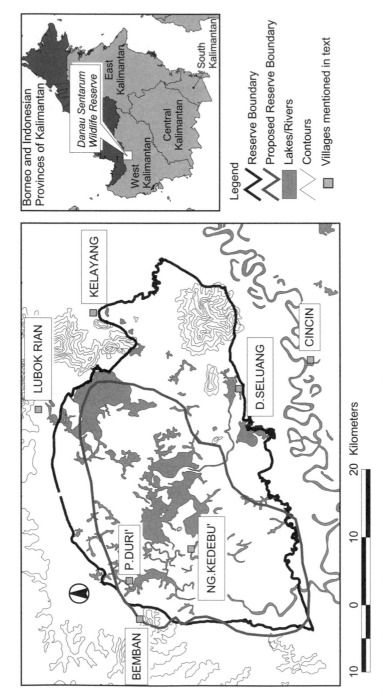

Figure 16-1. *Map of DSWR and Its Place in West Kalimantan*

Sentarum area prompted its protection, in the 1980s, as an 80,000-hectare wildlife reserve. It is an area of seasonally flooded blackwater lakes in the remote interior of Kalimantan, in the Indonesian part of Borneo, near the border with Malaysia. From a biodiversity standpoint, the reserve is of great value. It is the only large area of primary freshwater swamp of this type remaining in Borneo (Giesen 1996). Its flora are unique, and several type specimens are known from this area. Its fish fauna are very rich and include the rare and valuable red-color phase Asian arowana (*Scleropages formosus*). Other important species include the endemic proboscis monkey (*Nasalis larvatus*), orangutan (*Pongo pygmaeus*), and estuarine crocodile (*Crocodylus porosus*).

During the past few years, considerable effort has been devoted to expanding the reserve boundary to include the hills to its east; more recently, efforts have been made to expand it northward as well to include relatively intact but logged-over tall peat swamp forest. This expansion would result in a total protected area of 197,000 hectares.

This site was chosen for study for several reasons. In addition to those listed in Chapter 8—long-term experience on the site, access to numerous reports, presence of several ethnic groups with different management systems, histories, and levels of conflict—we wanted to compare our sustainability conclusions with the results of previous methods testing in the same location (see also Chapters 5, 8, and 12).

The primary *forest actors* (Colfer and others 1999c) in this area include Muslim Melayu fisherfolk who live in the seasonally flooded core of the reserve, and Christian and animist Iban swidden cultivators who live in the surrounding hills. These two groups occupy ecologically very different habitats and have significantly different natural resource management systems. Other important local stakeholders include residents of the larger, Melayu "mother villages" along the Kapuas River (the main river in West Kalimantan), traders, timber concessionaires, timber workers, the Conservation Project, and local government. The 1996 population in DSWR was estimated at 6,750 in the wet season, rising to 9,185 in the dry season, when fish are plentiful. The 1986 population was estimated at 6,170 (Giesen 1987), but the season is not specified. There are 55 villages in the reserve, of which 45 are permanent and 10 seasonal. Comparing Giesen's data and recent data from the Conservation Project, the number of villages has not increased significantly despite the population increase. Between 1986 and 1996, one seasonal village became permanent and two new seasonal villages emerged. The number of personnel from timber companies, the Conservation Project, and local government is minimal; yet their rights to manage local forests are much firmer, from an official standpoint, than are those of the local communities.

Because land tenure emerges as an important feature in our conclusions, a brief mention of this issue in DSWR is in order. Melayu and Iban have had traditional tenure regulations since time immemorial, and they continue to function to a large extent. Within these traditional systems, the areas of

DSWR have long been divided, by ethnic group, into sultanates (which form the basis for current counties [*kecamatan*]) and into large and small villages (with differing legal status at this time), each with its own identified territory and jurisdiction. Within communities, individuals and families also have rights based on traditional rules.

However, the certainty with which any of these entities and individuals can defend or protect their ownership of local areas has decreased significantly. Most of the area now considered part of the Danau Sentarum National Park (which DSWR recently became) was at some time given to a timber concessionaire to manage, despite the presence of communities within the concession. In the 1980s, the area was officially declared a wildlife reserve. By Indonesian law, settlements cannot exist within a wildlife reserve. Although moving people out of the reserve was not seriously considered by most would-be official managers, DSWR's official protection status and subsequent active management further weakened local people's certainty about their rights.

Study Sites

The Iban village of Bemban (officially part of the larger Melayu village, Pulau Duri') is located just outside the western reserve boundary, but its resource use areas cross the boundary. Wadley (1999b) reports that this area has been continuously farmed since at least the mid-1850s. In 1990, the Bemban longhouse contained 71 people in 13 "doors" or households (1990 Indonesian Census). Half are Protestant, half are Catholic. These Iban practice an "integral" form of shifting cultivation, which includes dry land and swamp rice farming, hunting, fishing, growing various cash crops and secondary subsistence crops, and extensive use of nearby forests for gathering other products. Bemban lays claim to one exclusive use zone (*wilayah kerja kampung*), four zones shared with other villages (*wilayah kerjasama*), and two traditionally protected forest areas (*hutan adat*) (Figure 16-2).[3] The Bemban claim area is within the boundaries of two logging concessions and also part of the reserve. P.T. Hutan Hebat arrived in 1986, and its activities were still evident in 1992–1993 (Colfer and others 1997b). In June 1996, a new logging camp was established in the south by a second logging concession and was still active a year later.

Nanga Kedebu', a Melayu fishing village, has been within the reserve boundary since 1981, when Danau Sentarum was proposed as a nature reserve, but this designation was made clear to Nanga Kedebu' residents only in 1992. The people of Nanga Kedebu' number approximately 108 in the rainy season, about 199 in the dry season (Colfer's 1992 census), when many people from the Kapuas River come to share in the (sometimes) abundant fish harvests. In the 1986 dry season, Giesen recorded 120 people in the village. Fishing and fish processing are the most important economic activities, but villagers also collect rattan, firewood, and timber from their surrounding

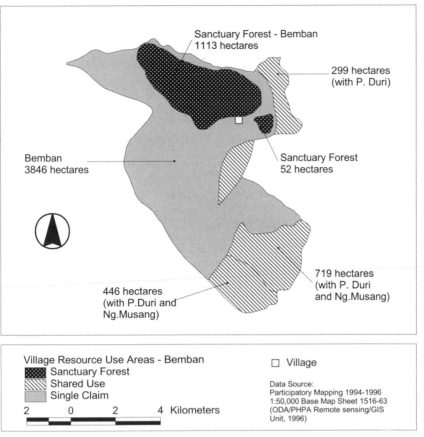

Figure 16-2. *Village Resource Use Areas in Bemban, 1994*

forests (Colfer and others 1993). Nanga Kedebu' comprises several use zones (Figure 16-3 shows their location and size): four shared use zones; two zones with conflicting claims; and claim to a sanctuary forest, known locally as Hutan Nung. All the neighbors are Melayu. Only a small part of the area claimed by Nanga Kedebu' falls within the boundaries of a logging concession; the rest falls within the reserve. The first logging company arrived in 1973 (Colfer and others 1997b), and P.T. Hutan Hebat made a brief appearance in 1980. No commercial logging activities remain today.

Danau Seluang, much smaller in total claim area than the other two communities, lies in the southeast corner of the reserve. Previously well outside the reserve boundary, this Melayu fishing village and part of its resource use area now lie within the proposed boundary (see also Figure 16-1). The current population varies from 216 in the rainy season to 312 in the dry season (Population Census, U.K. Department for International Development [DfID][4]

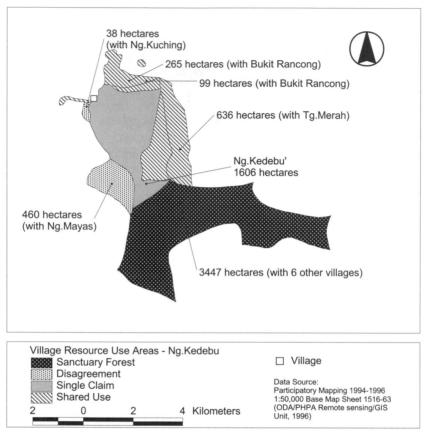

Figure 16-3. *Village Resource Use Areas in Nanga Kedebu', 1994*

Conservation Project, 1997). It is interesting to note that Giesen recorded a population of only 150 in 1986. Nearer to dry land than Nanga Kedebu', the area offers more opportunities for agricultural activities, for example, on the river levees. Participatory mapping shows that Danau Seluang claims one exclusive use and two shared use zones (one with Iban neighbors and one with Melayu neighbors) (see Figure 16-4). Inhabitants of Danau Seluang also use some forest areas on the nearby hill, Bukit Semujan. The area claimed by the village lies completely within the logging concession of P.T. Hutan Hebat. According to local reports, logging started in 1980 in the immediate area but has now stopped, and replanting is taking place (Colfer and others 1997b). Over the years, Danau Seluang has been more affected by logging activities than the other two villages.

Two of the villages (Bemban and Danau Seluang) are on the boundary of the DSWR, and one (Nanga Kedebu') is inside. All three have experienced

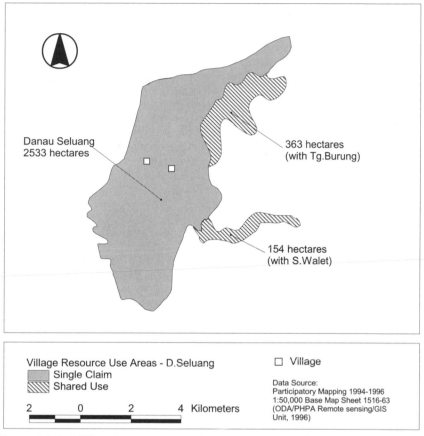

Figure 16-4. *Village Resource Use Areas in Danau Seluang, 1994*

commercial timber activities within their territories. In addition to our general interest in seeing whether forest cover had decreased substantially over the period 1973–1994, we were particularly interested in the results from Bemban, a village where shifting cultivation—a controversial lifestyle among many foresters and policymakers—forms the subsistence base.

Data

Although we recognize that many readers are unfamiliar with some of the language used in this section, its inclusion is important if remote sensing specialists are to find the study useful as well, and we hope they will. The remotely sensed data we selected for the change analysis included Landsat multispectral scanner (MSS) satellite imagery (April 13, 1973); Landsat thematic mapper (TM) satellite imagery (August 28, 1990); and black-and-white vertical aerial photography, scale 1:35,000 (January 2, 1994).

The Landsat imagery was available as digital data for both 1973 and 1990. The Landsat MSS imagery was purchased from the EROS Data Center in the United States. All four bands were available, but the radiometric quality of the image was poor. The Landsat TM imagery was purchased as seven bands of digital data from the Thailand Remote Sensing Center. The quality of the data was high, and little or no cloud cover obscured the image. The aerial photos, procured as 230-millimeter by 230-millimeter standard metric photo prints, were taken by the topographic mapping section of the Indonesian Army (DITTOP) in early January 1994. They were good-quality vertical photos with very little cloud cover. The photos exhibited 60% overlap and 40% sidelap, thus offering good stereoscopic (three-dimensional) viewing capabilities, revealing clear visual topographic features.

Nonremotely sensed data was also used in the study. We were fortunate to have at our disposal the GIS developed by the DfID Conservation Project for DSWR. The GIS contains a large amount of spatial and attribute data and was accessible through the GIS software packages, PC ARC/INFO and PC ARCVIEW (from Environmental Systems Research Institute [ESRI]).

Of particular interest to this study were the village work area boundaries, because they demarcated the areas where we conducted the change analysis. The boundaries of the community claim areas (*wilayah kerja*) were the end result of the community-based mapping activities carried out as part of the DfID Conservation Project. These boundaries were mapped in a participatory manner with local communities over a four-year period. Sketch maps, base maps, radar imagery, and global positioning systems (GPS) were used in the process. The hand-drawn boundaries were then transferred to the DSWR GIS. (For a detailed description of the community-based mapping in and around DSWR, see Dennis and others 1997a.)

Methods

To begin to address the question of forest cover and its change over time, we use data collected and interpreted by Dennis as part of the "land cover change mapping activity" of the DfID Conservation Project. The villages selected for this study—Bemban, Nanga Kedebu', and Danau Seluang—are located throughout the reserve (see Figure 16-1). Between mid-1994 and 1997, Dennis and colleagues mapped the different kinds of indigenous resource use claim areas using participatory techniques for DSWR and environs (Dennis and others 1997a). In this section, we explain some important decisions that we have made relating to working scale, data preprocessing and preparation, image classification and interpretation, and postclassification procedures.

Choosing a Working Scale

A common dilemma faced by those carrying out land cover change analyses over a long time period (in this case, 21 years) is the question of working scale. Ideally, change detection procedures should involve data acquired by the same (or similar) sensor and should be recorded using the same spatial resolution, viewing geometry, spectral bands, time of day, and period/season of acquisition (Richards 1992).

During the period in which we intended to assess change (1973 to 1994), the choice of sensors for the early period (1970s) was limited to Landsat MSS. By the 1990s, a wider choice of sensors was theoretically available. However, we were ultimately limited by the availability of cloud-free imagery and photography. In this case, we were forced to compare Landsat MSS at a spatial resolution of 80 meters by 80 meters (recommended maximum mapping scale 1:100,000) with Landsat TM at a spatial resolution of 30 meters by 30 meters (recommended maximum mapping scale 1:50,000) and 1:35,000 black-and-white aerial photography.

In addition to the differences in sensors and spatial resolution, the seasonal variability within the dataset added to the difficulties of maintaining a consistent level of interpretation. DSWR is a seasonally flooded forest area with a complex and fluctuating hydrological cycle (Klepper 1994; Giesen 1996). From week to week, the water levels fluctuate, submerging or exposing vegetation in the process. The Landsat MSS imagery, dated April 1973, showed high flood levels, thus much of the dwarf swamp forest and burnt swamp areas were submerged. These data contrasted sharply with the Landsat TM imagery, dated August 1990, which was at the height of the dry season, clearly showing areas of dwarf swamp forests and recent swamp forest burning. The aerial photography, dated January 1994, represents a typical picture of medium-to-high water levels.

Data Preprocessing and Preparation

The majority of processing was carried out by Dennis using the facilities of the Perlindungan Hutan dan Pelestarian Alam (PHPA, the directorate general for forest protection and nature conservation)/DfID Remote Sensing/ GIS Unit. The results of the interpretation were digitized by Puntodewo at CIFOR. The digital preprocessing was carried out using the PC ERDAS software (ERDAS 1991). Classifications and subsequent spatial analysis was achieved with PC ARC/INFO and ARCVIEW.

Ensuring that the three different data sets overlay exactly (co-register) is critical for valid change analysis. Without it, spurious results will occur. The Landsat TM was geometrically corrected using ground control points obtained from topographic base maps and GPS surveys. A first-order polynomial transformation and nearest–neighbor resampling procedure[5] was used

371

during the geocorrection. The resulting pixel root–mean–square error (RMSE) was 1.5, which implies good accuracy with the topographic base map. The Landsat MSS was then georegistered to the Landsat TM image using a second–order polynomial transformation with an RMSE of 1.5. The registration between images then was checked visually. The correspondence was deemed sufficient for the change analysis.

The digitized community claim boundaries from the GIS were superimposed on the images to clearly identify the area of interest. For each of the three villages (Bemban, Danau Seluang, and Nanga Kedebu'), the image areas relating to these boundaries were "cut" from the Landsat TM and MSS images. This subset of the original image was saved as a new file that was exported from the image processing system to the GIS for classification.

Image Classification and Interpretation

On–screen digitizing was chosen as the method of classification for the Landsat MSS and TM. We decided not to use conventional automated image classification in this study for two reasons. First, all three data sets were different spatial resolutions or scales. Second, the aerial photo interpretation would be a vector dataset produced from digitizing a hand–drawn interpretation. Therefore, extra care had to be taken that the classification produced for each year could be easily compared with any of the other years.

One classification technique that would produce three data sets of a format that could be easily compared within the GIS is on–screen digitizing. For each of the villages, we imported 1973 and 1990 subsetted images into the GIS. Within the GIS, we directly digitized (drew) land cover boundaries on the screen. (The classification scheme used is described in Annex 16-1.) For all the village areas classified, we had a large amount of field data and knowledge, including site photographs, to aid in interpretation. On–screen digitizing can be a lengthy process when classifying large areas, but our study areas were relatively small. The entire process of digitizing took about two days per village, including editing of the results. The final product of the on–screen digitizing was a vector file (a shape file in ARCVIEW terminology) that contained an associated database file with areas and codes for each of the land cover polygons.

Despite the great amount of user interaction in the classification process, problems were still encountered. The Landsat MSS image was one of the early products of the Landsat mission, and the quality of the image, even after image restoration, was not good. In addition, the limited number of spectral bands (four: green, red, and two near–infrared [IR]) meant that the results of the classification were variable. The inclusion of a mid-infrared band is generally favored for discrimination of natural vegetation types. Inaccurate classification was particularly apparent on the Bemban image in areas of heterogeneous vegetation and areas of steep topography. In all three areas, MSS image speckling caused by poor image quality led to some misclassification. In comparison, the classification of the Landsat TM was easier because of the excellent image quality and higher spatial and spectral resolution of the image than the MSS.

A mirror stereoscope was used when interpreting the air photos, and boundaries were drawn onto a plastic overlay. These boundaries were subsequently transferred to the base map at a scale of 1:50,000. The interpretation of the aerial photos was digitized by hand from the base map and converted to a digital file for inclusion in the GIS.

During the classification process, a great deal of time was spent producing three sets of classifications that could be compared within the GIS, despite the resolution differences.

Postclassification Procedures

Assessing the accuracy of the classification is a very important part of the process. Through field checking, interviews with villagers, and viewing many site photos we were able to make a qualitative estimation of the classification accuracy with considerable confidence. The classifications have a high level of accuracy, and the small errors we found are discussed in Results. The final classification results were all in vector digital format in the GIS. The GIS also contained contours, streams/rivers, villages, and community claim boundaries. Area estimates and final maps were generated simply and rapidly using the ARCVIEW GIS program.

Results

The forest cover classifications allow us to compare forest conditions under three (at least hypothetical) conditions: relatively pure, indigenous management (1972); indigenous management, supplemented by industrial timber/government management (1990); and the current situation, which involves a mix of indigenous management, co-management, and national management for both timber and conservation (1994). We say "at least hypothetical" because our recent findings conclude that almost all local stakeholders considered forest management to be primarily in the hands of local people (Chapter 12)—despite the fairly long-standing presence of industrial logging activities and Ministry of Forestry officials in the area. We also had the impression that the most obvious management still remained in the hands of local people during our 1996 visit.

Bemban

In the Bemban area, where swidden agriculture is practiced, we were particularly interested in the trend of change in cleared areas relative to swamp and hill forest areas (see Figure 16-5 and Tables 16-1, 16-2, and 16-3). These

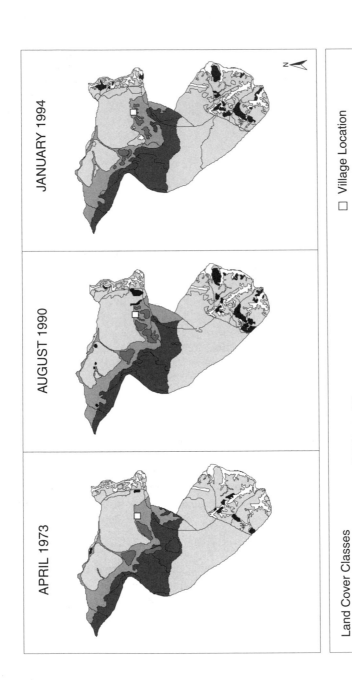

Figure 16-5. *Land Cover Change Map of Bemban*

Table 16-1. Forest Cover Change in Bemban, 1973, 1990, and 1994

| | Area (ha) | | Change 1973–1990 (%) | | Area (ha) | Change 1990–1994 (%) | |
Class	1973	1990	Cumulative	Annual	1994	Cumulative	Annual
Swamp forest	4,095	3,956	–3	–0.2	3,709	–6	–1
Hill forest	1,153	1,079	–6	–0.4	1,065	–1	–0.3
Lowland forest	126	120	–5	–0.3	69	–42	–10
Mosaic	831	873	+5	+0.3	1,001	+15	+4
Burnt swamp forest	89	260	+193	+11	442	+70	+18
Lake/river	425	431	NA	NA	433	NA	NA
Total area	6,719	6,719	NA	NA	6,719	NA	NA

Note: NA = not applicable.

Table 16-2. Direction of Change Matrix, 1973–1990 (hectares)

Class	Hill forest (1973)	Lowland forest (1973)	Swamp forest (1973)	Mosaic (1973)	Burnt swamp forest (1973)	Water (1973)	Total (1990)
Hill forest (1990)	1,022 (89%)		14 (0.3%)	43 (5%)	1 (1%)		1,079
Lowland forest (1990)		101 (80%)		18 (2%)			120
Swamp forest (1990)			3,841 (94%)	83 (9%)	22 (25%)	9 (25%)	3,956
Mosaic (1990)	131 (11%)	25 (20%)	37 (0.9%)	680 (82%)	66 (74%)		873
Burnt swamp forest (1990)			185 (4.5%)	7 (0.8%)		2	260
Water (1990)			17 (0.4%)			414	431
Total (1973)	1,153	126	4,095	831	89	425	6,719

Note: Percentage values show how much of that class remained as that type or became another type between 1973 and 1990. For example, 1,022 hectares (89%) of lowland forest remained that class from 1973–1990, but 11% became mosaic.

Table 16-3. Direction of Change Matrix, 1990–1994 (hectares)

Class	Hill forest (1990)	Lowland forest (1990)	Swamp forest (1990)	Mosaic (1990)	Burnt swamp forest (1990)	Water (1990)	Total (1994)
Hill forest (1994)	1,022 (95%)			43 (5%)			1,065
Lowland forest (1994)		62 (52%)		7 (0.8%)			69
Swamp forest (1994)	12		3,665 (93%)	18 (0.1%)	14 (5%)		3,709
Mosaic (1994)	45 (4%)	58 (48%)	85 (2%)	805 (92%)	8 (3%)		1,001
Burnt swamp forest (1994)			204 (5%)		238 (92%)		442
Water (1994)			2			431	433
Total (1990)	1,079	120	3,956	873	260	431	6,719

Note: Percentage values show how much of that class remained as that type or became another type between 1990 and 1994.

forest types are used in swidden cultivation in the area (a frequently believed "cause" of forest degradation).

Interpretation of the forest cover change estimates shows some interesting results. Swamp forest covers the largest area of any one forest type in Bemban territory. The area of swamp forest decreased 386 hectares between 1973 and 1994 (Table 16-1). This change was greater in the second time period (1990 and 1994), with a yearly decrease of 1% compared with 0.2% for 1973–1990.

Certain small changes may seem illogical and this is due to interpretation error. For example, in Table 16-2, we see that 14 hectares of swamp forest became hill forest by 1990. In other cases, such as swamp forest becoming water, we could have misinterpreted flooded burnt forest, which looks like open water on the image. Most of the spurious changes are small, and the major changes are what we concentrate on. It is also highly unlikely that we could achieve a 100% accurate classification through interpretation of remotely sensed imagery.

The predominant cause of change in swamp forest is burning (Tables 16-2 and 16-3). Burnt swamp forest increased significantly, almost fivefold, in the areas around the lakes in the southern area (Figure 16-5), from approximately 89 hectares in 1973 to 442 hectares in 1994. The annual increase in burning of swamp forest was greater between 1990 and 1994 (18% as opposed to 11%; Table 16-1). The most probable reasons for this increase include population growth from a nearby village, use of the area for fishing, and the presence of a new logging camp. Local people report that timber company employees use burning to enhance hunting success, because deer come to feed on new regrowth (also seen in other places in Borneo). Another interpretation involves the discovery in the early 1990s of Asian arowana, a valuable species of fish; some local fisherfolk reported that burning made it easier to catch them. The areas most affected by fire also coincide with shared claim areas, where tenure is likely to be less secure (Figure 16-5). It can be hypothesized that less secure tenure in these areas may encourage burning or discourage fire prevention.

The annual rate of swamp forest fires varies with length and intensity of the dry season. The Landsat TM image (1990) depicts a very dry season. Therefore, the annual percentage change should be interpreted with care, because the annual rate varies considerably due to yearly fluctuations in burning. According to village interviews (June 1999), the most recent El Niño year (1997–1998) was particularly bad for burning—though not compared with some other areas in Kalimantan (Dennis 1999; Hoffmann and others 1999). Forest fires constitute a complex issue in and around DSWR; the reasons for burning are both physical and anthropogenic and very difficult to pinpoint (Luttrell 1994). However, it is clear that the majority of fires in the swamp forest is not the result of clearing for cultivation. Having said that, some less frequently or severely inundated areas of swamp forest are cleared for cultivation, but cultivation on dry land continues to dominate.

The other main forest types that have decreased in area over the past 21 years are hill forest and lowland forest (Table 16-1). These forests are found in the areas most favorable for swidden agriculture. Based on our historical interviews (Colfer and others 1997c), which suggested that the village territory has not changed dramatically, we conclude that the area of hill forest within the Bemban claim area was approximately 1,153 hectares in 1973.[6] By early 1994, the area was reduced by 8% to 1,065 hectares. The general areas chosen for cultivation have varied little, with large areas of hill forest remaining intact. During the course of the village sketch-mapping sessions, we found that much of the remaining hill forest areas contained village burial sites and as such may not be disturbed. The remaining decline in hill forest can be accounted for by a greater percentage of areas that are a mosaic of swidden agriculture and secondary forest. Between 1973 and 1990, 89% of the hill forest remained intact, and 131 hectares were cleared for swidden agriculture (Table 16-2). The annual rate of conversion of hill forest to agriculture decreased slightly (−0.6% to −0.4%) from 1990 to 1994. From 1973 to 1994, small areas of mosaic reverted to forest (43 hectares) as they were left fallow.

Because the livelihood of the people of Bemban is based on swidden agriculture, we were interested in changes in cleared land vis-à-vis total land area. One should remember that this class encompasses several stages, from cleared and cultivated land to fairly well-advanced secondary forest regrowth. With remotely sensed data, we were unable to detect secondary forest, so it is likely that this stage is classified with one of the forest classes. During the course of field checking, we found that many areas that had been recently cleared according to the 1973 imagery were by 1997 in an advanced stage of regeneration. In 1994, for instance, cleared land (mosaic) accounted for 1,001 hectares, 16% of the total land area (excluding water) (Table 16-1). In 1973, cleared land accounted for 831 hectares, just over 13% of the land area. The greater amount of cleared land in 1994 may be due to improved access afforded to local people by the construction of a road within their territory, combined with more time for agriculture (because of less time spent in travel). Greater ease in marketing of farm produce also may have been a factor.

Apart from hill forest areas, other areas suitable for cultivation within the Bemban territory are provided by the dry lowland forest areas. The results of our analysis show a significant decrease in the area of lowland forest between 1973 and 1994, with the greatest decrease between 1990 and 1994 (Table 16-1). This decrease was due to a rise in swidden agriculture (Tables 16-2 and 16-3). Although the areas involved are relatively small, we see a decrease of 45% over the 21-year period. Certainly, for the people of Bemban, these lowland forest areas are more accessible than the hill forest. The building of the road in the early 1990s significantly improved access, with not atypical adverse effects on forest cover. On the basis of our change estimates, we would suggest that the amount of clearing was not dramatic in the past and probably was at appropriate levels for the Bemban swidden cultivation system. But the more recent rate of change—particularly for lowland forest—is indeed of concern for the future.

Nanga Kedebu'

In Nanga Kedebu', the village work area consists of swamp forest, open lakes, and rivers. Forest areas are used for the collection of timber and nontimber forest products such as rattan. A small area (three hectares) for cultivation exists behind the village (Figure 16-6). Of the three villages, this fishing community is the most striking example of stability of forest cover. Swamp forest has been reduced by only 0.7% over the 21-year period (see Table 16-4). In 1973, a few families lived in this location, and cleared forest was not detected on the imagery. Even by 1994, the inhabitants still had cleared considerably less than 1% of their land area. One obvious area directly behind the village fell victim to an inadvertent, uncontrolled fire not long before 1992. During the time of the research, the area was used for agricultural endeavors that frequently failed due to unpredictable flooding. Most of the forest area within the territory of Nanga Kedebu' is flooded a large part of the year, thus making that area unsuitable for any form of agriculture.

Again, looking at swamp forest, we see no change in the rate of reduction in Nanga Kedebu' territory between 1973 and 1990; two hectares per year is the average. The slight decrease in swamp forest was centered around two areas, one behind the village and another in the swamp forest to the south of the lake. Field checking showed that these areas represented attempts to clear for cultivation in the 1970s, but the areas have since been abandoned. Results suggest that indigenous management in this area has had remarkably little negative effect on the forest. The dominance of fishing over agriculture as the primary economic activity would seem to be an important factor, particularly in light of the increasing population reported by some observers (for example, Aglionby and Whiteman 1996; Colfer and others 1999d). It is also interesting that, despite the large areas of swamp forest with similar conditions to the other two villages, little swamp forest has been burned. Possible

Table 16-4. *Forest Cover Change in Nanga Kedebu', 1973, 1990, and 1994*

Class	Area (ha)		
	1973	1990	1994
Swamp forest	5,508	5,476	5,470
Cleared forest	0	30	31
Burnt swamp forest	2	5	5
Lake/river	1,547	1,546	1,548
Total	7,057	7,057	7,054

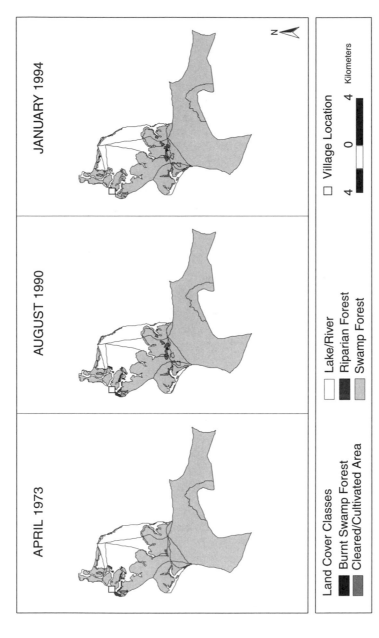

Figure 16-6. *Land Cover Change Map of Nanga Kedebu'*

contributing factors include the lack of a currently active timber company in the area, the absence or rarity of arowana or other valuable fish and wildlife resources, observation of the negative impacts of fires on the availability of nontimber forest products in neighboring communities, proximity to the headquarters of the Conservation Project (in the 1990s), and the presence of influential community members with an awareness of ecological issues.

Danau Seluang

The work area claimed by the village of Danau Seluang is made up of swamp forest, riverine forest, open water, and rivers. Fishing is the predominant activity, and the collection of nontimber forest products such as rattan and honey is of secondary importance. Changes in forest cover have been significant (Figure 16-7). This area has been somewhat infamous in recent years for its frequent forest fires; there was a staggering increase in burning from 54 hectares in 1973 to 278 hectares in 1994 (Table 16-5). Interviews with Danau Seluang inhabitants in 1996 about the causes of these fires focused on motivations of jealousy or revenge and on inadvertent wildfire caused by insufficiently extinguished cooking fires. The locations of the fires, in dwarf swamp forest along the fringes of the lake, would tend to support the "cooking fires" theory. Another possible source, reported to Dennis in 1997, was the burning of water hyacinths, which cause navigational problems for small boats. We remain dissatisfied with our explanations, despite considerable efforts on several occasions to understand the picture more fully. This community does seem marked by more conflict, both within and without (for example, mutual suspicion about fire setting and disagreements over community leadership, fishing regulations with neighboring fishing communities, and land use with neighboring agricultural communities; see also Chapters 8 and 12), than were the other two communities examined.

The area of swamp forest decreased by 187 hectares between 1973 and 1994. Losses in riparian forest have been significant. In fact, only 17% of the 1973 riparian forest area remains (Table 16-5). The amount of cleared/cultivated/regrowth area and the burnt swamp forest also have increased substantially (by factors of 4.6 and >5, respectively). Interestingly, most of the riparian area that has been lost (and the cleared/cultivated area that has been added) are, again, in a shared use area. The community sharing this area with Danau Seluang consists of shifting cultivators who use the river levees for agriculture, whereas Danau Seluang residents use the river for fishing and rarely cultivate. However, there is some history of conflict between these communities (for a similar story, see Box 12-1 in Chapter 12). The dramatically increased area of cleared/cultivated area still accounts for only 8% of Danau Seluang's total land area (rivers and lakes excluded). Swamp and riparian forest lost one hectare per year from 1973 to 1990 and from 1990 to 1994. Hill forest remained the same.

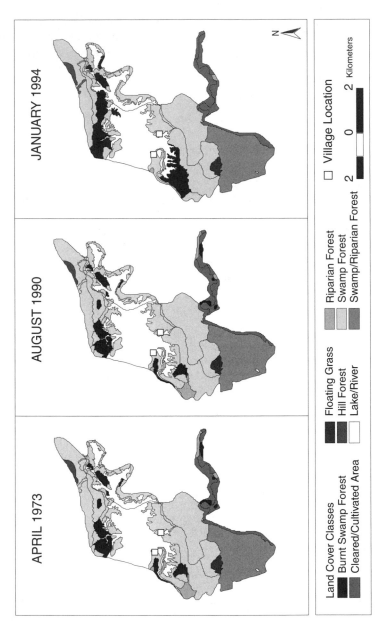

Figure 16-7. Land Cover Change Map of Danau Seluang

Table 16-5. *Forest Cover Change in Danau Seluang, 1973, 1990, and 1994*

Class	Area (ha)		Change 1973–1990 (%)		Area (ha)	Change 1990–1994 (%)	
	1973	1990	Cumulative	Annual	1994	Cumulative	Annual
Swamp forest	1,516	1,419	-6	-0.35	1,329	-6.0	-2
Riparian forest	206	67	-67	-4.0	36	-46.0	-15.0
Swamp/riparian forest	494	470	-5	-0.3	466	-0.8	-2.6
Hill forest	21	21	0	0	21	0	0
Cleared/cultivated area/regrowth	43	178	+314	+18.0	198	+11.0	+3.6
Burnt swamp forest	54	173	+220	+13.0	278	+61.0	+20.0
Floating grass	27	27	0	0	27	0	0
Lake/river	693	690	NA	NA	698	NA	NA
Total	3,054	3,045	NA	NA	3,053	NA	NA

Note: NA = not applicable.

Conclusions

In this study, we used remotely sensed data, supplemented by our ethnographic understanding and socioeconomic data, to assess the spatial and temporal changes in forest cover from 1973 to 1994 in the DSWR area. The forest cover change analyses indicate discernible trends and patterns. Communities on the DSWR periphery have experienced greater loss in forest cover than the community in the center of the reserve. Where the change in rate of forest loss has been noticeable (for example, the swamp forests in Danau Seluang and Bemban), the rate of loss has been more rapid recently than under purely indigenous management; fires have played a prominent role in this decline. Possible factors relating to fire include some support for Brookfield and others' (1995) conclusion that logging had played a significant role in making areas of Borneo more fire-prone, a concurrent increase in population, including outsiders with little stake in the area, and resource extraction (like burning to gain access to arowana or deer).

Despite the differences in sensor type and resolution available in this example, we believe that the level of accuracy of the classifications was sufficient to arrive at a realistic picture of past forest cover change. Although this analysis does not allow us to predict the future—which would of course be necessary were we to have a true measure of sustainability—it provides the next best thing: a picture of current conditions and the trajectory of change. Many factors are involved in assessing sustainability, and forest cover change is only one. The forest change analysis provided important insights into possible mechanisms behind the changes. We found, for instance, that shared use areas consistently showed decreased forest area, particularly burning. This finding is especially relevant to our C&I pertaining to security of intergenerational access to resources (see Section 3). We also found that forest loss in one of the fishing communities was as high as in a shifting cultivation community and that neither community had a dramatic rate of forest loss compared with many areas of the world.

Additional GIS analysis of these data will focus on assessing how specific areas are changing and the extent to which the landscape is becoming fragmented. It also would be valuable to add dates to the time series. In the interim, however, we will use this analysis as a template, seeking similar analyses in other areas of the world.

The results and analysis show that the integration of remote sensing and GIS can add significantly to the possibilities for assessing sustainability. These maps can provide some measure of sustainability against which to examine differences in social science results. The hints about sustainability provided in this chapter help us to place particular forest management units and villages on a continuum from comparatively sustainably managed to comparatively unsustainably managed areas. On a global basis, although the DSWR is considered forest rich by our definition, its prognosis for sustainability seems to vary considerably by area.[7] We share Brondizio and others' (1994) positive assessment of this approach for understanding and planning improved land-use systems, and we believe that, in combination with data on other contributing factors, it can prove valuable in assessing sustainability in general.

Annex 16-1. Forest Classification Scheme Used in Remote Sensing Interpretation

The following description relies heavily on Giesen's for vegetation descriptions (consult Giesen 1996 for further details). Some of the classes mentioned here are not described by Giesen; therefore, no species descriptions exist at this time. In addition, we describe the factors important in interpreting these classes from the remotely sensed data.

Swamp Forest

This class combines Giesen's *dwarf swamp forest, stunted swamp forest,* and *tall swamp forest.*

Dwarf swamp forest is predominantly characterized by *Barringtonia acutangula, Croton cf. ensifolius, Gardenia tentaculata, Ixora mentanggis, Pternandra teysmanniana,* and *Memecylon edule.*

Stunted swamp forest is characterized by two vegetation types: Kenarin-Menungau-Kamsia is probably the most widespread and is characterized by *Diospyros coriacea (kenarin), Vatica menungau (menungau),* and *Mesua hexapetalum (kamsia);* Kawi-Kamsia is the second type and is characterized by *Shorea balangeran (kawi).*

Tall swamp forest is characterized by two main vegetation types: Kelansau-Emang-Melaban is characterized by *Dryobalanops abnormis (kelansau), Hopea mengerawan (emang),* and *Tristaniopsis obovata (melaban);* the second type, Ramin-Mentangur-Kunyit, is characterized by *Gonystylus bancanus (ramin)* and *Calophyllum sclerophyllum (mentangur kunyit).*

Riparian forest appears to have many of the same species as the Kenarin-Menungau-Kamsia stunted swamp forest but is characterized by the presence of *Gluta renghas.*

The Landsat multispectral scanner (MSS) images were taken during high water level, and most areas containing dwarf swamp forest are flooded; however, this flooded forest is included in the swamp forest class. The 1990 image represents an extreme dry season, and the submerged vegetation class is not present. Subdivision of the swamp forest class is possible with the Landsat thematic mapper (TM) imagery. However, at the edge of homogeneous areas of tall, stunted, or dwarf swamp forest, a mixed class is often found. For the purposes of the land cover change classification, the three swamp forest classes from Landsat TM were recoded into one class. Discrimination of the three swamp forest classes was relatively easy using stereoscopic interpretation of the air photos.

Hill Forest

This class is represented by Giesen's *hill forest* and *heath forest*. It is located on hills and slopes; contours were used to define the boundary between dry land and hill forest. This class equates with the closed canopy >70% class of the Food and Agriculture Organization (FAO 1990).

Hill forest exhibits a preponderance of dipterocarp species, such as *Anisoptera grossivenia, Dipterocarpus gracilis, Shorea leprosula,* and *S. seminis.*

Interpretation of the Landsat MSS showed that spectrally, swamp forest and hill forest (shadow side) were difficult to separate. Contour lines were used to identify which areas of forest should be classified as hill, dry land, or swamp. After the hill forest areas had been selected based on elevation, two distinct spectral classes were evident: one on the back slope, which had a much lower spectral reflectance, and one on the sunlit slope, which displayed a bright reflectance similar to cleared land. The bright reflectance made it both visually and digitally difficult to discriminate between cleared and hill forest on the sunlit slope. Similar problems were encountered with the Landsat TM. By comparison, this class was relatively easy to discriminate on the air photos.

Lowland Forest

This class was found in areas between the low-lying flooded areas and the hill forest. Originally, these areas were included with the swamp forest class, but field checks showed that these areas were never flooded and were composed of slightly different species.

Mosaic of Cleared Forest/Cultivation/Regrowth

This class is characterized by clearings that are either active areas or shifting cultivation areas (originally cleared for cultivation but not currently in use for agriculture) and clearings caused by logging or fire.

Secondary scrub is generally dominated by ferns (*Pteridium aquilinum*), the shrublets *Melastoma malabathricum* and *Rhodomyrtus tormentosa,* the trees *Macaranga* spp., and a locally valuable hill variety, *Fagraea fragrans.*

Spectrally, this a mixed class. However, it tends to have a much higher spectral reflectance than the fragmented forest class. During visual and digital classification, this class was confused with the sunlit hill slopes. Field checks and the air photos helped to correct some of these errors.

Burnt Swamp Forest

This class is spectrally similar to newly cleared areas but located in swamp areas.

On the MSS, this class is not as highly reflective as newly cleared areas, maybe due to some flooding. On the TM images, this class has quite a broad range of spectral values, ranging from highly reflective (no woody vegetation) to less reflective with a mixture of burnt, woody, and nonwoody vegetation. Information from field checks improved interpretation of this class.

In the air photos, *burnt swamp forest* is easily identified by charred trees and the lack of woody vegetation. In flooded areas, burnt areas are identified by areas of water with patchy vegetation and single standing trees as distinct from clear open water.

Lakes/Rivers

Areas of open water.

Endnotes

1. Center for International Forestry Research scientists have been involved in efforts to develop modeling tools for anticipating future forest use scenarios. For example, Jerry Vanclay coordinated the development of the computer model, FLORES (Forest Land Oriented Resource Envisioning System; see Vanclay 1998), and Philippe Guizol and Herry Purnomo have been working on CORMAS (Common-Pool Resources and Multi-Agent Systems).

2. See Chapters 5, 8, and 12, also on the DSWR.

3. These represent the Indonesian classification system and do not exactly reflect Iban forest management. *Hutan adat,* for instance, covers a wide range of Iban forest categories, both sacred and nonsacred (personal communication from Reed Wadley, October 1, 1999).

4. Formerly the U.K. Overseas Development Administration (ODA).

5. Resampling is the process by which the pixel values in the original distorted input image are assigned in the geometrically corrected output image. In the nearest-neighbor resampling method, the output data file value is equal to the input pixel whose coordinates are closest to the retransformed coordinates of the output pixel.

6. In 1989, Bemban became officially part of Pulau Duri', a nearby Melayu community that shares borders; since then, our impression is that Pulau Duri' community members have increasingly been using Bemban's land (for small-scale food crop agriculture and, to a lesser extent, for a rubber planting project), especially in the southeastern portion of Bemban territory. Bemban willingness to cooperate derives partially from gratitude to Pulau Duri' residents for their help in 1986, when the Bemban longhouse burned to the ground, leaving its inhabitants destitute.

7. As we finalize this book, there is widespread interest and support for oil palm expansion in Indonesia and ongoing planting of oil palm just north of the reserve. This situation has been further exacerbated by a phenomenal increase, throughout forested areas of Indonesia, of illegal logging (December 2000).

CONCLUSION

Concluding Remarks and Next Steps

In this book, we have both examined (as case studies) and compared four main topics as related to forest management: gender and diversity, a conservation ethic, security of intergenerational access to resources, and rights and responsibilities to manage the forest cooperatively and equitably. To round out the analysis, we concluded with two comparative analyses from slightly different perspectives. The topics were selected on the basis of the most up-to-date research on human–forest interactions. Twenty-five qualified scholars and researchers of various disciplines from ten countries were involved in conducting the research in six countries over a six-year period. We used globally accepted social criteria and indicators (C&I) to guide our study of the causal links between human well-being and sustainable forest management and used various accepted social science methods to ensure the validity of our results. We further compared our results from the quick assessment tools developed by the Center for International Forestry Research (CIFOR) with the longer-term experience of the researchers in the countries of study.

Yet, the results do not provide us with clear evidence of straightforward, direct, causal links between the four issues identified above and sustainable forest management. They do provide us with a whole range of useful information—for forest managers, policymakers, and researchers. But the most useful information, in our view, is that which is embedded in its systemic context. The most significant similarities and conclusions from our research applied to *locations* and *ethnic groups,* not to our topics of study. The chapters about Kalimantan (1, 4, 5, 8, 12, and 16) and about Central Africa (2, 3, 9, and 10) give us a holistic understanding of the issues that link people and forests in those contexts. Those chapters indeed teach us something about gender and diversity, conservation ethics, security of intergenerational access to resources, and rights and responsibilities to manage the forest cooperatively and equitably. We see, for instance, that women have active roles related to forests in Cameroon, Brazil, and Indonesia (and how those roles link to other parts of people's lives). We see the different relations of different ethnic groups with forest resources (in Cameroon) and among themselves (in Indonesia). In the Brazil case, power differences among stakeholders rather than ethnicity emerge as central. The lifeway of forest-based ethnic groups, like the Benuaq in East Kalimantan or the Iban in West Kalimantan, carry elements that serve obvious conservation functions. The rationale for the involvement of Brazil's colonists in sustainable forest management is different, relating to their history of oppression and their commitment to their own plots of land. We were able to trace the complex links among access to resources, distribution of benefits, equity, and forest maintenance in Trinidad, Indonesia, Gabon, Cameroon, and, to a lesser extent, Brazil. And, by contextual analysis, we showed how and why the sharing of rights and responsibilities among stakeholders is important and needed—to harmonize different management systems and empower marginalized people. But taken out of context, these issues (gender and diversity, conservation ethics, access, and rights and responsibilities) mutate into a form less usable for management and policy-related decisions.

Something is inherently unsatisfying about the comparative chapters (6, 11, and 14) for at least two reasons. One is that the results are not particularly rich in information that can guide us in decisionmaking. Does it really help us a great deal to be able to say that local communities have statistically significantly higher rights in forest-rich than in forest-poor areas? Forest managers or policymakers are likely to be dealing with a particular forest of a given quality, characterized by a particular level of governmental or local management, when a decision needs to be made. It may not help much to know that forest-rich areas tend to be characterized by slightly greater rights for local folk.

The other reason that these comparative analyses are unsatisfying is that the issues have been taken out of context; the links between them and the other components of the systems of which they form a part are broken for the purposes of the analysis. Whereas the more descriptive cases make the importance of conservation ethics, access to resources, and rights and responsibilities clear by showing the links among the various parts of the cultural (and to a lesser extent, biophysical) systems, we cannot draw such links from the comparative analyses. The picture we receive from such analyses is misleading, suggesting both a simplicity and a stability that in fact do not exist.

One message we take home from this experience is that we must develop research methods and approaches—paradigms, if you will—that allow us to deal with complexity and change. We probably will have to focus on processes as much as on C&I-like structures. *Understanding* may prove more important than the chimera of "straight line" causation; *surprise* must become a source of insight about the systems we examine rather than a debilitating shock that throws our research off course.

We are not making an argument against all reductionist science, by any means; such science plays an important role in knowledge generation for humankind. The argument is rather that when looking at dynamic systems such as cultures and forests, we have to develop new approaches and methods—or acknowledge existing approaches and methods that address such systems more effectively.

Before discussing the specific approaches we have in mind, some important policy implications from these studies must be set out clearly.

- Formal forest managers now tend to pay little attention to the human systems functioning within and adjacent to forests. When noticed at all, such systems tend to be lumped into broad categories such as *communities* or *forest-dependent people*. Much more effective mechanisms must be developed to gain access to the knowledge, skills, and human resources that these currently marginalized human beings can bring to forest management. Effective management will require external acknowledgement of the diversity of human beings who make use of forests—men and women, old and young, rich and poor, different ethnic groups. Such recognition, of both positive or potentially positive contributions by communities and the diversity of knowledge and skills within communities, is important pragmatically and ethically if we truly seek to improve human well-being and forest management.

- Although the concept of a conservation ethic is difficult to define and measure, these studies have provided ample evidence that cognitive and emotional aspects of local cultural systems can contribute to maintaining or improving local environments. Certain groups have strong emotional and ethical links to their environments that are not shared by other groups, and that can serve as a basis for improved collaboration with outsiders also interested in conserving or more sustainably using a resource or an environment. Effective mechanisms for working collaboratively with such groups differ from those needed when working with groups without such concerns. Efforts need to be made to identify, support, and strengthen elements of local cultures with a current or potential conservation function.

- Our research has strengthened and supported the research of others showing the importance of security of access to resources. Lack of clarity and conflict about ownership and use rights as well as inequitable access and distribution of benefits from forests—sometimes related to the recent arrival of wealthy and powerful outsiders—are wreaking havoc with tropical forests. Simple privatization of land is not the answer. Care must be taken to develop solutions that fit with local contexts and needs. Formal forest managers and policymakers should redouble their efforts to resolve these resource conflicts and enhance equity in access to forest benefits.

- Similarly, the importance of clearly defining and agreeing on respective rights and responsibilities to manage forests in a cooperative and equitable manner has again been shown. The desirability of the trend toward more democratic practices, currently being demanded in many countries, is consistent with the findings of this research. People who have a voice— not the only voice, but a voice nonetheless—in the use of local resources are more likely to embrace stewardship than are those who have been as utterly disempowered as have many tropical forest dwellers. Local forest people must have more of a voice in formal forest management—for a more humane world, but also for a more environmentally secure one.

Let us move on to possible next steps. If we agree that the somewhat reductionist scientific approach that we initially took in this research is inadequate, then what should we do? We should keep using some of the methods that allow us to look at whole systems, such as participant observation, Galileo or CATPAC, and network analysis. We should expand our modeling work—making sure that whatever we do is grounded in field realities and not simply what some have called "high-tech masturbation"*—to include system dynamics modeling and efforts such as the Forest Land-Oriented Resource Envisioning System (FLORES).

FLORES is a complex model that is being developed as a forest equivalent to the commercial software, SimCity (FLORES 1999). It includes subcomponents that address wildlife, agriculture, silviculture, and household decisionmaking and has policy levers that can be manipulated to examine the land use implications of various policy changes. Early work has focused on Sumatra and Zimbabwe.

The Consultative Group on International Agricultural Research (CGIAR or the CG system) has started a new initiative on Integrated Natural Resource Management. Its purpose is to expand the horizons of the CG system beyond the component research that has characterized much of its work to date. Such interests include addressing some of the problems identified in this book, and trying to identify new approaches with promise.

The approach that has grown directly out of our own increasing conviction that human systems are in fact complex adaptive systems, intimately connected with each other and with biological systems in a self-organizing process of coadaptation, has taken form in a new research program at CIFOR: Local People, Devolution, and Adaptive Collaborative Management of Forests. This program evolved from our realization that the identification of C&I, though useful, was insufficient. If we wanted genuine improvement in forest conditions or human well-being, the conditions spec-

---

*I'm referring to the love affair that many scientists (especially western ones) have with high-tech solutions—often, ones that do not result in useful tools for conservation or management. To some, these scientists seem involved most fundamentally in stimulation of their intellect alone—particularly the writers of papers full of mathematical formulas that are unintelligible to most people.

ified as desirable in the C&I would have to be catalyzed in the real world. We did not believe it could be done by using a reductionist approach alone. Also, on the basis of research reported in this book, we realized that the situation was pretty dire. Something needed to be done *quickly.*

A core group of about 15 people, working with cooperating teams in 12 countries, has developed an approach that incorporates attention to the dynamic aspect of systems, examining the appropriateness of collaborative management involving communities in these countries. The program involves four main elements:

1. Research on issues of relevance to adaptive and collaborative management (for example, devolution, institutions, social learning, conflict resolution, power differentials, and gender)
2. A series of sites where participatory action research is being undertaken to catalyze local community action to improve human well-being and environmental quality
3. A parallel set of activities to document and evaluate the process as it progresses
4. Modeling and synthesizing activities at the site, national, and international scales

Some progress has been made on the first element (Wollenberg 1997, 1999; Edmunds and Wollenberg 1999; Ramirez 1999; Campbell and others 2000c; Kajembe and Monela 2000; Kayambazinthu 2000; Wollenberg and others in press) and the fourth element (FLORES 1999; Purnomo and others 2000; work in progress by Godwin Kowero, Ravi Prabhu, and Herry Purnomo), but work on the second and third elements has only recently begun. This approach is based on a more realistic assessment of the nature of the phenomena we are dealing with—adaptive interacting systems—than was the research reported in this book.

This book is about people managing forests. It represents an effort, based on a combination of short- and long-term fieldwork, to approach forest management from a grounded theoretical perspective. Unlike conventional books on forest management, we have focused on kinds of forest management that have evolved in a naturalistic fashion—adaptive and interrelated with other systems, human and "natural." The recurring and largely unacknowledged geographical overlap between systems of formal timber management and community management should be clear. Continuing to pretend that local or indigenous management does not exist is like hiding one's head in the sand. Both formal and informal management must be acknowledged, and the necessary negotiations to clarify respective rights and responsibilities must begin.

We also have made a case for expanding the realm of legitimate, forest-related, scientific inquiry to include more systems-oriented research. Our qualitative studies have long led us in this direction, and our experience with the research reported here further strengthened our conviction: reductionist approaches are insufficient for understanding and coping with these inter-related and dynamic systems.

Nancy Langston has written a fascinating but discouraging historical account of what happened ecologically to the forests of Oregon's Blue Mountains—an area "blessed" by various kinds of management over the years. In her conclusion (Langston 1995, 298), she says

> Each time, when faced with complexity, foresters eventually denied it and retreated into their certainty. In their minds they partly acknowl-edged this complexity and partly denied it, just as the same forester could simultaneously have both a knowledge of ecology's interconnec-tions and a faith in his own ability to redesign a more efficient forest. Foresters were not, as many environmentalists claim, greedy or stupid. Like everyone else, they needed to hold onto a story that made their lives make sense. Their work was based on the faith that they were making the forests better. If they let themselves see the evidence in front of them—that the forests were dying, not getting better—they would have to give up the vision that made sense of their lives. Instead, they blinded themselves to the consequences of their actions, ignored the doubts that crept in, and condescended to people who challenged their version of the forest.
>
> Everyone does this, not just foresters....

This book is a clarion call for us not to repeat the mistakes that Langston describes for the Blue Mountains, that Fairhead and Leach (1996) describe for the West African nation of Guinea, and that we have seen ourselves in Indonesia's forests. The mistakes are based on two important kinds of errors. First, we lack awareness about the impact of our own images of how the landscape should look and function on our scientific work. We believe we have far greater objectivity than in fact we do. Second, we fail to recognize the dynamic and systemic nature of the phenomena we seek to understand, trying to simplify by extracting bits of reality from their contexts, dissecting them, and assuming wrongly that when we put them back into the system, they will function better.

We have looked at the nature of the links between human well-being and sustainable forest management in this book, with the intention of providing useful insights for formal forest managers and policymakers in their decision-making. Understanding these links—links within complex systems that do not lend themselves readily to cause-and-effect analysis—is, we believe, a pre-requisite if we really want to create and sustain conditions whereby Earth and its people can thrive. We conclude with a simple reminder of the urgency of the human and environmental problems that currently plague the world's for-ests and forest peoples—and the fervent hope that the kinds of approaches we suggest will contribute to their solution fast enough to make a difference.

References

Afshar, H., and C. Dennis. 1992. *Women and Adjustment Policies in the Third World*. London, U.K.: Macmillan.

Agarwal, B. 1990. Social security and the family: Coping with seasonality and calamity in rural India. *Journal of Peasant Studies* 17: 341–412.

Aglionby, J.C., and A. Whiteman. 1996. *The Utilisation of Economic Data for Conservation Management Planning: A Case Study from Danau Sentarum Wildlife Reserve*. Unpublished report. Pontianak, Indonesia: Conservation Project, Indonesia–U.K. Tropical Forest Management Programme.

Aguiar, R. 1994. *Direito do Meio Ambiente e Participação Popular*. Brasília, Brazil: Ministério do Meio Ambiente e da Amazónia Legal/IBAMA.

Ahl, V., and T.F.H. Allen. 1996. *Hierarchy Theory: A Vision, Vocabulary, and Epistemology*. New York: Columbia University Press.

Alexandre, P., and J. Binet. 1963. *Le Groupe dit Pahouin (Fang-Boulou-Beti)*. Paris, France: Vendôme, Presses Universitaires de France.

Allen, T.F.H., and T.W. Hoekstra. 1992. *Toward a Unified Ecology*. New York: Columbia University Press.

———. 1994. Toward a definition of sustainability. In *Sustainable Ecological Systems: Implementing an Ecological Approach to Land Management*, W.W. Covington and L.F. DeBano, technical coordinators. General Technical Report RM-247. Fort Collins, CO: USDA Forest Service, Rocky Mountain Forest and Range Experiment Station, 98–107.

Allen, T.F.H., and Starr, T.B. 1982. *Hierarchy: Perspectives for Ecological Complexity*. University of Chicago Press, Chicago, Illinois.

Allen, T.F.H., J.A. Tainter, and T.W. Hoekstra. 1999. Supply-side sustainability. *Systems Research and Behavioral Science* 16: 403–27.

———. Forthcoming. *Supply-Side Sustainability*. New York: Columbia University Press.

Anderson, A., P. May, and M.J. Balick. 1991. *The Subsidy from Nature: Palm Forests, Peasantry, and Development on an Amazon Frontier*. New York: Columbia University Press.

Ang-Lopez, M.J., R.H. Asong, R. Alcala, and L. Lopez Roderiquez. 1996. Insights on women in some environmental conservation projects in Antigua. In *Participatory and*

Community-Based Approaches in Upland Development: A Decade of Experience and a Look at the Future, edited by J.M. Marco and E.A. Nune Jr. Laguna, The Philippines: Philippines Uplands Resource Centre, 57–61.

Appell, G. 1986. Kayan land tenure and the distribution of devolvable usufruct in Borneo. *Borneo Research Bulletin* 18: 119–30.

Ascher, W. 1993. *Political Economy and Problematic Forestry Policies in Indonesia: Obstacles to Incorporating Sound Economics and Science.* Durham, NC: Duke University, Center for Tropical Conservation.

Atkinson, J.M., and S. Errington (eds.). 1990. *Power and Difference: Gender in Island Southeast Asia.* Stanford, CA: Stanford University Press.

AtKisson, A. 1999. *Believing Cassandra: An Optimist Looks at a Pessimist's World.* White River Junction, VT: Chelsea Green.

Bahuchet, S. 1993. Afrique centrale. In *Situation des Populations Indigènes des Forêts Denses et Humides,* edited by S. Bahuchet and P. de Maret. Brussels, Belgium: European Commission, 389–441.

Bahuchet, S., and de Maret, P. 1993. *Situation des Populations Indigènes des Forêts Denses et Humides.* Brussels, Belgium: European Commission.

Bailey, C., and Zerner, C. 1992. Local management of fisheries resources in Indonesia: Opportunities and constraints. In *Contributions to Fishery Development Policy in Indonesia,* edited by R. Pollnac, C. Bailey, and Alie Poernomo. Jakarta, Indonesia: Central Research Institute for Fisheries (CRIFI) Publication, 38–56.

Bailey, C., R. Pollnac, and S. Malvestuto. 1990. The Kapuas River Fishery: Problems and Opportunities for Local Resource Management. Paper presented at the International Association for the Study of Common Property. September 27–30, Duke University, Durham, NC.

Bailey, G.N., C. Ioakim, G.P.C. King, C. Turner, M-F. Sanchez-Goni, D. Sturdy, N.P. Winder, and S.E. van der Leeuw. 1998. Northwestern epirus in the Paleolithic. In *The Archaeomedes Project: Understanding the Natural and Anthropogenic Causes of Land Degradation and Desertification in the Mediterranean Basin,* edited by S.E. van der Leeuw. Luxembourg: Office for Official Publications of the European Communities, 71–112.

Bailiff, I., J. Bell, P.V. Castro, and others. 1998. Environmental dynamics in the Vera Basin. In *The Archaeomedes Project: Understanding the Natural and Anthropogenic Causes of Land Degradation and Desertification in the Mediterranean Basin,* edited by S.E. van der Leeuw. Luxembourg: Office for Official Publications of the European Communities, 115–72.

Balee, W. 1992. People of the fallow: A historical ecology of foraging in lowland South America. In *Conservation of Neotropical Forests: Working from Traditional Resource Use,* edited by K.H. Redford and C. Padoch. New York: Columbia University Press, 35–57.

Banana, A.Y., and W. Gombya-Ssembajjwe. 1996. Successful Forest Management: The Importance of Security of Tenure and Rule Enforcement in Ugandan Forests. International Forestry Resources and Institutions (IFRI) Research Program Working Paper. Bloomington, IN: Indiana University.

Banuri, T., and F.A. Marglin (eds.). 1993. *Who Will Save the Forests? Knowledge, Power, and Environmental Destruction.* London, U.K.: Zed Books.

Barbier, E.B., J.C. Burgess, and C. Folke. 1994. *Paradise Lost? The Ecological Economics of Biodiversity.* London, U.K.: Earthscan Publications Ltd.

Barnett, G.A., and J.K. Woelfel (eds.). 1998. *Readings in the Galileo System: Theory, Methods, and Applications.* Dubuque, IA: Kendall-Hunt.

Barr, C.M. In press. Will HPH reform lead to sustainable forest management? Questioning the assumptions of the "sustainable logging" paradigm in Indonesia. In *Which Way*

Forward? Policy, People, and Forests in Indonesia, edited by C.J.P. Colfer and I. Resosudarmo. Washington, DC: Resources for the Future.

Bass, S., B. Dalal-Clayton, and J. Pretty. 1995. *Participation in Strategies for Sustainable Development.* Environmental Planning Issues Series no. 7 (May). London, U.K.: International Institute for Environment and Development.

Bazerman, M. 1998. *Judgment in Managerial Decision Making* (Fourth edition). New York: John Wiley & Sons.

Beard, J.S. 1946. *The Natural Vegetation of Trinidad.* Oxford, U.K.: Colonial Forest Service.

Becker, B. 1997. *Sustainability Assessment: A Review of Values, Concepts, and Methodological Approaches.* Issues in Agriculture Series no. 10. Washington, DC: Consultative Group on International Agricultural Research.

Behan, R.W. 1988. A plea for constituency-based management. *American Forests* 94: 46–48.

Behrens, C.A. 1990. Applications of satellite image processing to the analysis of Amazonian cultural ecology. In National Aeronautics and Space Administration, *Conference Proceedings, Applications of Space Age Technology in Anthropology.* Hancock County, MS: John C. Stennis Space Center.

Belbin, L. 1993. *PATN: Pattern Analysis Package.* Canberra, Australia: Commonwealth Scientific and Industrial Research Organization, Division of Wildlife and Ecology.

Berkes, F., and C. Folke. 1994. Investing in cultural capital for a sustainable use of natural capital. In *Investing in Natural Capital: The Ecological Economics Approach to Sustainability,* edited by A.M. Jansson, M. Hammer, C. Folke, and R. Costanza. Washington, DC: Island Press.

Bernstein, J.H., E. Roy, and A. Bantong. Not dated. The Use of Plot Surveys for the Study of Ethnobotanical Knowledge: A Brunei Dusun Example. Unpublished manuscript.

Berry, S. 1988. Property rights and rural resource management: The case of tree crops in West Africa. *Cahier des Sciences Humaines* 24: 3–16.

Besley, T. 1995. Property rights and investment incentives: Theory and evidence from Ghana. *Journal of Political Economy* 103: 903–37.

Bikie, H., O. Ndoye, and W.D. Sunderlin. 1999. Crise Économique, Systèmes de Production, et Changement du Couvert Forestier dans la Zone Forestière Humide du Cameroun. Unpublished paper. Bogor, Indonesia: Center for International Forestry Research.

Blaney, S., S. Mbouity, and J-M. Nkombe. 1997. *Complexe d'Aires Protégées de Gamba (Gabon): Caractéristiques Socio-économiques des Populations des Départements de Ndougou, de la Bassa-Banio, et de la Mougoutsi (Mouenda).* Report to World Wide Fund for Nature (WWF—Programme pour le Gabon). Libreville, Gabon: WWF Gabon.

Bopda, A. 1993. Le secteur vivrier sud-camerounais face à la crise de l'economie cacaoyère. *Travaux de l'Institut de Géographie de Reims* 83/84: 109–22.

Borrini-Feyerabend, G., with D. Buchan. 1997. *Beyond Fences: Seeking Social Sustainability in Conservation.* Gland, Switzerland: International Union for the Conservation of Nature.

Brenan, J.P.M. 1978. Some aspects of the phytogeography of tropical Africa. *Annals of the Missouri Botanical Garden* 65: 437–38.

Brocklesby, M.A., and B. Ambrose-Oji. 1997. *Neither the Forest nor the Farm. Livelihoods in the Forest Zone: The Role of Shifting Agriculture on Mount Cameroon.* Rural Development Forestry Network Paper 21d. London, U.K.: Overseas Development Institute.

Brocklesby, M.A., P. Etuge, G. Ntube, J. Alabi, M. Anje, V. Bau Bau, and J. Molua. 1997. *CIFOR Cameroonian Test of Social Methods for Assessing Criteria and Indicators for Sustainable Forest Management.* CIFOR Report, Mt. Cameroon Project, Limbe. Bogor, Indonesia: Center for International Forestry Research.

Brondizio, E.S., E.F. Moran, P. Mausel, and Y. Wu. 1994. Land use change in the Amazon estuary: Patterns of Caboclo settlement and landscape management. *Human Ecology* 322: 249–78.

Brookfield, H., L. Potter, and Y. Byron. 1995. *In Place of the Forest: Environmental and Socio-economic Transformation in Borneo and the Eastern Malay Peninsula.* Tokyo, Japan: United Nations University.

Browder, J.O. 1988. Public policy and deforestation in the Brazilian Amazon. In *Public Forest Policies, and the Misuse of Forest Resources,* edited by R. Repetto and M. Gillis. New York: Cambridge University Press, 247–97.

Brown, K., and F. Ekoko. 2000. Forest Encounters: Local Actors and Forest Cover Change in the Humid Forest Zone of Cameroon. Draft paper. Norwich, U.K.: University of East Anglia.

Brown, K., and S. Lapuyade. 1999. "I Have Become the Man of the Family": Gendered Visions of Social, Economic, and Forest Change in Southern Cameroon. Draft paper. Norwich, U.K.: University of East Anglia.

Bruce, J. 1989. *Community Forestry: Rapid Appraisal of Tree and Land Tenure. Forests, Trees, and People.* Community Forestry Note 5. Rome, Italy: Food and Agriculture Organization.

Bruun, O., and A. Kalland (eds.). 1996. *Asian Perceptions of Nature: A Critical Approach.* Surrey, U.K.: Curzon Press.

Buck, L.E., C.G. Geisler, J.W. Schelhas, and E. Wollenberg (eds.). In press. *Biological Diversity: Balancing Interests through Adaptive Collaborative Management.* Boca Raton, FL: CRC Press.

Budiono, E. 1993. Aspek-aspek Ekologis pada Masyarakat Dayak Bentian dalam Hubungannya dengan Pelestarian Hutan [Ecological Aspects of the Bentian Dayaks, in their Relationship with Forest Sustainability]. Unpublished paper. East Kalimantan, Indonesia: Skripsi Sarjana Fakultas Kehutanan Unmul, Samarinda.

Burford de Oliveira, N. 1997. Ancient Gods in the Path of Change: Indications of the Sustainability of a Community Managed Forest in Tenganan, Bali. Internal document. Bogor, Indonesia: Center for International Forestry Research.

———. 1999. *Community Participation in Developing and Applying Criteria and Indicators of Sustainable and Equitable Forest Management.* Project report. Bogor, Indonesia: Center for International Forestry Research.

Burford de Oliveira, N., P. Shiembo, A.M. Tiani, and M. Vabi. 1998. Developing and Testing Criteria and Indicators for the Sustainability of Community Managed Forests in the SOLIDAM Zone, Central Province, Cameroon. Draft paper. Bogor, Indonesia: Center for International Forestry Research.

Burford de Oliveira, N., with B. Ritchie, C. McDougall, H. Hartanto, and T. Setyawati. 2000. *Developing Criteria and Indicators of Community Managed Forests.* Bogor, Indonesia: Center for International Forestry Research.

Burgess, P., with Elias, P.M. Laksono, R.J. Watling, and W.M. Wan Razali. 1995. *Final Report: Indonesia Test, March 5–April 2, 1995.* Project on Testing Criteria and Indicators for Sustainable Management of Forests. Bogor, Indonesia: Center for International Forestry Research.

Caballe, G. 1999. *Test de PCIV en République Centrafricaine.* Libreville, Gabon: African Timber Organization.

Campbell, B.M., W. Kozanayi, C. Lovell, A. Mandondo, N. Nemarundwe, and B. Sithole. 2000a. *Managing Common Pool Resources in Catchments: Are There Any Ways Forward?* Bogor, Indonesia: CIFOR, Centre for Ecology and Hydrology, and Institute for Environmental Studies (University of Zimbabwe).

Campbell, B.M., and F. Matose, with D. Kayambazinthu and G. Kajembe. 2000b. *Institutions and Natural Resources in the Miombo Region.* Bogor, Indonesia: Center for International Forestry Research.

Campbell, B.M., B. Sithole, and N. Nemarundwe. 2000c. Zimbabwe case study: Empowering communities in Zimbabwe—new configurations of power. In *Empowering Communities to Manage Natural Resources: Case Studies from Southern Africa,* Volume 2, edited by S. Shackleton and B. Campbell. Council for Scientific and Industrial Research (CSIR) Report ENV-P-C-2000–025. Pretoria, South Africa: CSIR and World Wide Fund for Nature (Southern Africa), 182–95.

Canan, P., and M. Hennessy. 1981. *Moloka'I Data Book: Community Values and Energy Development.* Honolulu, HI: University of Hawaii Urban and Regional Planning Program.

Canuday, J.F. 1996. Fight vs. mining, logging: Lumads ready for total war. *Philippine Daily Inquirer,* October 8, p. 17.

Carter, J. 1996. *Recent Approaches to Participatory Forest Resource Assessment.* Rural Development Forestry Study Guide 2. London, U.K.: Rural Development Forestry Network, Overseas Development Institute.

Carter, J., M. Stockdale, F. Sanchez Roman, and A. Lawrence. 1995. Local people's participation in forest resource assessment: An analysis of recent experience, with case studies from Indonesia and Mexico. *Commonwealth Forestry Review* 74: 333–42.

Cary, J.W. 1995. *An Analysis of Perceptions of High Country Landscapes: A Test of Comparative Quantitative Methods and an Artificial Neural Network Technique.* Report of results from a pilot study for Manaaki Whenua Landcare Research New Zealand, Ltd. May. Lincoln, New Zealand.

Cary, J.W., and W.E. Holmes. 1982. Relationships among farmers' goals and farm adjustment strategies: Some empirics of a multidimensional approach. *The Australian Journal of Agricultural Economics* 26: 114–30.

Cary, J.W., A.W. Kenelly, and R.C. Bell. 1989. *A Comparison of Methods for Tracking Consumer Perceptions of Meat.* Parkville, Victoria, Australia: University of Melbourne, School of Agriculture and Forestry and Department of Psychology.

CFCA (Caribbean Forest Conservation Association). 1994. Report of a Seminar on "National Parks and Protected Areas." Chaguaramas, Republic of Trinidad and Tobago: Institute of Marine Affairs.

Chalmers, W.S. 1992. *Trinidad and Tobago, National Forestry Action Programme. Report of the Country Mission Team.* Rome, Italy: Food and Agriculture Organization, Forestry Division.

Chambers, R. 1988. Sustainable rural livelihoods: A key strategy for people, environment, and development. In *The Greening of Aid, Sustainable Livelihoods in Practice,* edited by C. Conroy and M. Litvinoff. London, U.K.: Earthscan Publications in association with International Institute for Environment and Development, 1–17.

———. 1994. Foreword. In *Beyond Farmer First: Rural People's Knowledge, Agricultural Research and Extension Practice,* edited by I. Scoones and J. Thompson. London, U.K.: Intermediate Technology Publications Ltd., xiii–xvi.

Chambers, R., A. Pacey, and L.A. Thrupp (eds.). 1989. *Farmer First: Farmer Innovation and Agricultural Research.* London, U.K.: Intermediate Technology Publications Ltd.

Chandrasekharan, D. 1997. Proceedings of the Electronic Conference, Addressing Natural Resource Conflicts through Community Forestry, January to May 1996. Rome, Italy: Food and Agriculture Organization, Forest, Trees and People Programme.

Chatterji, A.P., with A. Schwarz and Arabari Communities. 1996. *Community Forest Management in Arabari: Understanding Sociocultural and Subsistence Issues.* New Delhi, India: Society for Promotion of Wastelands Development.

REFERENCES

Chayanov, A.V. 1986. *The Theory of Peasant Economy,* edited by D. Thorner, B. Kerblay, and R.E.F. Smith. Madison, WI: University of Wisconsin Press.

Child, B., S. Ward, and T. Tavengwa. 1997. *Natural Resource Management by the People: Zimbabwe's CAMPFIRE Programme.* International Union for the Conservation of Nature-Regional Office for Southern Africa (IUCN-ROSA) Environmental Issues Series no. 2. Harare, Zimbabwe: IUCN-ROSA.

Christiansen, H. Not dated. Iban Plant Classification and Plant Names. Unpublished manuscript.

CIFOR (Center for International Forestry Research). 1998. CIFOR research recognises critical role of women in sustainable forest management. *CIFOR News* 20: 1–2.

———. 1999. *CIFOR Generic C&I Template.* CIFOR C&I Toolbox Series no. 2. Bogor, Indonesia: CIFOR.

Clay, J.W. 1988. *Indigenous Peoples and Tropical Forests: Models of Land Use and Management from Latin America.* Cambridge, MA: Cultural Survival, Inc.

Coakes, S. 1996. The Social Aspect: Participation of Stakeholders in Forest Management. Paper presented at the International Conference on Certification and Labelling of Products from Sustainably Managed Forests. May 26–31, Brisbane, Australia.

Coleman, J.S. 1988. Social capital in the creation of human capital. *American Journal of Sociology* 94: S95–120.

Colfer, C.J.P. 1981. Women, men, and time in the forests of Kalimantan. *Borneo Research Bulletin* 13: 75–85.

———. 1982. Women of the forest: An Indonesian example. In *Women in Natural Resources: An International Perspective,* edited by F. Stock and D. Ehrenreich. Moscow, ID: University of Idaho Press, 153–82.

———. 1983a. Change and indigenous agroforestry in East Kalimantan. *Borneo Research Bulletin* 15: 3–20, 70–86.

———. 1983b. On communication among "unequals." *International Journal of Intercultural Communication* 7: 263–83.

———. 1985a. On circular migration: From the distaff side. In *Labour Circulation and the Labour Process,* edited by G. Standing. Geneva, Switzerland: Croom Helm Ltd., 182–218.

———. 1985b. Female status and action in two Dayak communities. In *Women in Asia and the Pacific: Toward an East-West Dialogue,* edited by M. Goodman. Honolulu, HI: University of Hawaii Press, 183–211.

———. 1991. Indigenous rice production and the subtleties of culture change. *Agriculture and Human Values* 8: 67–84.

———. 1995. Who Counts Most in Sustainable Forest Management? CIFOR Working paper no. 7. Bogor, Indonesia: Center for International Forestry Research.

———. 2000a. Are Women Important in Sustainable Forest Management? Paper presented at the Twenty-first International Union of Forestry Research Organizations Congress,. Kuala Lumpur, Malaysia, August 7–12 (proceedings in press).

———. 2000b. Cultural Diversity in Forest Management. Keynote Address, Twenty-first International Union of Forestry Research Organizations Congress, Kuala Lumpur, Malaysia, August 7–12 (proceedings in press).

Colfer, C.J.P., with R.G. Dudley. 1993. *Shifting Cultivators of Indonesia: Managers or Marauders of the Forest?* Community Forestry Case Study no. 6. Rome, Italy: Food and Agriculture Organization.

Colfer, C.J.P., and R.L. Wadley. 1996. Assessing "Participation" in Forest Management: Workable Methods and Unworkable Assumptions. Working paper no. 12. Bogor, Indonesia: Center for International Forestry Research.

Colfer, C.J.P., B. Newton, and Herman. 1989. Ethnicity: An important consideration in Indonesian agriculture. *Agriculture and Human Values* 6: 52–67.

Colfer, C.J.P., R.L. Wadley, B. Suriansyah, and E. Widjanarti. 1993. Use of forest products in three communities: A preliminary view. In *Conservation Sub-Project Quarterly Report and Attachments*, edited by R.G. Dudley and C.J.P. Colfer. Study no. 7. Bogor, Indonesia: Asian Wetland Bureau (now Wetlands International).

Colfer, C.J.P., with R. Prabhu and E. Wollenberg. 1995. *Principles, Criteria and Indicators: Applying Ockham's Razor to the People-Forestry Link*. CIFOR Working paper no. 8. Bogor, Indonesia: Center for International Forestry Research.

Colfer, C.J.P., J. Woelfel, R.L. Wadley, and E. Harwell. 1996a. *Assessing People's Perceptions of Forests in Danau Sentarum Wildlife Reserve*. CIFOR Working paper no. 13. Bogor, Indonesia: Center for International Forestry Research.

Colfer, C.J.P., R.L. Wadley, and E. Widjanarti. 1996b. Using indigenous organizations from West Kalimantan. In *Indigenous Organizations and Development*, edited by P. Blunt and D.M. Warren. London, U.K.: Intermediate Technology Publications, Inc., 228–38.

Colfer, C.J.P., with N. Peluso and S.C. Chin. 1997a. *Beyond Slash and Burn: Building on Indigenous Management of Borneo's Tropical Rain Forests*. New York: New York Botanical Garden Press.

Colfer, C.J.P., R.L. Wadley, E. Harwell, and R. Prabhu. 1997b. *Intergenerational Access to Resources: Developing Criteria and Indicators*. CIFOR Working paper no. 18. Bogor, Indonesia: Center for International Forestry Research.

Colfer, C.J.P., R.L. Wadley, J. Woelfel, and E. Harwell. 1997c. From heartwood to bark: Indonesian gender issues in sustainable forest management. *Women in Natural Resources* 18(4): 7–14.

Colfer, C.J.P., A. Salim, B. Tchikangwa, A.M. Tiani, M.A. Sardjono, and R. Prabhu. 1998a. Social Criteria and Indicators: Assessing Human Well-Being in and around Industrial Timber Enterprises. CIFOR Working paper. Bogor, Indonesia: Center for International Forestry Research.

Colfer, C.J.P., A. Salim, A.M. Tiani, B. Tchikangwa, M.A. Sardjono, and R. Prabhu. 1998b. Whose Forest Is This, Anyway? C&I on Access to Resources. Paper presented at the International Union of Forestry Research Organizations Conference. August 24–28, Melbourne, Australia. Proceedings in press.

Colfer, C.J.P., and others. 1999a. *The BAG (Basic Assessment Guide for Human Well-Being)*. CIFOR C&I Toolbox Series no. 5. Bogor, Indonesia: Center for International Forestry Research.

Colfer, C.J.P., and others. 1999b. *The Grab Bag: Supplementary Methods for Assessing Human Well-Being*. CIFOR C&I Toolbox Series no. 6. Bogor, Indonesia: Center for International Forestry Research.

Colfer, C.J.P., with R. Prabhu, M. Günter, C. McDougall, N. Miyasaka Porro and R. Porro. 1999c. *Who Counts Most? Assessing Human Well-Being in Sustainable Forest Management*. CIFOR C&I Toolbox Series no. 8. Bogor, Indonesia: Center for International Forestry Research.

Colfer, C.J.P., R.L. Wadley, and P. Venkateswarlu. 1999d. Understanding local people's use of time: A pre-condition for good co-management. *Environmental Conservation* 26: 41–52.

Colfer, C.J.P., R.L. Wadley, A. Salim, and R.G. Dudley. 2000a. Understanding patterns of resource use and consumption: A prelude to co-management. *Borneo Research Bulletin* 38 (Special Issue on Danau Sentarum Wildlife Reserve, edited by W. Giesen).

Colfer, C., R.A. Dennis, and G. Applegate. 2000b. *The Underlying Causes and Impacts of Fires in Southeast Asia: Site 8, Long Segar, East Kalimantan Province, Indonesia*. Bogor, Indonesia: Center for International Forestry Research.

Collinson, M. (ed.) 2000. *A History of Farming System Research*. Wallingford, U.K.: CAB International-Food and Agriculture Organization.

Conklin, H.C. 1957. *A Report on an Integral System of Shifting Cultivation in the Philippines*. Rome, Italy: Food and Agriculture Organization.

Copus, A.K., and J.R. Crabtree. 1996. Indicators of socio-economic sustainability: An application to remote rural Scotland. *Journal of Rural Studies* 12(1): 41–54.

Crawford, C.S., L.M. Ellis, D. Shaw, and N.E. Umbreit, 1999. Restoration and monitoring in the middle Rio Grande bosque: Current status of flood pulse related efforts. In *Rio Grande Ecosystems: Linking Land, Water, and People*, edited by D. Finch, J. Whitney, J. Kelly, and S. Loftin. Proceedings RMRS-P-7. Fort Collins, CO: USDA Forest Service, Rocky Mountain Research Station, 158–63.

Croll, E., and D. Parkin (eds.). 1992. *Bush Base: Forest Farm: Culture, Environment and Development*. London, U.K.: Routledge.

Crumley, C.L. 2000. From garden to globe: linking time and space with meaning and memory. In *The Way the Wind Blows: Climate, History, and Human Action*, edited by R.J. McIntosh, J.A. Tainter, and S. Keech McIntosh. New York: Columbia University Press, 193–208.

Curran, P.J., G.M. Foody, R.M. Lucas, and M. Honzak. 1995. A methodology for remotely sensing the stages of regeneration in tropical forests. In *TERRA 2: Understanding the Terrestrial Environment—Remote Sensing Data Systems and Networks*, edited by P.M. Mather. New York: John Wiley & Sons, 189–202.

Davis-Floyd, R., and C. Sargent (eds.). 1997. *Childbirth and Authoritative Knowledge: Cross-Cultural Perspectives*. Berkeley, CA: University of California Press.

Davison, J., and V.H. Sutlive. 1991. The children of Nising: Images of headhunting and male sexuality in Iban ritual and oral literature. In *Female and Male in Borneo: Contributions and Challenges to Gender Studies*, edited by V.H. Sutlive and G.N. Appell. Monograph Series no. 1. Williamsburg, VA: Borneo Research Council, 153–230.

Deere, C. 1990. *Household and Class Relations: Peasants and Landlords in Northern Peru*. Berkeley, CA: University of California Press.

Dennis, R.A. 1999. *A Review of Fire Projects in Indonesia (1982–1998)*. Bogor, Indonesia: Center for International Forestry Research.

Dennis, R.A., A. Erman, and R. Tarigan. 1997a. *Community-Level Mapping in and Around the Danau Sentarum Wildlife Reserve, West Kalimantan. Indonesia*. Tropical Forest Management Programme Project Report. Jakarta, Indonesia: U.K. Department for International Development.

Dennis, R.A., C.J.P. Colfer, and A. Puntodewo. 1997b. *Assessing the Biophysical Aspects of Sustainable Forest Management Using Remote Sensing: an Indonesian Example*. Draft. CIFOR C&I Project/DfID Indonesian Tropical Forestry Management Project, Conservation Sub-Project. Bogor, Indonesia: Center for International Forestry Research.

Dennis, R.A., A. Puntodewo, and C.J.P. Colfer. 1999. Fishermen, farmers, forest change and fire. *GIS Asia Pacific* (February/March): 26–30.

Diamond, I., and G. Orenstein (eds.). 1990. *Reweaving the World: the Emergence of Ecofeminism*. San Francisco, CA: Sierra Club Books.

Diaw, M.C. 1997. Si, Nda Bot, *and* Ayong: *Shifting Cultivation, Land Use, and Property Rights in Southern Cameroon*. Rural Development Forestry Network paper 21e. London, U.K.: Overseas Development Institute.

Diaw, M.C., P.R. Oyono, F. Sangkwa, C. Bidja, S. Efoua, and J. Nguiebouri. 1998. *Social Science Methods for Assessing Criteria and Indicators of Sustainable Forest Management: A Report of the Tests Conducted in Cameroon Humid Forest Benchmark and in the Lobe and Ntem River Basins—Part I*. Bogor, Indonesia: Center for International Forestry Research.

Djuweng, S. 1992. Kampung Loboh Laman Banua: Konsep dan Praktek Pengusahaan Teritorial pada Suku Dayak Simpang [The Banua Community of Loboh Laman: Territorial Land Use Concepts and Practices among the Simpang Dayaks]. Paper presented at the Second Biennial Conference, Borneo Research Council. Kota Kinabalu, Sabah, Malaysia.

Dove, M.R. 1980. Sexual Equality and Economic Development among Tropical Forest Horticulturalists. Paper presented to the American Anthropological Association. December 3–7, Washington, DC.

————. 1981. Subsistence Strategies in Rain Forest Swidden Agriculture. Ph.D. dissertation. Stanford, CA: Stanford University.

————. 1985. Swidden Agriculture in Indonesia: The Subsistence Strategies of the Kalimantan Kantu'. Berlin, Germany: Mouton Publishers.

————. 1993. A revisionist view of tropical deforestation and development. Environmental Conservation 20: 17–24, 56.

————. 1996. Center, periphery and biodiversity: A paradox of governance and a developmental challenge. In Intellectual Property Rights and Indigenous Knowledge, edited by S. Brush and D. Stabinsky. Washington, DC: Island Press, 41–67.

Dove, M.R., and D.M. Kammen. 1997. The epistemology of sustainable resource use: Managing forest products, swiddens, and high-yielding variety crops. Human Organization 56: 91–101.

Drake, R.A. 1991. The cultural logic of textile weaving practices among the Ibanic people. In Female and Male in Borneo: Contributions and Challenges to Gender Studies, edited by V.H. Sutlive and G.N. Appell. Monograph Series no. 1. Williamsburg, VA: Borneo Research Council, 271–94.

Drijver, C.A. 1992. People's participation in environmental projects. In Bush Base, Forest Farm: Culture, Environment and Development, edited by E. Croll and D. Parkin. London, U.K.: Routledge, 131–45.

Drouineau, D., R. Nasi, F. Legault, and M. Cazet. 1999. L'Aménagement Forestier au Gabon—Historique, Bilan Perspectives. FORAFRI Document 19. Libreville, Gabon: Capitalisation et Synthèse des Recherches sur les Ecosystèmes forestiers humides d'Afrique (FORAFRI).

Dubisch, J. 1971. Dowry and the Domestic Power of Women in a Greek Island Village. Paper presented at the 70th Annual Meeting, American Anthropological Association. November, New York.

Dubois, O. 1998. Capacity to Manage Role Changes in Forestry: Introducing the "4Rs" Framework. IIED Forest Participation Series no. 11. London, U.K.: International Institute for Environment and Development, Forestry and Land Use Programme.

Dubois, O. 1999. Short Note on Possible Uses of the "4Rs" Framework. London, U.K.: International Institute for Environment and Development, Forestry and Land Use Programme.

Dudley, R.G. 1996a. The Fishery of the Danau Sentarum Wildlife Reserve, West Kalimantan, Indonesia: Fishery Analysis. Report to Wetlands International. Bogor, Indonesia: Wetlands International.

————. 1996b. The Fishery of the Danau Sentarum Wildlife Reserve, West Kalimantan, Indonesia: Management Considerations. Report to Wetlands International. Bogor, Indonesia: Wetlands International.

Dudley, R.G., and C.J.P. Colfer. 1993. Conservation Sub-Project Quarterly Report and Attachments. June. Bogor, Indonesia: Asian Wetland Bureau.

Due, J., and C. Gladwin. 1991. Impacts of structural adjustment programs on African women farmers and female-headed households. American Journal of Agricultural Economics 73: 1431–39.

REFERENCES

Dvorâk, K.A. (ed.) 1993. *Social Science Research for Agricultural Technology Development: Spatial and Temporal Dimensions.* Tucson, AZ: CAB International and International Institute of Tropical Agriculture with support from the Rockefeller Foundation.

Eba'a-Atyi, R. 1998. *Cameroon's Logging Industry: Structure, Economic Importance and Effects of Devaluation.* CIFOR Occasional Paper 14. Bogor, Indonesia: Center for International Forestry Research.

Economist Intelligence Unit. 1997. *Trinidad and Tobago.* Country report. 4th Quarter. London, U.K.: Economist Intelligence Unit.

Edmunds, D. 1998. *Notes for CIFOR Gender Analysis Training.* Bogor, Indonesia: Center for International Forestry Research.

Edmunds, D., and E. Wollenberg. 1999. A Strategic Approach to Multistakeholder Negotiation. Unpublished report. Bogor, Indonesia: Center for International Forestry Research.

Ekwoge, H., J-M. Mbani, and L. Tita. 1997. *Stakeholder Analysis for Wildlife Management—Mokoko Area.* First draft. Limbe, Cameroon: Mt. Cameroon Project.

Elmhirst, R.J. 1997. *Gender, Environment and Culture: A Political Ecology of Transmigration in Indonesia.* London, U.K.: University of London.

Elson, D. 1995. Male bias in macro-economics: the case of structural adjustment. In *Male Bias in the Development Process,* edited by D. Elson. Manchester, U.K.: Manchester University Press, 164–90.

Engel, P.G.H., A. Hoeberichts, and L. Umans. In press. Accommodating multiple actors in local forest management: A focus on facilitation, actors and practices. *International Journal of Agricultural Resources, Governance and Ecology.*

ERDAS. 1991. *ERDAS Version 7.5 Field Guide.* Atlanta, GA: ERDAS.

ECC (European Communities Commission). 1996. *Forests in Sustainable Development. Guidelines for Forest Sector Development Co-Operation.* Brussels, Belgium: ECC.

Everitt, B.S. 1993. *Cluster Analysis.* London, U.K.: Edward Arnold.

Eyebe, A., O. Ndoye, and M. Ruiz Pérez. 1999. *Importance des Produits Forestiers Non Ligneux pour les Communautés Rurales et Urbaines du Cameroun. Quelques Freins à l'Eclosion du Secteur.* Bogor, Indonesia: Center for International Forestry Research.

Fairhead, J., and M. Leach. 1994/95. *Whose Forest? Modern Conservation and Historical Land Use in Guinea's Ziama Reserve.* Rural Development Forestry Network paper 18c. London, U.K.: Overseas Development Institute.

———. 1996. Rethinking the forest-savanna mosaic: Colonial science and its relics in West Africa. In *The Lie of the Land,* edited by M. Leach and R. Mearns. Oxford, U.K.: James Currey, for The International African Institute.

FAO (Food and Agriculture Organization). 1989. *Household Food Security and Forestry.* Rome, Italy: FAO Forests, Trees, and People Programme.

———. 1990. *Forest Resources Assessment 1990. Survey of Tropical Forest Cover and Study of Change Processes.* FAO Forestry paper 130. Rome, Italy: FAO.

———. 1993. *Forest Resources Assessment 1990. Tropical Countries.* FAO Forestry paper 112. Rome, Italy: FAO.

———. 1995a. *Proceedings of Expert Meeting on Harmonisation of Criteria and Indicators for Sustainable Forest Management,* February 13–16. Rome, Italy: FAO.

———. 1995b. *Forestry Statistics Today for Tomorrow.* Rome, Italy: FAO.

———. 1995c. *State of the World's Forests.* Rome, Italy: FAO.

———. 1995d. *National Forestry Action Plan Update.* Rome, Italy: FAO.

———. 1997. *State of the World's Forests 1997.* Rome, Italy: FAO.

Farmer, A., and J. Tiefenthaler. 1995. Fairness concepts and the intrahousehold allocation of resources. *Journal of Development Economics* 47: 179–89.

Fearnside, P. 1993. Deforestation in Brazilian Amazonia: The effect of population and land tenure. *Ambio* 22: 537–45.

Federal Ministry for Environment, Youth, and Family. 1996. *Testing of Criteria and Indicators of Sustainable Forest Management within the International CIFOR Project.* Vienna, Austria: Federal Environment Agency.

Feyerabend, P.F. 1962. Explanation, reduction, and empiricism. In *Minnesota Studies in the Philosophy of Science* (Volume 3), edited by H. Feigl and G. Maxwell. Minneapolis, MN: University of Minnesota Press, 28–97.

Filer, C., with N. Sekhran. 1998. *Loggers, Donors and Resource Owners.* London, U.K and Boroko, Papua New Guinea: International Institute for Environment and Development and National Research Institute.

Finnish Ministry of Agriculture and Forestry. 1996. *National Implementation of Criteria and Indicators for Sustainable Forest Management in Finland.* Helsinki, Finland: Department of Forestry.

Firth, R. 1966. *Malay Fishermen: Their Peasant Economy.* New York: W.W. Norton.

FLORES. 1999. Workshop CD. Forest Land Oriented Resource Envisioning System (FLORES) Model Design Workshop. Central and West Sumatra, Indonesia, January 22–February 4. Bukittinggi, Sumatra, Indonesia: Center for International Forestry Research.

Foldy, J., and J.K. Woelfel. 1990. Cognitive processes as damped harmonic oscillators. *Quality and Quantity* 24: 1–16.

Ford Foundation. 1998. *Forestry for Sustainable Rural Development: a Review of Ford Foundation-Supported Community Forestry Programs in Asia.* New York: Ford Foundation.

Forestry Division. 1995a. *Annual Report of the Forestry Division 1995.* Port of Spain, Republic of Trinidad and Tobago: Forestry Division.

———. 1995b. *Forest Resources Management Plan for the South-East Conservancy.* Port of Spain, Republic of Trinidad and Tobago: Forestry Division.

Fortmann, L., and J.W. Bruce. 1988. *Whose Trees? Proprietary Dimensions of Forestry.* Boulder, CO: Westview Press.

Foster, G. 1965. Peasant society and the image of limited good. *American Anthropologist* 67: 293–315.

Foucault, M. 1980. *Power/Knowledge.* New York: Pantheon Books.

Fox, J. 1996. Response to Cabarle and Lynch's paper, *Conflict and Community Forestry: Legal Issues and Responses.* March 11. Food and Agriculture Organization E-mail Conference, January to March.

Frans, S. 1992. Pola Pengusahaan Tanah dan Permasalahan pada Masyarakat "Dayak Banuaka" di Kabupaten Kapuas Hulu, Kalimantan Barat [Patterns of Land Use and Problems of Banuaka' Communities in Kapuas Hulu District, West Kalimantan]. Paper presented at the Second Biennial Conference, Borneo Research Council. Kota Kinabalu, Sabah, Malaysia.

Freeman, J.D. 1970. *Report on the Iban.* London, U.K.: Athlone Press.

FSC (Forest Stewardship Council). 1994a. *Forest Stewardship Council Ratification Documents.* July. Oaxaca, Mexico: FSC.

———. 1994b. *FSC Principles and Criteria for Natural Forest Management.* Oaxaca, Mexico: FSC.

———. 1998. Forest Stewardship Council's ten principles. Reprinted in R. Prabhu, C.J.P. Colfer, and R. Dudley. 1999. *Guidelines for Developing, Testing and Selecting Criteria and Indicators for Sustainable Forest Management.* CIFOR C&I Toolbox Series no. 1. Bogor, Indonesia: Center for International Forestry Research.

Furukawa, H. 1994. *Coastal Wetlands of Indonesia: Environment, Subsistence, and Exploitation,* translated by Peter Hawkes. Kyoto, Japan: Kyoto University Press.

Gabriel, K.R. 1971. The biplot graphical display of matrices with application to principal component analysis. *Biometrika* 58: 453–67.

Gadsby, E.L., and P.D. Jenkins. 1992. *Report on Wildlife and Hunting in the Proposed Etinde Forest Reserve.* Report to the Limbe Botanic Garden and Rainforest Genetic Conservation Project. Limbe, Cameroon: Mt. Cameroon Project.

Gale, R.P., and S.M. Cordray. 1994. Making sense of sustainability: Nine answers to "what should be sustained?" *Rural Sociology* 59: 311–32.

Gami, N. 1995. *Etude du Milieu Humain—Parc National d'Odzala-Congo.* Unité de Recherche en Nutrition et Alimentation Humaines DGRST-Congo. Congo rapport intermédiaire. May. Village Mbandza, Congo: Conservation et Utilisation Rationnelle des Ecosystèmes Forestiers en Afrique Centrale (ECOFAC), Organisation Africaine du Bois (OAB).

———. 1998. *Test des Critères et Indicateurs de Gestion Durable des Forêts au Gabon. Rapport d'Evaluation de Certains Principes, Critères, et Indicateurs en Sciences Sociales et Humaines.* Libreville, Gabon: Organisation Africaine du Bois (OAB).

Gami, N., and G. Mavah. 1997. Résumé Général des Données de l'Etude Carto-graphique et Socio-économique de PROECO 003: axes, Ouesso–Liouesso et Ouesso–Sembé (Forêt du Nord Congo-Brazzaville). Report to Deutsche Gesellschaft für Technische Zusammenarbeit (GTZ), PROECO Project. Brazzaville, Republic of Congo: GTZ, PROECO project.

Gartlan, S. 1992. Practical constraints on sustainable logging in Cameroon. In *Conservation of West and Central African Rainforests*, edited by K. Cleaver, M. Munasinghe, M. Dyson, N. Egli, A. Peuker, and F. Wencelius. Washington, DC: The World Bank, 141–5.

Gatuslao, R.M. 1988. The Higaonons' war. *Midweek*, December 21: 3–5.

Geschiere, P., and F. Nyamnjoh. 1998. Witchcraft as an issue in the "politics of belong-ing": Democratization and urban migrants' involvement with the home village. *African Studies Review* 41(3): 69–91.

Getz, W., L. Fortmann, D. Cumming, J. du Toit, J. Hilty, R. Martin, M. Murphree, N. Owen-Smith, A. Starfield, and M. Westphal. 1999. Sustaining natural and human cap-ital: Villagers and scientists. *Science* 283: 1855–6.

Giesen, W. 1987. *Danau Sentarum Wildlife Reserve: Inventory, Ecology and Management Guidelines.* Bogor, Indonesia: World Wide Fund for Nature and Perlindungan Hutan dan Pelestarian Alam (PHPA, agency for forest protection and nature conservation).

———. 1996. *Habitat Types and Their Management. Danau Sentarum Wildlife Reserve, West Kalimantan, Indonesia.* Bogor, Indonesia: Wetlands International Indonesia Pro-gramme/Perlindungan Hutan dan Pelestarian Alam (PHPA).

WRI/IUCN/UNEP (World Resources Institute/International Union for the Conserva-tion of Nature/U.N. Environment Programme). 1992. *Global Biodiversity Strategy.* Washington, DC: WRI/IUCN/UNEP.

Goheen, M. 1996. *Men Own the Fields, Women Own the Crops.* Madison, WI: University of Wisconsin Press.

Gomes, M.E., and A.D. Kanner. 1995. The rape of the well-maidens: Feminist psychol-ogy and the environmental crisis. In *Ecopsychology*, edited by T. Roszak, M.E. Gomes, and A.D. Kanner. San Francisco, CA: Sierra Club Books, 111–21.

Gonzalez, I.C.C. 1999. *Indigenous Management of Forest Resources in East Kalimantan, Indo-nesia. The Role of Secondary Forests.* M.Sc. thesis, Tropical Forestry. Wageningen, the Netherlands: Wageningen Agricultural University.

Goodland, R. 1991. Tropical Deforestation: Solutions, Ethics, and Religions. World Bank Sector Policy and Research Staff Environment Working paper no. 43. Washing-ton, DC: The World Bank.

Goodman, D., and Redclift, M. 1991. *Refashioning Nature: Food, Ecology, and Culture.* New York: Routledge.

Goodman, M. (ed.) 1985. *Women in Asia and the Pacific: Towards an East-West Dialogue.* Honolulu, HI: University of Hawaii, Women's Studies Program.

Green, L. 1986. The theory of participation: A qualitative analysis of its expression in national and international health policies. *Advances in Health Education and Promotion* 1(A): 211–36.

Green, S.F., G.P.C. King, V. Nitsiakos, and S.E. van der Leeuw. 1998. Landscape perceptions in Epirus in the late 20th century. In *The Archaeomedes Project: Understanding the Natural and Anthropogenic Causes of Land Degradation and Desertification in the Mediterranean Basin*, edited by S.E. van der Leeuw. Luxembourg: Office for Official Publications of the European Communities, 330–59.

Grigsby, W. 1995. *The Nature of Land: Tenure in an Uncertain Environment*. Doctoral dissertation. Seattle, WA: Washington State University.

Günter, M. 1998. Report on the Implementation and Testing of Social Criteria and Indicators for Sustainable Forest Management and Development in Small Island States. Internal Paper. Port of Spain, Republic of Trinidad and Tobago: United Nations–Economic Commission on Latin America and the Caribbean (UN–ECLAC) Environment Unit.

Guyer, J.I. 1978. The food economy and French colonial rule in central Cameroon. *Journal of African History* 19: 577–97.

————. 1984. *Family and Farm in Southern Cameroon*. African Research Studies no. 15. Boston, MA: Boston University, African Studies Center.

Guyer, J.I. (ed.) 1995. *Money Matters: Instability, Values, and Social Payments in the Modern History of West Africa*. Portsmouth, NH: Heinemann.

Hames, R. 1991. Wildlife conservation in tribal societies. In *Biodiversity: Culture, Conservation and Ecodevelopment*, edited by M.L. Oldfield and J.B. Alcorn. Boulder, CO: Westview Press, 172–99.

Harms, R. 1984. *River of Wealth, River of Sorrow: The Central Zaire Basin in the Era of the Slave Trade: 1500–1891*. New Haven, CT: Yale University Press.

————. 1987. *Games Against Nature: An Eco-Cultural History of the Nunu of Equatorial Africa*. Cambridge, U.K.: Cambridge University Press.

Harris, M. 1968. *The Rise of Anthropological Theory: A History of Theories of Culture*. New York: Thomas Y. Crowell.

Harrisson, T. 1970. *The Malays of Southwest Sarawak Before Malaysia*. London, U.K.: Macmillan.

Harwell, E. 1997. *Law and Culture in Resource Management: An Analysis of Local Systems for Resource Management in the Danau Sentarum Wildlife Reserve, West Kalimantan, Indonesia*. Consultant's report for Wetlands International. Bogor, Indonesia: Wetlands International.

————. 2000a. Remote sensibilities: Discourses of technology and the making of Indonesia's natural disaster. *Development and Change* 31: 307–40.

————. 2000b. The Un-Natural History of Culture: Ethnicity, Tradition, and Territorial Conflicts in West Kalimantan, Indonesia, 1800–1997. Doctoral dissertation, Yale University, New Haven, CT.

Hawkesworth, M. 1997. Confounding gender. *Signs* 22: 651–85.

Head, S., and R. Heinzman (eds.). 1990. *Lessons of the Rainforest*. San Francisco, CA: Sierra Club Books.

Hecht, S.B., and A. Cockburn. 1989. *The Fate of the Forest: Developers, Destroyers and Defenders of the Amazon*. New York: Verso.

Helsinki Process. 1993. *Ministerial Conference on the Protection of Forests in Europe*. June 16–17. Helsinki, Finland: Maa-ja metsatalousministerio.

Herbert, F. 1984. *Dune*. New York: Putnam Publishing Group.

Heuveldop, J. 1994. *Assessment of Sustainable Tropical Forest Management*. Hamburg, Germany: Kommissionsverlag Max Wiedebusch.

Hobley, M. 1996. *Participatory Forestry: The Process of Change in India and Nepal.* Rural Development Forestry Study Guide 3. London, U.K.: Rural Development Forestry Network, Overseas Development Institute.

Hobley, M., and K. Shah. 1996. *What Makes a Local Organisation Robust? Evidence from India and Nepal.* ODI Natural Resource Perspectives no. 11. July. London, U.K.: Overseas Development Institute.

Hoekstra, T.W., T.F.H. Allen, J. Kay, and J.A. Tainter. 2000. Criteria and indicators for ecological and social system sustainability, with system management objectives. In *North American Test of Criteria and Indicators of Sustainable Forestry* (Volume 1), edited by S. Woodley, G. Alward, L.I. Gutierrez, T. Hoekstra, B. Holt, L. Livingston, J. Loo, A. Skibicki, C. Williams, and P. Wright. Washington, DC: USDA Forest Service, 117–25.

Hoffmann, A.A., A. Hinrichs, and F. Siegert. 1999. Fire Damage in East Kalimantan in 1997/98 Related to Land Use and Vegetation Classes: Satellite Radar Inventory Results and Proposal for Further Actions. Integrated Forest Fire Management—Sustainable Forest Management Project Report no.1a. Samarinda, East Kalimantan, Indonesia: Ministry of Forestry and Estate Crops, Deutsche Gesellschaft für Technische Zusammenarbeit (GTZ) GmbH and Kreditanstalt für Wiederaufbau (KfW).

Howell, S. 1984. *Society and Cosmos: Chewong of Peninsular Malaysia.* Chicago, IL: University of Chicago Press.

IBAMA (Instituto Brasileiro do Meio Ambiente). 1995. *Ministério do Meio Ambiente, dos Recursos Hídricos e da Amazônia Legal.* Portaria # 48/95. July 10. Brasília, Brazil: Instituto Brasileiro do Meio Ambiente e dos Recursos Naturais Renováveis.

Ibo, J., and E. Leonard. 1997. *Forests, Farmers and the State: Participatory Forest Management in the Ivory Coast.* Forest Participation Series no. 7. London, U.K.: International Institute for Environment and Development.

IITA/HFS (International Institute of Tropical Agriculture, Humid Forest Station). *IITA/HFS Medium Term Plan 1990–95.* Nkolbisson, Cameroon: IITA/HFS.

Ingles, A.W., A. Musch, and H. Qwist-Hoffman. 1999. *The Participatory Process for Supporting Collaborative Management of Natural Resources: An Overview.* Rome, Italy: Food and Agriculture Organization, Forests, Trees, and People Programme.

Isham, J., D. Narayan, and L. Pritchett. 1995. Does participation improve performance? Establishing causality with subjective data. *The World Bank Economic Review* 9: 175–200.

ISNAR (International Service for National Agricultural Research). 1997. *Gender Analysis for Management of Research in Agriculture and Natural Resources.* The Hague, the Netherlands: ISNAR.

ITTO (International Tropical Timber Organization). 1992a. *ITTO Guidelines for the Sustainable Management of Natural Tropical Forests.* Policy Development Series 1. Yokohama, Japan: ITTO.

———. 1992b. *Criteria for the Measurement of Sustainable Tropical Forest Management.* Yokohama, Japan: ITTO.

———. 1997. *Report of the ITTO Expert Panel on Criteria and Indicators.* Yokohama, Japan: ITTO, Appendix 4, Part III.

ITW (Initiative Tropenwald). 1994. *Assessment of Sustainable Tropical Forest Management,* compiled by Jochen Heuveldop. Hamburg, Germany: Kommissionsverlag Max Wiedebusch.

IUCN (International Union for the Conservation of Nature). 1995a. Monitoring and Assessment of Local Strategies for Sustainability: A Guide for Fieldworkers Carrying Out Monitoring and Assessment at Community Level. Draft. Gland, Switzerland: IUCN.

———. 1995b. *Assessing Progress towards Sustainability.* Gland, Switzerland: IUCN International Assessment Team Strategies for Sustainability Programme.

Jackson, B. 1997. Workshop on Tools and Methods for Monitoring and Evaluating Collaborative Management of Natural Resources in Southern Africa. International Union for the Conservation of Nature (IUCN) Workshop report. Gland, Switzerland: IUCN.

Jackson, C. 1996. Rescuing gender from the poverty trap. *World Development* 24: 489–504.

Joiris, D.V. 1996. L'esprit, l'igname, et l'eléphant: Essai d'interprétation symbolique d'un rituel chez les pygmées Baka du sud Cameroun. In *L'Alimentation en Forêt Tropicale. Interactions Bioculturelles et Perspectives de Développement. Bases Culturelles des Choix Alimentaires et Stratégies de Développement*, edited by C.M. Hladik, A. Hladik, H. Pagezy, O.F. Linares, G.J.A. Koppert, and A. Froment. Paris, France: U.N. Educational, Scientific, and Cultural Organisation, 961–72.

Jolliffe, T. 1986. *Principal Component Analysis.* New York: Springer Verlag.

Jordan, B. 1991. *Technology and Social Interaction: Notes on the Achievement of Authoritative Knowledge in Complex Settings.* Palo Alto, CA: Xerox Palo Alto Research Center, Institute for Research on Learning and Work Practice and Technology System Sciences Laboratory.

———. 1997. Authoritative knowledge and its construction. In *Childbirth and Authoritative Knowledge: Cross-Cultural Perspectives*, edited by R. Davis-Floyd and C. Sargent. Berkeley, CA: University of California Press, 55–79.

Jordan, B., with R. Davis-Floyd. 1993. *Birth in Four Cultures: A Cross-Cultural Investigation of Childbirth in Yucatan, Holland, Sweden, and the United States.* Prospect Heights, IL: Waveland Press.

Kaimowitz, D., Erwidodo, O. Ndoye, P. Pacheco, and W. Sunderlin. 1998. Considering the impact of structural adjustment policies on forest in Bolivia, Cameroon and Indonesia. *Unasylva* 194(49): 57–64.

Kajembe, G.C., and G.C. Monela. 2000. Tanzania case study: Empowering communities to manage natural resources: Where does the new power lie? A case study of Duru-Haitemba, Babati, Tanzania. In *Empowering Communities to Manage Natural Resources: Case Studies from Southern Africa* (Volume 2), edited by S. Shackleton and B. Campbell. Council for Scientific and Industrial Research (CSIR) Report ENV-P-C-2000–025. Pretoria, South Africa: CSIR and World Wide Fund for Nature (Southern Africa), 153–68.

Karsenty, A., L. Mendouga Mébenga, and A. Pénelon. 1997. Spécialisation des espaces ou gestion intégrée des massifs forestiers. *Bois et Forêts des Tropiques* 251: 43–53.

Kaskija, L. 1995. Punan Malinau: The Persistence of an Unstable Culture. Doctoral dissertation. Uppsala, Sweden: Uppsala University.

———. 1999. *Stuck at the Bottom: Opportunity Structures and Punan Malinau Identity.* Draft report to CIFOR. Bogor, Indonesia: Center for International Forestry Research.

Katz, E., A. Lammel, and M. Goloubinoff (eds). In press. *Entre Ciel et Terre: Climat et Sociétés.* Paris, France: L'Harmattan-IRD.

Kayambazinthu, D. 2000. Malawi case study: Empowering communities to manage natural resources: Where does the power lie? In *Empowering Communities to Manage Natural Resources: Case Studies from Southern Africa* (Volume 2), edited by S. Shackleton and B. Campbell. Council for Scientific and Industrial Research (CSIR) Report ENV-P-C-2000–025. Pretoria, South Africa: CSIR and World Wide Fund for Nature (Southern Africa), 46–71.

Kearney, M. 1996. *Reconceptualizing the Peasantry: Anthropology in Global Perspective.* Boulder, CO: Westview Press.

Kemf, E. (ed.) 1993. *The Law of the Mother: Protecting Indigenous Peoples in Protected Areas.* San Francisco, CA: Sierra Club Books.

Klepper, O. 1994. *A Hydrological Model of the Upper Kapuas River and the Kapuas Lakes.* Consultancy report for the Asian Wetland Bureau/Perlindungan Hutan dan Pelestarian Alam (PHPA), for the U.K.–Indonesia Tropical Forest Management Project, Subproject 5: Conservation. Bogor, Indonesia: Asian Wetland Bureau.

REFERENCES

Kundstadter, P., E.C. Chapman, and S. Sabhasri (eds.). 1978. *Farmers in the Forest: Economic Development and Marginal Agriculture in Northern Thailand*. Honolulu, HI: East-West Center.

Kwenzi Mikala, J.T. 1997. Le Gabon: le pays et les hommes. In *L'Esprit de la Forêt,* edited by L. Perrois. Bordeaux, France: Somogy, Musée d'Aquitaine, 17–30.

Laburthe-Tolra, P. 1981. *Les Seigneurs de la Forêt: Essai sur le Passé Historique, l'Organisation Sociale et les Normes Éthiques des Anciens Beti du Cameroun*. Paris, France: Publications de la Sorbonne.

Lammerts van Bueren, E., and E. Blom. 1997. *Hierarchical Framework for the Formulation of Sustainable Forest Management Standards: Principles, Criteria and Indicators*. Leiden, the Netherlands: The Tropenbos Foundation.

Langston, N. 1995. *Forest Dreams, Forest Nightmares*. Seattle, WA: University of Washington Press.

Laporte, N. 1999. *Assessing Land Cover Change in the Dense Humid Forest of Cameroon: Comparison of Satellite Image Maps and Household Surveys*. Report to CIFOR. Bogor, Indonesia: Center for International Forestry Research.

Leach, M. 1994. *Rainforest Relations: Gender and Resource Use Among the Mende of Gola*. Edinburgh, U.K.: Edinburgh University Press.

Leach, M., and R. Mearns (eds.). 1996. *The Lie of the Land: Challenging Received Wisdom on the African Environment*. Oxford, U.K.: James Currey, for The International African Institute.

LEI (Lembaga Ekolabel Indonesia). 1997. From the Web page of Lembaga Ekolabel Indonesia at http://www.iscom.com/~ekolabel/buku1.html (accessed October 2, 2000).

Lele, S. 1993. Sustainability: A plural, multi-dimensional approach. Unpublished manuscript. Bogor, Indonesia: Center for International Forestry Research.

Leopold, A. 1970. *A Sand County Almanac*. New York: Ballatine Books.

Leplaideur, A. 1985. *Les Systèmes Agricoles en Zone Forestière: Paysans du Centre et du Sud Cameroun*. Montpellier, France: Institut pour la Recherche Agronomique Tropicale/Centre de Coopération Internationale pour la Recherche Agronomique en Développement.

Lewis, H.T. 1989. Ecological and technological knowledge of fire: Aborigines versus park rangers in northern Australia. *American Anthropologist* 91: 940–61.

Lightfoot, C., S. Feldman, and M. Zainul Abedin. 1991. *Households, Agroecosystems, and Rural Resources Management*. Manila, the Philippines: International Center for Living Aquatic Resource Management.

Liverman, D., E.F. Moran, R.R. Rindfuss, and P. Stern. 1998. *People and Pixels. Linking Remote Sensing and Social Science*. Washington, DC: National Academy Press.

Long, N., and M. Villareal. 1994. The interweaving of knowledge and power in development interfaces. In *Beyond Farmer First: Rural People's Knowledge, Agricultural Research and Extension Practice*, edited by I. Scoones and J. Thompson. London, U.K.: Intermediate Technology Publications Ltd., 41–51.

Lueck, D. 1995. Property rights and the economic logic of wildlife institutions. *Natural Resources Journal* 35: 625–70.

Luttrell, C. 1994. *Forest Burning in Danau Sentarum*. Preliminary report for AWB/PHPA, June 1994. Indonesia-U.K. Tropical Forest Management Project, Sub-project: Conservation. Bogor, Indonesia: Asian Wetland Bureau/Perlindungan Hutan dan Pelestarian Alam.

Lynch, O.J., and J.B. Alcorn. 1994. Tenurial rights and community-based conservation. In *Natural Connections: Perspectives in Community-Based Conservation*, edited by D. Western and R.M. Wright. Washington, DC: Island Press, 373–92.

Madrah, T. 1997. *Lemu, Ilmu Magis Suku Dayak Benuaq dan Tunjung [Lemu, Benuq and Tunjung Magic]*. Jakarta, Indonesia: Puspa Swara.

Madrah, T., and Karaakng. 1997. *Tempuutn. Mitos Dayak Benuaq dan Tunjung [Lemu, Benuaq and Tunjung Myths]*. Jakarta, Indonesia: Puspa Swara.

Malingreau, J.P. 1991. Remote sensing for tropical forest monitoring: an overview. In *Remote Sensing and Geographical Information Systems for Resource Management in Developing Countries*, edited by A.S. Belward and C.R. Valenzuela. Dordrecht, the Netherlands: Kluwer Academic, 253–78.

Malvestuto, S. 1989. *Research and Development Strategies for the Kapuas River Fishery in West Kalimantan, Indonesia: A Project Proposal*. Submitted to U.S. Agency for International Development, Jakarta, Indonesia.

Mangel, M., and 41 other authors. 1996. Principles for the conservation of wild living resources. *Ecological Applications* 6(2): 338–62.

Mashman, V. 1991. Warriors and weavers: A study of gender relations among the Iban of Sarawak. In *Female and Male in Borneo: Contributions and Challenges to Gender Studies*, edited by V.H. Sutlive and G.N. Appell. Monograph Series no. 1. Williamsburg, VA: Borneo Research Council, 231–70.

Mayaux, P., F. Achard, and J.P. Malingreau. 1998. Global tropical forest area measurements derived from coarse resolution maps at a global level: A comparison with other approaches. *Environmental Conservation* 25: 37–52.

Mbot, J.E. 1997. Quand l'esprit de la forêt s'appelait jachère. In *L'Esprit de la Forêt: Terres du Gabon*, edited by L. Perrois. Brussels, Belgium: Union Européenne, Programme culturel Bantu, 33–51.

McCay, B., and J.M. Acheson (eds.). 1987. *The Question of the Commons: the Culture and Ecology of Communal Resources*. Tucson, AZ: University of Arizona Press.

McDougall, C. 1998. *Final Test of The BAG, Bulungan, East Kalimantan*. Internal document. Bogor, Indonesia: Center for International Forestry Research.

McIntosh, R.J., J.A. Tainter, and S. Keech McIntosh. 2000. Climate, history, and human action. In *The Way the Wind Blows: Climate, History, and Human Action*, edited by R.J. McIntosh, J.A. Tainter, and S. Keech McIntosh. New York: Columbia University Press, 1–42.

Meillassoux, C. 1981. *Maidens, Meals, and Money: Capitalism and the Domestic Economy* (translation of *Femmes, greniers, et capitaux*). New York: Cambridge University Press.

Mengin-Lecreulx, P., with A. Anvo, C. Huttel, H. van Haaften, and N.K. Anatole. 1995. *Final Report: Test Côte d'Ivoire, June 2–30*, translated by Guy Ferlin, edited by R. Prabhu and L.C. Tan. Bogor, Indonesia: Center for International Forestry Research.

Milligan, G.W., and M.C. Cooper. 1985. An examination of procedures for determining the number of clusters in a data set. *Psychometrika* 50: 159–79.

Ministry of Agriculture, Land, and Marine Resources. 1992. *Land Rationalisation and Development Programme*. Port of Spain, Republic of Trinidad and Tobago: Ministry of Agriculture, Land and Marine Resources.

Mintz, S. 1985. From plantations to peasantries in the Caribbean. In *Caribbean Contours*, edited by S. Mintz and S. Price. Baltimore, MD: The Johns Hopkins University Press, 127–53.

Miyasaka Porro, N. 1997. *Changes in Peasant Perceptions of Development and Conservation*. Master's thesis. Gainesville, FL: University of Florida.

Mojena, R. 1977. Hierarchical grouping method and stopping rules: An evaluation. *Computer Journal* 20: 359–63.

Momberg, F., K. Atok, and M. Sirait. 1996. *Drawing on Local Knowledge: A Community Mapping Training Manual: Case Studies from Indonesia*. Jakarta, Indonesia: Ford Foundation, Yayasan Karya Sosial Pancur Kasih, and WWF.

Moran, E. 1990. Ecosystem ecology in biology and anthropology: A critical assessment. In *The Ecosystem Approach in Anthropology: From Concept to Practice*, edited by E. Moran. Ann Arbor, MI: University of Michigan Press, 3–40.

REFERENCES

Muhtaman, D.R., C.A. Siregar, and P. Hopmans. 2000. *Criteria and Indicators for Sustainable Plantation Forestry in Indonesia.* Report. Bogor, Indonesia: Center for International Forestry Research.

Munyanziza, E., and K.F. Wiersum. 1999. Indigenous knowledge of miombo trees in Morogoro, Tanzania. *Indigenous Knowledge and Development Monitor* 7(2): 10–13.

Murombedzi, J. 1999. Land expropriation, communal tenure, and common property resource management in southern Africa. *The Common Property Resource Digest* 50 (October): 1–4.

Murphy, Y., and R. Murphy. 1974. *Women of the Forest.* New York: Columbia University Press.

Myers, N. 1980. *Conversion of Tropical Moist Forests.* Report prepared by Norman Myers for the Committee on Research Priorities in Tropical Biology of the National Research Council. Washington, DC: National Academy of Sciences.

———. 1991. *Population, Resources, and the Environment: The Critical Challenges.* New York: U.N. Family Planning Agency.

———. 1993. *Rainforests.* Emmaus, PA: Rodale Press.

Nagle, J. 1991. *Report on Forest Economics.* Trinidad and Tobago NFAP Report. Rome, Italy: Food and Agriculture Organization.

Narayan, D., and L. Srinivasan (eds.). 1994. *Participatory Development Tool Kit: Training Materials for Agencies and Communities.* Washington, DC: The World Bank.

NASA (National Aeronautics and Space Administration). 1990. *Proceedings, Applications of Space-Age Technology in Anthropology Conference.* Hancock County, Mississippi: John C. Stennis Space Center, South Mississippi.

Nasi, R., J-M. Bouvard, P. Hecketsweiler, R. Ondo, N. Gami, and S. Dondyas. 1998. *Initiative de l'Organisation Africaine du Bois sur les Critères et Indicateurs pour la Gestion Durable des Forêts Africaines. Test du Gabon.* Libreville, Gabon: Organisation Africaine du Bois.

Ndoye, O. 1998. *Non-Timber Forest Product Markets and Potential Degradation of the Forest Resource in Central Africa: The Role of Research in Finding a Balance Between Welfare Improvement and Forest Conservation.* Bogor, Indonesia: Center for International Forestry Research.

Ndoye, O., and D. Kaimowitz. 1998. Macro-economics, Markets, and the Humid Forests of Cameroon, 1967–1997. Draft paper. Bogor, Indonesia: Center for International Forestry Research.

Ndoye, O., and D. Russell. 1993. The Cocoa Crisis in Southern Cameroon: Implications for Environmental and Agricultural Development in the Forest Zone. Paper presented at the annual meeting of the African Studies Association. December 4–7, Boston, MA.

Ndoye, O., M. Ruiz Pérez, and A. Eyebe. 1997/98. *The Markets of Non-Timber Forest Products in the Humid Forest Zone of Cameroon.* Rural Development Forestry Network Paper 22c. London, U.K.: Overseas Development Institute.

———. 1998. Non-Timber Forest Product Markets and Potential Degradation of the Forest Resource in Central Africa. The Role of Research in Finding a Balance between Welfare Improvement and Forest Conservation. Paper presented at the International Expert Workshop on Non-Wood Forest Products (NWFPs) for Central Africa. May 10–15, Limbe Botanic Garden, Limbe, Cameroon.

Nelson, R., and B.N. Holben. 1986. Identifying deforestation in Brazil using multiresolution satellite data. *International Journal of Remote Sensing* 7: 429–48.

Newton, B. 1977. Perceptions of Sex Roles at the University of Hawaii. Paper presented at Women in Communication Convention. October, Honolulu, HI.

Newton, B., E. Buck, and J. Woelfel. 1984. Metric multidimensional scaling of viewers' perceptions of TV in five countries. *Human Organization* 42: 162–70.

Ngo, M. 1996. A new perspective on property rights: Examples from the Kayan of Kalimantan. In *Borneo in Transition: People, Forests, Conservation and Development*, edited by C. Padoch and N. Peluso. New York: New York Botanical Garden Press, 137–49.

Nguinguiri, J-C. 1998. *Les Approches Participatives dans la Gestion des Ecosystèmes Forestiers d'Afrique Centrale.* Report to Conférence de responsables de recherche agronomique africains/Capitalisation et Synthèse des Recherches sur les Ecosystèmes forestiers humides d'Afrique (CORAF/FORAFRI). Pointe-Noire, Congo: CORAF/FORAFRI.

———. 1999. *Les Approches Participatives dans la Gestion des Ecosystèmes Forestiers d'Afrique Centrale: Revue des Initiatives Existantes.* CIFOR Occasional Paper no. 23. Bogor, Indonesia: Center for International Forestry Research.

Norgaard, R.B. 1992. Sustainability and the Economics of Assuring Assets for Future Generations. World Bank Policy Research Working paper 832. Washington, DC: The World Bank.

NRI (Natural Resources Institute) 1992. *Natural and Human Resource Studies and Land Use Options, Department of Nyong and So'o, Cameroon.* Chatham, U.K.: NRI/Overseas Development Administration.

Nurse, M.C., C.R. McKay, J.T. Young, and C.A. Asanga. 1995. Biodiversity conservation through community forestry, in the montane forests of Cameroon. In *From the Field, Rural Development Forestry Network paper 18d.* London: Overseas Development Institute, 14–19.

OIBT (Organisation Internationale des Bois Tropicaux). 1992. Critères de Mesure de l'Amenagement Durable des Forêts Tropicales, Serie C. Bulletins Techniques: Politique forestière 3. Yokohama, Japan: OIBT.

Oksa, J. 1993. The benign encounter: The great move and the role of the state in Finnish forests. In *Who Will Save the Forests? Knowledge, Power and Environmental Destruction*, edited by T. Banuri and F.A. Marglin. London, U.K.: Zed Books, 114–41.

Ondo, R. 1997. *Enjeux de la Forêt: Rapports Entreprises Forestières/Populations Autochtones.* Study completed for the Compagnie Equatoriale des Bois (CEB). Libreville, Gabon: CEB.

Ostrom, E. 1990. *Governing the Commons: The Evolution of Institutions for Collective Action.* Cambridge, U.K.: Cambridge University Press.

———. 1994. *Neither Market Nor State: Governance of Common-Pool Resources in the Twenty-First Century.* Washington, DC: International Food Policy Research Institute.

———. 1999. *Self Governance and Forest Resources.* CIFOR Occasional Paper no. 20. Bogor, Indonesia: Center for International Forestry Research.

Oxford Dictionary of Current English. 1987. New York: Oxford University Press.

Oyono, P.R., and M.C. Diaw with F. Sangkwa, J. Nguiiebouri, C. Bidja, and S. Efoua. 1998. *Methodes des Science Sociales sur les Critères et Indicateurs de Gestion Durable des Forets, Part II.* Report of the social science methods tests conducted in the International Institute for Tropical Agriculture (IITA) Humid Forest Benchmark in Cameroon and in the Ntem and Lobe Basins. Yaoundé, Cameroon and Bogor, Indonesia: IITA and Center for International Forestry Research.

Padoch, C., and N.L. Peluso (eds.). 1996. *Borneo in Transition: People, Forests, Conservation, and Development.* Kuala Lumpur, Malaysia: Oxford University Press.

Padoch, C., and C. Peters. 1993. Managed forest gardens in West Kalimantan, Indonesia. In *Perspectives on Biodiversity: Case Studies of Genetic Resource Conservation and Development*, edited by C.S. Potter, J.I. Cohen, and D. Janczewski. Washington, DC: AAAS Press, 167–76.

Pagezy, H. 1996. Aspects psychoculturels de l'exploitation des ressources naturelles dans la région du lac Tumba (Congo-Démocratique). In *Bien Manger et Bien Vivre*, edited by A. Froment, C. de Garine, Binam Bikoi, and J.F. Loung. Paris, France: ORSTOM/ l'Harmatan, 447–58.

Palmer, C. 1993. Folk management, "soft evolutionism," and fishers' motives: Implications for the regulation of the lobster fisheries of Maine and Newfoundland. *Human Organization* 52: 414–20.

Panday, D.N., S. Chadha, A. Chatterjee, A. Swarz, and M. Poffenberger. 1997. *Participatory Mapping for Joint Forest Management Inventory, Planning, and Monitoring: Methods Manual* (Volume 3). Berkeley, CA and New Delhi, India: Asia Forest Network.

Participatory Rural Appraisal Handbook. 1990. World Resources Institute, Natural Resources Management Support Series no. 1. Washington, DC: Prepared jointly by National Environment Secretariat, Edgerton University, Clark University and the Center for International Development and Environment of the World Resources Institute.

Pearce, D.W., G.D. Atkinson, and W.R. Dubourg. 1994. The economics of sustainable development. *Annual Review of Energy and Environment* 19: 457–74.

Pedersen, P. 1996. Nature, religion and cultural identity: The religious environmentalist paradigm. In *Asian Perceptions of Nature: A Critical Approach*, edited by O. Bruun and A. Kalland. Surrey, U.K.: Curzon Press, 258–76.

Peluso, N. 1994. *The Impact of Social and Environmental Change on Forest Management: A Case Study from West Kalimantan, Indonesia.* FAO Community Forestry Case Study Series no. 8. Rome, Italy: Food and Agriculture Organization.

Pemda Kaltim. 1990. *Sejarah Pemerintahan di Kalimantan Timur dari Masa ke Masa.* Samarinda, Indonesia: Pemda Kaltim.

Perez, L.M. 1996. Analysis of social elements in forestry certification. *Proceedings, University of British Columbia (UBC)–Universiti Pertanian Malaysia (UPM) Conference on the Ecological, Social, and Political Issues of the Certification of Forest Management.* Vancouver, BC, Canada: Faculties of Forestry, UBC and UPM, 147–65.

Peters, C. 1993. *Forest Resources of the Danau Sentarum Wildlife Reserve: Observations on the Ecology, Use and Management Potential of Timber and Non-Timber Products.* Field Report I to Asian Wetland Bureau (AWB). Bogor, Indonesia: AWB.

———. 1994. *Forest Resources of the Danau Sentarum Wildlife Reserve: Strategies for the Sustainable Exploitation of Timber and Non-Timber Products.* Field Report II to Asian Wetland Bureau (AWB). Bogor, Indonesia: AWB.

Pierce, A. 1996. Issues Pertaining to the Certification of Non-Timber Forest Products. A Forest Stewardship Council discussion paper. Oaxaca, Mexico: Forest Stewardship Council.

Pimbert, M.P., and J.N. Pretty. 1995. Parks, People and Professionals: Putting "Participation" into Protected Area Management. Discussion paper. Geneva, Switzerland: U.N. Research Institute for Social Development, International Institute for Environment and Development, and World Wide Fund for Nature–International.

Poffenberger, M. (ed.) 1990. *Keepers of the Forest: Land Management Alternatives in Southeast Asia.* West Hartford, CT: Kumarian Press.

———. 1996. *Communities and Forest Management: With Recommendations to the Intergovernmental Panel on Forests.* Cambridge, U.K.: International Union for the Conservation of Nature–The World Conservation Union.

Poffenberger, M. 1998. *Stewards of Vietnam's Upland Forests.* Research Network report 10. Berkeley, CA: University of California, Center for Southeast Asian Studies, Asia Forest Network.

Poffenberger, M., and B. McGean (eds.). 1993a. *Community Allies: Forest Co-Management in Thailand.* Research Network report 2. Berkeley, CA: University of California, Center for Southeast Asian Studies, Asia Forest Network.

———. 1993b. *Upland Philippine Communities: Guardians of the Final Forest Frontiers.* Research Network report 4. Berkeley, CA: University of California, Center for Southeast Asian Studies, Asia Forest Network.

Poffenberger, M., B. McGean, N.H. Ravindranath, and M. Gadgil. 1992a. *Field Methods Manual* (Volume 1). New Delhi, India: Society for Promotion of Wastelands Development.

Poffenberger, M., B. McGean, A. Khare, and J. Campbell. 1992b. *Field Methods Manual* (Volume 2). New Delhi, India: Society for Promotion of Wastelands Development.

Poffenberger, M., C. Josayma, P. Walpole, and K. Lawrence. 1995. *Transitions in Forest Management: Shifting Community Forestry from Project to Process.* Research Network Report no. 6. August. Berkeley, CA: University of California, Center for Southeast Asia Studies, Asia Forest Network.

Poffenberger, M., with P. Bhattacharya, A. Khare, A. Rai, S.B. Roy, N. Singh, and K. Singh. 1996. *Grassroots Forest Protection: East Indian Experiences.* Research Network report no. 7. March. Berkeley, CA: University of California, Center for Southeast Asia Studies, Asia Forest Network.

Pokam, J., and W.D. Sunderlin. 1999. L'impact de la Crise Economique sur les Populations, les Migrations, et le Couvert Forestier du Sud-Cameroun. Unpublished paper. Bogor, Indonesia: Center for International Forestry Research.

Polunin, N.V.C. 1983. Do traditional marine "reserves" conserve? A view of the Indonesian and New Guinean evidence. In *Contending with Global Change: Study No. 2, Traditional Marine Resource Management in the Pacific Basin, an Anthology,* edited by K. Ruddle and R.E. Johannes. Jakarta, Indonesia: U.N. Educational, Scientific, and Cultural Organisation/Regional Office in Science and Technology in Southeast Asia, 191–212.

Porro, R., and N. Miyasaka Porro. 1998. *Methods for Assessing Social Science Criteria and Indicators for the Sustainable Management of Forests: Brazil Test.* Report. Bogor, Indonesia: Center for International Forestry Research.

Posey, D. 1992. Interpreting and applying the "reality" of indigenous concepts: What is necessary to learn from the natives? In *Conservation of Neotropical Forests: Working from Traditional Resource Use,* edited by K.H. Redford and C. Padoch. New York: Columbia University Press, 21–34.

Posey, D., and W. Balee. 1989. *Resource Management in Amazonia: Indigenous and Folk Strategies.* New York: New York Botanical Garden.

Prabhu, R. 1995. A Conceptual Framework for a System to Evaluate the Sustainability of Forest Ecosystem Management: A Discussion Paper. Internal document. Bogor, Indonesia: Center for International Forestry Research.

Prabhu, R., C.J.P. Colfer, P. Venkateswarlu, L.C. Tan, R. Soekmadi, and E. Wollenberg. 1996. *Testing Criteria and Indicators for the Sustainable Management of Forests: Phase I Final Report.* CIFOR Special publication. Bogor, Indonesia: Center for International Forestry Research.

Prabhu, R., W. Maynard, R. Eba'a Atyi, C.J.P. Colfer, G. Shepherd, P. Venkateswarlu, and F. Tiayon. 1998. *Testing and Developing Criteria and Indicators for Sustainable Forest Management in Cameroon: The Kribi Test, Final Report.* Special publication. Bogor, Indonesia: Center for International Forestry Research.

Prabhu, R., C.J.P. Colfer, and R.G. Dudley. 1999. *Guidelines for Developing, Testing, and Selecting Criteria and Indicators for Sustainable Forest Management: A C&I Developer's Reference.* CIFOR C&I Toolbox Series 1. Bogor, Indonesia: Center for International Forestry Research.

Prabhu, R., H.J. Ruitenbeek, T.J.B. Boyle, and C.J.P. Colfer. In press. Between voodoo science and adaptive management: The role and research needs for indicators of sustainable forest management. In Proceedings of the International Union of Forestry Research Organizations Conference, August 24–28, Melbourne.

Prakash, S., and M. Thompson. 1994. *Institutions and Transactions: The Risk and Fairness Approach to Environmental Economics.* Unpublished proposal submitted to Center for International Forestry Research (CIFOR). November. Bogor, Indonesia.

Prescott-Allen, R. 1995. Towards a Barometer of Sustainability for Zimbabwe. Draft report to International Union for the Conservation of Nature, Gland, Switzerland. July. Gland, Switzerland: IUCN.

Pretty, J.N. 1994. Alternative systems of inquiry for sustainable agriculture. *IDS Bulletin* 25(2): 37–48.

Pretty, J., and H. Ward. 1999. Social Capital and the Environment. Unpublished paper. Colchester, U.K.: University of Essex, Centre for Environment and Society and Department of Government. (Also submitted to World Development.)

Princet, M. 1994. *Perception des Parcs Nationaux en Afrique Centrale.* Montpellier, France: Société d'Eco-Aménagement for the European Economic Community.

Puri, R.K. 1997. Hunting Knowledge of the Penan Benalui of East Kalimantan, Indonesia. Doctoral dissertation. Honolulu, HI: University of Hawaii.

Purnomo, H., and others. 2000. *Criteria and Indicators Modification and Adaptation Tool (CIMAT), Version 2.* Bogor, Indonesia: Center for International Forestry Research.

Putnam, R.D. 1993. The prosperous community: Social capital and public life. *The American Prospect* 13: 35–42.

Rainforest Alliance. 1993. *Smart Wood Certification Program: Generic Guidelines for Assessing Natural Forest Management.* Revised draft. October. New York: Rainforest Alliance.

Ramirez, R. 1999. Stakeholder analysis and conflict management. In *Conflict and Collaboration in Natural Resource Management,* edited by D. Buckles. Ottawa, Canada: International Development and Research Centre Community-Based Natural Resource Management Program, 101–26.

Razavi, S. 1999. Gendered poverty and well-being: Introduction. *Development and Change* 30: 409–33.

Reboul, C. 1977. Determinants sociaux de la fertilité des sols. *Actes de la Recherche en Sciences Sociales* 17/18: 85–112.

Redford, K.H., and J.A. Mansour. 1996. *Traditional Peoples and Biodiversity Conservation in Large Tropical Landscapes.* Arlington, VA: America Verde Publications, The Nature Conservancy.

Redford, K.H., and C. Padoch (eds.). 1992. *Conservation of Neotropical Forests: Working from Traditional Resource Use.* New York: Columbia University Press.

Reed, D. (ed.) 1996. *Structural Adjustment, the Environment, and Sustainable Development.* London, U.K.: Earthscan.

Resolve. 1994. *The Role of Alternative Conflict Management in Community Forestry.* Rome, Italy: Food and Agriculture Organization, Community Forestry Programme.

Rew, A., C. Barnes, G. Clarke, and B. Vosey. 1996. *Final Report of the Consultancy to Prepare a Strategy for Participatory Biodiversity Conservation* (Volumes 1 and 2). Report for Mount Cameroon Project (MCP). Swansea, Wales, U.K.: Centre for Development Studies.

Rhoades, R.F. 1984. *Breaking New Ground: Agricultural Anthropology.* Lima, Peru: International Potato Center.

Richards, J.A. 1992. *Remote Sensing Digital Image Analysis.* Cambridge, U.K.: Springer-Verlag.

Richardson, C.W., R.G. Lee, and M.L. Miller. 1996. Thinking about ecology: Cognition of Pacific Northwest forest managers across diverse institutions. *Human Organization* 55: 314–23.

Riezebos, E.P., A.P. Vooren, and J.L. Guillaumet (eds.). 1994. *Le Parc National de Taï, Côte d'Ivoire.* La Fondation Tropenbos Series 8. Wageningen, the Netherlands: The Tropenbos Foundation.

Ritchie, B., C. McDougall, M. Haggith, and N. Burford de Oliveira. 2000. *An Introductory Guide to Criteria and Indicators for Sustainability in Community Managed Forest Landscapes.* Bogor, Indonesia: Center for International Forestry Research.

Riwut, T. 1979. *Kalimantan Menbangun [Kalimantan Developing]*. Jakarta, Indonesia: P.T. Jayakarta Agung Offset.

Rocheleau, D. 1988. Women, trees, and tenure: Implications for agroforestry research and development. In *Whose Trees? Proprietary Dimensions of Forestry*, edited by L. Fortmann and J.W. Bruce. Boulder, CO: Westview Press, 254–72.

———. 1999. Confronting complexity, dealing with difference: Social context, content, and practice in agroforestry. In *Agroforestry in Sustainable Agricultural Systems*, edited by L. Buck, J.P. Lassoie, and E. Fernandes. Boca Raton, FL: CRC Press LLC, 191–235.

Rocheleau, D., B. Thomas-Slaytor, and E. Wangari. 1996. *Feminist Political Ecology*. London, U.K.: Routledge.

Roe, E. 1994. *Narrative Policy Analysis: Theory and Practice*. Duke University Press, Durham, NC.

Rogers, S.C. 1978. Woman's place: A critical review of anthropological theory. *Comparative Studies in Society and History* 20: 123–62.

Roosevelt, A. 1989. Resource management in Amazonia before the conquest: Beyond ethnographic projection. *Advances in Economic Botany* 7: 30–62.

Rose, C.M. 1994. *Property and Persuasion: Essays on the History, Theory, and Rhetoric of Ownership*. Boulder, CO: Westview Press.

Roseberry, W. 1989. Peasants and the world. In *Economic Anthropology*, edited by S. Plattner. Stanford, CA: Stanford University Press, 108–26.

Roseman, M. 1991. *Healing Sounds from the Malaysian Rainforest*. Berkeley, CA: University of California Press.

Roszak, T. 1995. The spirit of the goddess. In *Ecopsychology: Restoring the Earth, Healing the Mind*, edited by T. Roszak, M.E. Gomes, and A.D. Kanner. San Francisco, CA: Sierra Club Books, 288–300.

Roszak, T., M.E. Gomes, and A.D. Kanner. 1995. *Ecopsychology: Restoring the Earth, Healing the Mind*. San Francisco, CA: Sierra Club Books.

Rudel, T. K. 1995. When do property rights matter? Open access, informal social controls, and deforestation in the Ecuadorian Amazon. *Human Organization* 54: 332–9.

Ruf, F. 1995. *Booms et Crises de Cacao: Les Vertiges de l'Or Brun*. Montpellier and Paris, France: Centre de Coopération Internationale pour la Recherche Agronomique en Développement—Systèmes Agraires, Karthala, and Ministère de la cooperation.

Ruitenbeek, H.J. 1996. Distribution of ecological entitlements: implications for economic security and population movement. *Ecological Economics* 17: 49–64.

———. 1998. *A Model Summary for Sustainable Forest Management*. April. Research note prepared for Center for International Forestry Research (CIFOR), Bogor, Indonesia: CIFOR.

Ruiz Pérez, M., O. Ndoye, and A. Eyebe. 1999. A Study of Non-Timber Forest Products Markets in the Humid Forest Zone of Cameroon. Unpublished paper. Bogor, Indonesia: Center for International Forestry Research.

Russell, D. 1991. Food Supply and the State: The Social Organization and History of the Rice Trade in Kisangani, Zaire. Unpublished doctoral dissertation, Anthropology. Boston, MA: Boston University.

———. 1993. *Resource Management in the Central African Forest Zone: A Handbook for IITA's Humid Forest Station*. Nkolbisson, Cameroon: International Institute for Tropical Agriculture–Humid Forest Systems.

Sader, S.A., T.A. Stone, and A.T. Joyce. 1990. Remote sensing of tropical forests: an overview of research and applications using non-photographic sensors. *Photogrammetric Engineering and Remote Sensing* 56: 1343–51.

417

Salas, M. 1994. "The technicians only believe in science and cannot read the sky": The cultural dimension of the knowledge conflict in the Andes. In *Beyond Farmer First: Rural People's Knowledge, Agricultural Research and Extension Practice*, edited by I. Scoones and J. Thompson. London, U.K.: Intermediate Technology Publications Ltd., 57–69.

Salick, J. 1992. Amuesha forest use and management: An integration of indigenous use and natural forest management. In *Conservation of Neotropical Forests: Working from Traditional Resource Use*, edited by K.H. Redford and C. Padoch. New York: Columbia University Press, 305–32.

Salim, A., and C.J.P. Colfer with C. McDougall. 1999. *The Scoring and Analysis Guide for Assessing Human Well-Being.* CIFOR C&I Toolbox Series no. 7. Bogor, Indonesia: Center for International Forestry Research.

Sanday, P. 1974. Female status in the public domain. In *Woman, Culture, and Society*, edited by M.R. Rosaldo and L. Lamphere. Stanford, CA: Stanford University Press, 189–206.

Sandin, B. 1980. *Iban Adat and Augury.* Penang, Malaysia: Universiti Sains Malaysia.

Sankar, S., P.C. Anil, M. Amruth, and P. Hopmans. 2000. *Criteria and Indicators for Sustainable Plantation Forestry in India.* Report. Bogor, Indonesia: Center for International Forestry Research.

Sardjono, M.A. 1990. Die Lembo-Kultur in OstKalimantan—ein Modell fuer die Entwicklung agroforstlicher Landnutzung in den Feuchttropen. Dissertation. Hamburg, Germany: University of Hamburg.

Sardjono, M.A., with E. Rositah, A. Wijaya, and A.M. Angie. 1997. A Test of Social Science Assessment Methods Concerning Indicators and Criteria for Sustainable Forest Management in East Kalimantan. Internal document. Bogor, Indonesia: Center for International Forestry Research.

Sarin, M. 1997. *Meeting of the Group on Gender and Equity in Joint Forest Management. September 29 to October 1, 1997. Report on Proceedings.* New Delhi, India: Society for the Promotion of Wastelands Development.

———. 1998. Putting Community Institutions under the Microscope: Selective Inclusion, Participation, and Rights under Joint Forest Management. Preliminary draft. Prepared for the Workshop on Shared Resources Management, Institute of Rural Management Anand, Anand, India.

Sather, C. 1990. Trees and tree tenure in Paku Iban society: The management of secondary forest resources in a long-established Iban community. *Borneo Review* 1: 16–40.

Savyasaachi. 1993. An alternative system of knowledge: Fields and forests in Abujhmarh. In *Who Will Save the Forests? Knowledge, Power, and Environmental Destruction*, edited by T. Banuri and F.A. Marglin. London, U.K.: Zed Books, 52–79.

Schmink, M. 1999. *Conceptual Framework for Gender and Community-Based Conservation.* Case Studies Series on Gender, Community Participation, and Natural Resource Management no. 1. Gainesville, FL: University of Florida, Center for Latin American Studies.

Schmink, M., and Wood, C. 1992. *Contested Frontiers in Amazonia.* New York: Columbia University Press.

Schotveld, A., and C.W. Stortenbeker. 1994. *Evaluating Sustainable Forest Management.* Report of Deskundigenwerkgroep Duurzaam Bosbeheer (DDB). Utrecht, The Netherlands: Ministry of Environment.

Scott, J.C. 1976. *The Moral Economy of the Peasant: Rebellion and Subsistence in Southeast Asia.* New Haven, CT: Yale University Press.

———. 1985. *Weapons of the Weak: Everyday Forms of Peasant Resistance.* New Haven, CT: Yale University Press.

———. 1986. Gender: A useful category for historical analysis. *American Historical Review* 91: 1053–75.

Selener, D. 1997. *Participatory Action Research and Social Change.* Ithaca, New York: Cornell University, The Cornell Participatory Action Research Network.

Sen, G. 1994. Development, population and the environment: A search for balance. In *Population Policies Reconsidered: Health, Empowerment and Rights,* edited by G. Sen, A. Germain, and L. Chen. Harvard Series on Population and Development Studies. Boston, MA and New York: International Women's Health Coalition, 63–74.

Senado Federal. 1996. *Meio Ambiente: Legislação. Dispositivos da Constituição Federal, Atos Internacionais, Código Florestal, Código de Mineração, Legislação Federal* (Volumes 1 and 2). Brasília, Brazil: Subsecretaria de Ediqões Técnicas.

Sewell, L. 1999. *Sight and Sensibility: The Ecopsychology of Perception.* New York: Penguin/ Putnam.

SGS. 1997. Sustainable Forest Management: A Practical Guide. First draft for consultation and review. London, U.K.: Oxford Centre for Innovation and International Institute for Environment and Development.

Shand, E. 1994. *Evaluation of Commercial Forest Plantation Resources—Trinidad and Tobago.* Report on situation and management of forest plantations. Rome, Italy: Food and Agriculture Organization.

Shaner, W.W., P.F. Philipp, and W.R. Schmehl. 1982. *Farming Systems Research and Development: Guidelines for Developing Countries.* Boulder, CO: Westview Press.

Shanin, T. 1990. *Defining Peasants: Essays Concerning Rural Societies, Expolary Economies, and Learning from Them in the Contemporary World.* Cambridge, U.K.: B. Blackwell.

Shanley, P. 1999. Extending ecological research to meet local needs: A case from Brazil. In *The Non-Wood Forest Products of Central Africa: Current Research Issues and Prospects for Conservation and Development,* edited by T. Sunderland, L. Clark and P. Vantomme. Rome, Italy: Food and Agriculture Organization.

Shanley, P., I. Hohn, and A. Valente da Silva. 1996. *Receitas sem Palavras: Plantas Medicinais da Amazonia.* Belem, Brazil: Editora Supercores.

Shanley, P., M. Cymerys, and J. Galvao. 1998. *Frutiferas da Mata na Vida Amazonica.* Belem, Brazil: Editora Supercores.

Shearer, R.R. 1997. Book Reviews—Ecofeminism. *Signs* 22: 496–501.

Shepherd, G. 1991. The communal management of forests in the semi-arid and sub-humid regions of Africa: Past practices and prospects for the future. *Development Policy Review* 9: 151–76.

Sigot, A., L.A. Thrupp, and J. Green (eds.). 1995. *Towards Common Ground: Gender and Natural Resource Management in Africa.* Nairobi, Kenya, and Washington, DC: African Centre for Technology Studies and World Resources Institute.

Silva, A. 1994. *Direito Ambiental Constitucional.* São Paulo, Brazil: Malheiros Editores Ltda.

Sirait, M., N. Podger, A. Flavelle, and J. Fox. 1994. Mapping customary land in East Kalimantan, Indonesia: A tool for forest management. *Ambio* 23(7): 411–17

Siskind, J. 1973. *To Hunt in the Morning.* New York: Oxford University Press.

Skole, D.L., and C.J. Tucker. 1993. Tropical deforestation and habitat fragmentation in the Amazon: Satellite data from 1978 to 1988. *Science* 260: 1905–10.

Smith, C.L. 1994. Connecting cultural and biological diversity in restoring Northwest salmon. *Fisheries* 19: 20–26.

Smith, C.L., and B.S. Steel. 1995. Core-periphery relationships of resource-based communities. *Journal of the Community Development Society* 26(1): 52–70.

SODEFOR (Société de Développement des Forêts). 1994. *Plan d'Amènagement de la Fôret Classée de la Bossematié (22,200 ha) 1995–2014.* Abengourou, Côte d'Ivoire: Kreditanstalt für Wiederaufbau (KFW)/SODEFOR/Deutsche Gesellschaft für Technische Zusammenarbeit (GTZ) GmbH.

Soedjito, H. 1994. *Keanekaragaman Sumberdaya Alam dan Budaya Kawasan Konservasi Kayan Mentarang, Potensi dan Tantangan bagi Pembangunan Propinsi Kalimantan Timur [Natural Resource and Cultural Variety in the Kayan Mentarang Nature Reserve, Potential and Constraints to the Development of East Kalimantan]*. Jakarta, Indonesia: Puslitbang-Lembaga Ilmu Pengetahuan Indonesia-World Wide Fund for Nature (WWF Indonesia Programme).

Soil Association. 1994. *Responsible Forestry Standards for the United Kingdom.* Bristol, U.K.: Soil Association.

Songan, P. 1993. A naturalistic inquiry into participation of the Iban peasants in the land development project in the Kalaka and Saribas Districts, Sarawak, Malaysia. *Borneo Research Bulletin* 25: 101–21.

Southern Appalachian Assessment. 1996. *Report 4 of 5: Southern Appalachian Man and the Biosphere Cooperative. Social/cultural/economic technical report.* Washington, DC: U.S. Department of Agriculture.

Sparr, P. (ed.) 1994. *Mortgaging Women's Lives: Feminist Critiques of Structural Adjustment.* London, U.K.: Zed Books.

Stamp, P. 1989. *Technology, Gender, and Power in Africa.* Ottawa, Canada: International Development and Research Centre.

Stanfield, D., and N. Singer. 1993. *Land Tenure and the Management of Land Resources in Trinidad and Tobago* (Parts I and II). Madison, WI: University of Wisconsin, Land Tenure Center.

Statistisches Bundesamt. 1996. *Länderbericht Karibische AKP-Staaten.* Wiesbaden, Germany: Statistisches Bundesamt.

Stevens, P. 1997. *Measuring the Sustainability of Forest Village Ecosystems: Concepts and Methodologies—A Turkish Example.* Forestry and Forest Products Technical Report no. 103. Canberra, Australia: Commonwealth Scientific and Industrial Research Organization.

Sunderlin, W., and Pokam, J. 1998. *Economic Crisis and Forest Cover Change in Cameroon: The Roles of Migration, Crop Diversification, and Gender Division of Labour.* Bogor, Indonesia: Center for International Forestry Research.

Sutherland, A. 1994. Managing bias: Farmer selection in tools for the field. In *Methodologies Handbook for Gender Analysis in Agriculture*, edited by H.S. Feldstein and J. Jiggins. West Hartford, CT: Kumarian, 15–20.

Sutlive, V.H., and G.N. Appell (eds.). 1991. *Female and Male in Borneo: Contributions and Challenges to Gender Studies.* Monograph Series no. 1. Williamsburg, VA: Borneo Research Council.

Tainter, J.A. 1988. *The Collapse of Complex Societies.* Cambridge, U.K.: Cambridge University Press.

———. 1995. Sustainability of complex societies. *Futures* 27: 397–407.

———. 1996. Complexity, problem solving, and sustainable societies. In *Getting Down to Earth: Practical Applications of Ecological Economics*, edited by R. Costanza, O. Segura, and J. Martinez-Alier. Washington, DC: Island Press, 61–76.

———. 1997. Cultural conflict and sustainable development: Managing subsistence hunting in Alaska. In *Sustainable Development of Boreal Forests. Proceedings of the 7th International Conference of the International Boreal Forest Research Association.* Moscow, Russia: All-Russian Research and Information Center for Forest Resources, 155–61.

———. 1999. Rio Grande Basin and the modern world: understanding scale and context. In *Rio Grande Ecosystems: Linking Land, Water, and People*, edited by D. Finch, J. Whitney, J. Kelly, and S. Loftin. Proceedings RMRS-P-7. Fort Collins, CO: USDA Forest Service, Rocky Mountain Research Station, 7–11.

———. 2000. Global change, history, and sustainability. In *The Way the Wind Blows: Climate, History, and Human Action,* edited by R.J. McIntosh, J.A. Tainter, and S.K. McIntosh. Columbia University Press, New York, 331–56.

Tainter, J.A., and G.J. Lucas. 1983. Epistemology of the significance concept. *American Antiquity* 48: 707–19.

Tarapoto Process. 1998. Excerpted (forest management unit level only) in Prabhu, R., C.J.P. Colfer, and R. Dudley. *Guidelines for Developing, Testing, and Selecting Criteria and Indicators for Sustainable Forest Management.* CIFOR C&I Toolbox Series no. 1. Bogor, Indonesia: Center for International Forestry Research.

Taussig, M. 1980. *The Devil and Commodity Fetishism in South America.* Chapel Hill, NC: The University of North Carolina Press.

Tchamou, N. 1993. Field notes for IITA Humid Forest Station. Nkolbisson, Cameroon: IITA Humid Forest Station.

Tchikangwa, N.B., with S. Sikoua, M. Metomo, and M.F. Adjudo. 1998. Test des Méthodes en Sciences Sociales de Vérification des Critères et Indicateurs d'Aménagement Durable des Forêts: Périphérie est de la Réserve du Dja (Sud-Cameroun). Internal document. Bogor, Indonesia: Center for International Forestry Research.

Tchoungi, R., S. Gartlan, J.A. Mope Simo, F. Sikod, A. Youmbi, M. Ndjatsana, and J. Winpenny. 1995. Structural Adjustment and Sustainable Development in Cameroon: A WWF Study. Working paper 83. London, U.K.: Overseas Development Institute.

Terborgh, J. 1999. *Requiem for Nature.* Washington, DC: Island Press.

Thenkabail, P.S., and C. Nolte. 1995. *Land Use and Vegetation Pattern in the Forest Margin Benchmark Area of Southwest Cameroon Determined Using Near-Time Satellite Images from SPOT HRV.* Nkolbisson, Cameroon: Office National de Développement des Forêts/ Centre de Télédétection et de Cartographie Forestière.

Thoreau, H.D. 1957 (orig. 1854). *Walden,* edited by Paul Sherman. Cambridge, MA: The Riverside Press.

Tiani, A.M., with B.E. Mvogo, A. Oyono, and D.N. Kenmegne. 1997. A Test of Social Science Assessment Methods (near Mbalmayo, Cameroon). Internal document. Bogor, Indonesia: Center for International Forestry Research.

Tiki, M.T., M. Ndjatsana, N.O. Besong, A.M. Tiani, Z. Mayna, W.C. Youmbi, N.L.P. Ngalie, and E. Adzeyuf. 1996. *La Gestion des Ressources Naturelles à Eyek 1 et Bitsok-Adjap (Zone SOLIDAM): Etude de Cas Réalisée avec la MARP.* Geneva, Switzerland: World Wide Fund for Nature.

Tostes, A. 1994. *Sistema de Legislaçio Ambiental.* CECIP. Rio de Janeiro, Brazil: Vozes.

Townsend, J.G., with U. Arrevillaga, J. Bain, S. Cancino, S. Frenk, S. Pacheco, and E. Perez. 1995. *Women's Voices from the Rainforest.* London, U.K.: Routledge.

Tsagué, A. 1995. *Etude de la Filière des Produits de la Cueillette du Prélèvement à la Commercialisation.* Dimako, Cameroon: Aménagement Pilote Intégré.

Tsing, A.L. 1993. *In the Realm of the Diamond Queen.* Princeton, NJ: Princeton University Press.

Tucker, C. 1999. Private versus common property forests: forest conditions and tenure in a Honduran community. *Human Ecology* 27: 201–30.

Tucker, C.J., B.N. Holben, and T.E. Goff. 1984. Intensive forest clearing in Rondonia, Brazil, as detected by satellite remote sensing. *Remote Sensing of Environment* 15: 255–62.

Umans, L. 1995. Meaning and practice of participation: The case of Nepal. (Forestry and People in South-East Asia). *BOS NiEuWSLETTER* 32(3): 42–50.

UNDP (United Nations Development Programme). 1985. *Evaluation and Development of Wildlife Resources, Trinidad and Tobago. Project Findings and Recommendations.* Port of Spain, Republic of Trinidad and Tobago: U.N. Development Programme.

UNICEF (United Nations Children's Fund). 1994. *The State of the World's Children— 1994.* Geneva, Switzerland: UNICEF.

Upton, C., and S. Bass. 1995. *The Forest Certification Handbook*. London, U.K.: Earthscan Publications Ltd.

Vanclay, J.K. 1998. FLORES: For exploring land use options in forested landscapes. *Agroforestry Forum* 9(1): 47–52.

van der Leeuw, S.E. 1998. Introduction. In *The Archaeomedes Project: Understanding the Natural and Anthropogenic Causes of Land Degradation and Desertification in the Mediterranean Basin*, edited by S.E. van der Leeuw. Luxembourg: Office for Official Publications of the European Communities, 2–22.

Van Dorp, M. 1995. Shaking the Tree: An Economic Geographical Analysis of the Foreign Impact on Forestry in Cameroon. M.Sc. dissertation. Amsterdam, the Netherlands: University of Amsterdam, Faculty of Economics.

van Haaften, H. 1995. Final Report/Diary. Prepared for CIFOR's project on Testing Criteria and Indicators for Sustainable Forest Management, Côte d'Ivoire. Bogor, Indonesia: Center for International Forestry Research.

Vayda, A.P. 1983. Progressive contextualization: Methods for research in human ecology. *Human Ecology* 11: 265–81.

———. 1996. *Methods and Explanations in the Study of Human Actions and Their Environmental Effects*. Special publication. Bogor, Indonesia: Center for International Forestry Research/World Wide Fund for Nature.

Vayda, A.P., C.J.P. Colfer, and M. Brotokusumo. 1980. Interactions between people and forests in East Kalimantan. *Impact of Science on Society* 30: 179–90.

von Bothmer, K-H. 1994. *Evaluation of Commercial Forest Plantation Resources—Trinidad and Tobago. Report on Concession Policy and Its Economic and Financial Implications*. Rome, Italy: Food and Agriculture Organization.

Wadley, R.L. 1996. Variation and changing tradition in Iban land tenure. *Borneo Research Bulletin* 27: 98–108.

———. 1997. Circular Labor Migration and Subsistence Agriculture: A Case of the Iban in West Kalimantan, Indonesia. Doctoral dissertation. Tempe, Arizona: Arizona State University, Department of Anthropology.

———. 1999a. Disrespecting the dead and the living: Iban ancestor worship and the violation of mourning taboos. *Journal of the Royal Anthropological Institute* 5: 595–610.

———. 1999b. The History of Population Displacement and Forced Settlement in and Around Danau Sentarum Wildlife Reserve, West Kalimantan, Indonesia: Implications for Co-Management. Paper presented at the Displacement, Forced Settlement and Conservation Conference. September 9–11, University of Oxford, U.K., Refugee Studies Programme.

———. Not dated. Agroforestry and Wildlife: A Report on Indigenous Forest Management and Conservation along the Danau Sentarum Periphery, West Kalimantan, Indonesia. Unpublished manuscript. Leiden, the Netherlands.

Wadley, R.L, C.J.P. Colfer, and I.G. Hood. 1996. The Role of Sacred Groves in Hunting and Conservation among the Iban of West Kalimantan, Indonesia. Paper presented at the 95th Annual Meeting of the American Anthropological Association. November 20–24, San Francisco, CA.

———. 1997. Hunting primates and managing forests: The case of Iban forest farmers in Indonesian Borneo. *Human Ecology* 25: 243–71.

———. In preparation. Sacred forest, hunting and conservation: An Iban case from West Kalimantan, Indonesia. In *Sacred Places and Biodiversity Conservation*, edited by L. Sponsel and G.N. Appell.

Waldrop, M.M. 1992. *Complexity: The Emerging Science at the Edge of Order and Chaos*. New York: Touchstone Books (Simon and Schuster).

Walker, A., and R. Sillans. 1962. *Rites et Croyances des Peuples du Gabon*. Paris, France: Présence Africaine.

Watts, J., and G.M. Akogo. 1994. Biodiversity assessment and development. Towards participatory forest management on Mount Cameroon. *Commonwealth Forestry Review* 73: 221–30.

Weber, J. 1977. Structures agraires et évolution des milieux ruraux: Le cas de la région cacaoyère du Centre-Sud Cameroun. Cahiers ORSTOM. Séries. *Sciences Humaines* 14(2): 113–39.

Webster's Third New International Dictionary. 1993. Springfield, MA: Merriam-Webster, Inc.

Weiner, A. 1976. *Women of Value, Men of Renown*. Austin, TX: University of Texas Press.

Weinstock, J.A. 1983. *Kaharingan and the Luangan Dayak: Religion and Identity in Central-East Borneo*. Ithaca, NY: Cornell University.

Wells, M. 1997. *Indonesia ICDP Study: Interim Report of Findings*. Jakarta, Indonesia: The World Bank.

Whitten, T., R.E. Soeriaatmadja, and S. Afiff. 1996. *The Ecology of Java and Bali*. Singapore: Periplus Editions.

Wickham, T. 1996. *Community-Based Participation in Wetland Conservation: Activities and Challenges of the Danau Sentarum Wildlife Reserve Conservation Project, Danau Sentarum Wildlife Reserve, West Kalimantan, Indonesia*. Report to Wetlands International. Bogor, Indonesia: Wetlands International.

Wickramasinghe, A. 1994. *Deforestation, Women, and Forestry*. Utrecht, the Netherlands: Institute for Development Research.

Widjono, R.A. 1992. *Hak Pemilikan Tanah Suku Dayak Benuaq di Kalimantan Timur [Land Ownership among the Benuaq Dayak of East Kalimantan]*. Paper presented at the Second Biennial Conference, Borneo Research Council (July). Kota Kinabalu, Sabah, Malaysia.

Wilde, V.L., and A. Vainio-Mattila. 1995. *Gender Analysis and Forestry*. Forests, Trees, and People Programme International Training Package. Rome, Italy: Food and Agriculture Organization.

Winthrop, R.H. 1991. *Dictionary of Concepts in Cultural Anthropology*. Westport, CT: Greenwood Press.

Woelfel, J.K., and J. Danes. 1980. Multidimensional scaling models for communication research. In *Multivariate Techniques in Communication Research*, edited by P. Monge and J. Capella. New York: Academic Press, 333–64.

Woelfel, J.K., and E.L. Fink. 1980. *The Measurement of Communication Processes: Galileo Theory and Method*. New York: Academic Press.

Woelfel, J.K., and G.A. Barnett. 1982. Multidimensional scaling in Riemann space. *Quality and Quantity* 16: 461–91.

———. 1992. Procedures for controlling reference frame effects in the measurement of multidimensional processes. *Quality and Quantity* 26: 367–81.

Woelfel, J.K., D.L. Kincaid, B. Newton, and J. Lee, 1986. The effect of compound messages on the global characteristics of Galileo spaces. *Quality and Quantity* 20: 133–45.

Woelfel, J.K., R.A. Holmes, M. Cody, and E. Fink. 1988a. A multidimensional scaling based procedure for designing persuasive messages and measuring their effects. In *Readings in the Galileo System: Theory, Methods, and Applications*, edited by G.A. Barnett and J. Woelfel. Dubuque, IA: Kendall-Hunt, 235–42.

Woelfel, J.K., R.A. Holmes, B. Newton, and D.L. Kincaid. 1988b. An experimental measure of the mass of occupation names. In *Readings in the Galileo System: Theory, Methods, and Applications*, edited by G.A. Barnett and J. Woelfel. Dubuque, IA: Kendall-Hunt, 313–32.

Woelfel, J.K., G.A. Barnett, and R. Pruzek. 1989. Rotation to simple processes: The effect of alternative rotation rules on observed patterns in time-ordered measurements. *Quality and Quantity* 23: 3–20.

Wolf, E. 1967. The vicissitudes of the closed corporate peasant community. *American Eth-nologist* 13: 325–9.

Wollenberg, E. 1997. Decision-Making Among Diverse Interests: The Use of Future Sce-narios in Local Forest Management Policy: A Proposed Methodology. Paper presented at International Seminar on Community Forestry at a Crossroads: Reflections and Future Directions in the Development of Community Forestry. July 17–19, Bangkok, Thailand.

———. 1999. The Social Nature of Forest Boundaries: Entitlement, Identity, and Reci-procity Among Kenyah Forest Users in East Kalimantan, Indonesia. Paper presented at the 95th meeting of the Association of American Geographers. March 23–27, Hono-lulu, HI.

Wollenberg, E., and C.J.P. Colfer. 1996. Social sustainability in the forest: A progress report of a project aiming to test criteria and indicators for the social dimensions of sustainable forest management. *Tropical Forest Update* 6(2): 9–11.

Wollenberg, L., A. Nawir, G. Limberg, I. Tjitradjaja, Z. Lubis, and C. Eghenter. 1996. *Incomes and Incentives for Conservation.* Bogor, Indonesia: Center for International For-estry Research.

Wollenberg, E., J. Anderson, and D. Edmunds. In press. Pluralism and the less powerful: experience in accommodating multiple interests in local forest management. Prepared for a special issue of *International Journal of Agricultural Resources, Governance, and Ecology.*

Woodley, S., G. Alward, L.I. Gutierrez, T. Hoekstra, B. Holt, L. Livingston, J. Loo, A. Skibicki, C. Williams, and P. Wright. 2000. *North American Test of Criteria and Indicators of Sustainable Forestry* (Volume 1). Washington, DC: USDA Forest Service.

World Bank. 1995. *Trinidad and Tobago: National Parks and Watershed Management Project.* Washington, DC: The World Bank.

———. 1996. *The World Bank Participation Sourcebook.* Washington, DC: The World Bank, Environment Department.

World Commission on Environment and Development. 1987. *Our Common Future.* Oxford, U.K.: Oxford University Press.

World Resources Institute/U.N. Environment Programme/U.N. Development Pro-gramme. 1992. *World Resources 1992–93: A Guide to the Global Environment. Toward Sustainable Development.* Oxford, U.K.: Oxford University Press.

Yonta, M. 1995. *Enquête Socio-economique Menée dans les Villages Situés autour et dans la Reserve Forestière de Mbalmayo.* Projet d'Amenagement et de Régénération des Forêts. Yaoundé, Cameroon: Office National de Développement des Forêts/Overseas Devel-opment Agency Plan d'Amenagement.

Young, G.L. 1992. Between the atom and the void: Hierarchy in human ecology. *Advances in Human Ecology* 1: 119–47.

Young, K. 1988. Gender and development: A relational approach. In *Gender Analysis for Management of Research in Agriculture and Natural Resources.* The Hague, the Nether-lands: International Service for National Agricultural Research, 101–9.

Young, T.R. 1994. *Evaluation of Commercial Forest Plantation Resources—Trinidad and Tobago.* Report of the Legislation Consultant Regarding Concession Allocation and other Timber Sales Systems. Rome, Italy: Food and Agriculture Organization.

Zerner, C. 1994. Through a green lens: The construction of customary environmental law and community in Indonesia's Maluku Islands. *Law and Society Review* 28: 1079–122.

Zweede, J., J. Kressin, R. Mesquita, J.N.M. Silva, V.M. Viana, and C.J.P. Colfer. 1997. *Final Report: Test Brazil, 22 October–21 November 1995,* edited by R. Prabhu and L.C. Tan. CIFOR Project on Testing Criteria and Indicators for the Sustainable Manage-ment of Forests. Bogor, Indonesia: Center for International Forestry Research.

Index

In page references, n. indicates an endnote, and italicized page numbers indicate figures or tables.